To Rise From Earth

United States

To Rise From Earth

An Easy-to-Understand Guide to Spaceflight

Second Edition

Wayne Lee

Space Mission Design Engineer
NASA Jet Propulsion Laboratory

Facts On File, Inc.

To Rise From Earth: An Easy-to-Understand Guide to Spaceflight

First Edition 1995
Second Edition 2000

Checkmark Books
An Imprint of Facts On File, Inc.
11 Penn Plaza
New York NY 10001

Library of Congress Cataloging-in-Publication Data

Lee, Wayne
To rise from earth: an easy-to-understand guide to spaceflight / Wayne Lee. – 2nd ed.
p. cm.
Includes index.
ISBN 0-8160-4091-5. – ISBN 0-8160-4092-3 (alk. paper)
1. Astronautics–Popular works. 2. Space flight–Popular works.
I. Title
TL793.L3137 1999
629.4–dc21 99-15971

The opinions expressed in this book are the author's. These opinions do not necessarily represent the official views of the National Aeronautics and Space Administration. The use of NASA materials and official NASA insignia in this book does not constitute an official endorsement by NASA or any of its associated contractor organizations, the government of the United States, the Jet Propulsion Laboratory, or the California Institute of Technology.

All photographs, except where otherwise noted, were provided courtesy of the National Aeronautics and Space Administration and belong to the public domain. Some photographs of Mars were provided courtesy of the United States Geological Survey for NASA and also belong to the public domain. Additional credits appear on page 307.

This book was produced by the *To Rise From Earth* Project in collaboration with the Flat Satellite Society. This is version 5-Alpha.

Cover design by Nora Wertz

Printed on acid-free paper in Hong Kong

Creative 10 9 8 7 6 5 4 3 2 1

“For while we cannot guarantee that we shall one day be first, we can guarantee that any failure to make this effort will make us last. We take an additional risk by making it in full view of the world.”

President John F. Kennedy

“Far better it to dare mighty things, to win glorious triumphs, even though checkered with failures, than to rank with those poor spirits who neither enjoy nor suffer much because they live in the gray twilight that knows not victory nor defeat.”

President Theodore Roosevelt

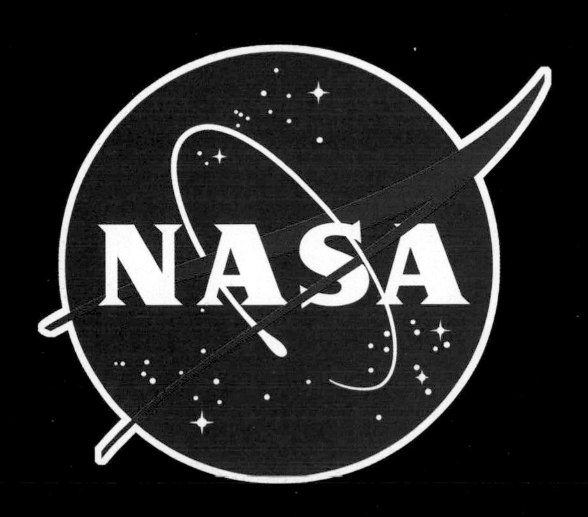

This book is dedicated to the success of the American civil space program into the twenty-first century and beyond.

HUGHES

Acknowledgments

No book is ever the product of just one person's hard work. Many individuals contributed their valuable time and suggestions toward improving the quality of this publication. However, I would first like to thank my wife Marguerite for patiently waiting out the endless hours that I spent working on this second edition.

Much of the artistic appeal of this book comes from the many photographs graciously provided by the National Aeronautics and Space Administration. In particular, I would like to thank Kim Grosso, Becky Fryday, Gloria Vale, and Mike Gentry of NASA Media Services at the Johnson Space Center in Houston, Texas, for helping me select many of the photographs. Kim Willis at JSC, from the Lockheed Martin Corporation, also deserves many thanks for selecting most of the Earth views that appear in chapters 2 and 3. At the Jet Propulsion Laboratory, Jurrie Van der Woude and Ed "Skip" McNevin deserve many thanks for helping me locate photographs from NASA's interplanetary missions. Jurrie (JPL) and Mike (JSC) were the masters of remembering the serial numbers for "lost" negatives. Other artwork and photographs were graciously provided by the NASA Kennedy Space Center media relations staff, General Dynamics Space Systems Division, McDonnell Douglas Aerospace, the Lockheed Martin Corporation, Trimble Navigation, the Washington Post, the United States Geological Survey, the Computer Support Corporation, and Dave Smith and Maria Zuber of the Mars Orbiter Laser Altimeter Team.

My original proofreaders spent endless hours reviewing the first ninety-two-page draft back in 1992 and made many valuable comments and suggestions that ultimately improved the book's overall quality. They are Dr. George Botbyl, Kathleen Crean, Barbie Kozel, Caroline McDonnell, Dr. Stephen Nichols, Glenn Peterson, Calina Seybold, John Sonntag, and Chris and Lisa Maria Tuason.

That original draft was written for the 1992 NASA Space Life Sciences Training Program (SLSTP). The program brings some of the nation's top college undergraduates to the Kennedy Space Center every summer. Dr. Linda Chamberlin and Dr. Gary Coulter from the SLSTP program deserve special thanks, as they provided encouragement from within the NASA community for the formulation of this project. In addition, I am grateful to the entire 1992 SLSTP class and staff members for serving as a test audience for the book. One member of that class, Nancy Balter, read the entire book while on vacation and then offered her "flat satellite" idea to NASA based on what she had learned. Although these types of satellites are impractical, her suggestion epitomizes the fulfillment of the project's goals.

After the 1992 SLSTP, the process of converting the book from an academic draft to a publishable work would not have been possible without the support of Dr. Wallace T. Fowler and Dr. Byron D. Tapley of the University of Texas Aerospace Engineering Department. They, along with the Texas Space Grant Consortium, provided the mechanism to conduct the proof-of-concept beta test for the *To Rise From Earth* Project. During this time, Danielle Miller, Shirley Pollard, and Peggy Trunnell at the Texas Space Grant Consortium Office at the University of Texas kept track of the administrative details of the project.

My two agents, Michael Larsen and Elizabeth Pomada, provided invaluable assistance toward turning this book from a draft into reality. The importance of their contribution cannot be understated, as they allowed a first-time author to secure a contract on the first try. In addition, many thanks also go out to my editor James Chambers and the Facts On File staff for their infinite patience in allowing me to micromanage them in the production of this book.

One of the biggest thanks and debts of gratitude go out to my mother and father. They provided me with the funds to purchase the computer equipment used to start this book. Also, an indirect thank you is owed to Michelle Kokel, who inadvertently set off the cascaded chain of events in 1987 that led to the writing of this book. She was also a proofreader.

The final acknowledgment goes out to the thousands of men and women who have worked to make the United States the undisputed leader in space exploration. This book would not have been possible without their efforts and sacrifices. ❑

USA
NASA
Endeavour

Table of Contents

USA
USA

To Rise From Earth

Entering Space

Chapter 1: An Introduction to Orbits and Rockets

"We choose to go to the Moon," President Kennedy daringly announced to a cheering Houston crowd at a September 1962 pep rally. "We chose to go to the Moon in this decade," reiterated the young president, "and do other things not because they are easy, but because they are hard, because there is new knowledge to be gained, and new rights to be won, and they must be won for the progress of all mankind." The blistering summer heat and humid Texas air attacked the president, but he was determined to deliver because the message was urgent.

Barely sixteen months earlier, Kennedy had appeared in front of a joint session of Congress to issue the boldest challenge in the history of civilization. Simply stated, he asked America to send humans to the surface of the Moon and return them safely to Earth before the end of the decade. The challenge to the nation became known as Project Apollo.

Consider the magnitude of what Kennedy asked for. At the time of the challenge, America's venture into the final frontier amounted to a single suborbital flight by Alan Shepard lasting a paltry 15 minutes and covering less than 120 miles. That single voyage represented a mere flea's jump as compared to a trip to the Moon. But America of the 1960s, despite an initial slow reaction to realizing the importance of spaceflight during the cold war era, rose to the occasion and responded magnificently. With "a giant leap for mankind" in 1969, astronauts Neil Armstrong and Buzz Aldrin completed Kennedy's challenge for America by walking on the Moon. Less than sixty-six years after the world's first airplane flight at Kitty Hawk, North Carolina, America had been transformed from the home of the Wright Brothers to, in the words of author Tom Wolfe, the land of the "right stuff."

Shortly after Armstrong's immortal giant leap, Wernher von Braun, the German rocket scientist who keyed America's success in space, predicted that humans would walk on the planet Mars by the year 1980. Even before 1980 came and went, it became painfully evident that America was playing the role of the hare in the classic fable about the race between the tortoise and hare. After the challenge of the Moon faded, so did the public's interest in sending humans to other worlds. Although the technology existed for a Mars trip, the national will did not, and nobody from Earth has set foot on another world since the Apollo Moon landings ended in December 1972.

Today, there exist tribal civilizations scattered in remote corners of the world that remain ignorant about modern technology such as computers, television, and laser surgery. But, some of them know about Apollo. They know that humans once walked on the Moon. Future historians will undoubtedly regard the Apollo Moon landings as humanity's most significant accomplishment in the twentieth century despite tremendous advancements in the fields of medicine, transportation, communications, and world peace. For the very first time, the invisible chains of gravity holding humans to the surface of the Earth had been broken. For the very first time, the future had finally arrived and humans began to push back on the final frontier, into the unknown expanse of the cosmos. The full benefits of beginning this voyage may not be revealed for many years to come.

Unfortunately, we did not have the luxury of hindsight during the early 1970s. Subsidizing expensive, long-term investments into the nation's future simply became unacceptable, and we began to demand an immediate, tangible return on our tax dollars. Under pressure to deliver this return, America embarked on a radical new

Previous Page:

NASA astronaut Gene Cernan prepares to climb aboard the lunar rover during America's final Moon landing of the twentieth century. The *Apollo 17* crew visited the Moon in December 1972. Nobody has been back since.

Opposite Page:

With a roar heard from miles away, Space Shuttle *Columbia* lifts off on the start of a 14-day science mission in July 1994. During that flight, seven astronauts carried out scientific experiments designed by over 200 scientists representing 13 different countries. The shuttle is the most complicated vehicle ever built by humans and represents America's first attempt at turning spaceflight into "routine business." Since the beginning of the shuttle program in 1981, there have been nearly 100 launches.

USA

transportation system called the Space Shuttle, a 99-ton spacecraft with wings that races into space like a rocket and lands on a runway like an ordinary plane. The shuttle first flew in 1981 and immediately began to prove itself as a versatile vehicle orbital platform capable of allowing astronauts to launch satellites, to retrieve and repair satellites in space, and to conduct scientific experiments in weightlessness. It still flies today on a semi-routine basis, but never ventures more than 300 miles from the Earth at any one time.

Compared to the spirit of Armstrong's giant leap, 300 miles represents only a minuscule fraction of the distance to the Moon. Astronauts have not played a significant part in exploring other worlds since 1972. Today, they continue to watch from the sidelines as emotionless robotic space probes have temporarily taken their place in the arena of solar system exploration. These electronic servants cannot convey the elation of discovery or the thrill of adventure, but their electronic eyes have photographed and gathered scientific data on every planet in the solar system except for Pluto. Why should we care about these distant worlds in our cosmic backyard?

Take a look at Jupiter, largest of all the planets. This gargantuan world possesses no solid surface and is essentially a swirling inferno of hydrogen gas 320 times more massive than the Earth. What could Earth possibly share with this turbulent ball of gas? In 1994, space probe *Galileo*, on its way to Jupiter, and the *Hubble Space Telescope* recorded a cataclysmic collision between Jupiter and comet Shoemaker-Levy. Scientific evidence points to a similar collision on Earth 65 million years ago that extinguished the dinosaurs and 90% of all the living species at the time.

On Venus, robotic spacecraft found a hostile climate where a thick, perpetual layer of carbon dioxide clouds has trapped heat from sunlight to the point where the surface temperature could melt lead. This scenario represents a classic example of global warming and the greenhouse effect gone mad. Data also indicates that despite Venus' infernal present, the past showed signs of a cooler climate with oceans, much like today's Earth.

On Mars, space probes found volcanoes larger than Mount Everest, canyons longer than the United States, and weather conditions drier than the Sahara and colder than Antarctica. Yet they also found evidence of torrential flooding millions of years ago, evidence that Mars once experienced a warm, wet past like conditions on Earth today. Today, Mars lies trapped in a global ice age.

Will Earth continue as probably the sole oasis of life in the solar system? Or will Earth suffer the fate of Venus or Mars? By studying the planets and comparing them to Earth, scientists hope to gain a better understanding of our planet, its history, and the evolution of life. Spaceflight provides the means to capture this scientific opportunity. But, it also allows us the chance to think about the Earth

Opposite Page:

Although humans have yet to explore Mars in person, NASA has sent several robotic probes to the Martian surface. In November 1976, this picture from the surface of Mars was beamed back to Earth by the *Viking 2* lander from a place called Utopia Planitia. Many scientists believe that simple forms of life may have once evolved on Mars. In late 1996, NASA resumed its Mars exploration program with the *Mars Global Surveyor* and the *Mars Pathfinder* missions. *Pathfinder* landed on the surface and deployed a robotic micro-rover to collect surface data.

Orbit Firsts

First Accomplishment	Date
Artificial Satellite *Sputnik 1* Launched by USSR	4 October 1957
Animal in Space Laika (dog on *Sputnik 2*) Launched by USSR	3 November 1957
American Satellite *Explorer 1* Launched by USA	31 January 1958
Weather Satellite *TIROS 1* Launched by USA	1 April 1960
Human in Orbit Yuri Gagarin Launched by USSR	12 April 1961
American in Orbit John Glenn (*Friendship 7* mission) Launched by USA	20 February 1962
Television Satellite *Telstar 1* Launched by USA	10 July 1962
Woman in Orbit Valentina Tereshkova Launched by USSR	16 June 1963
Humans in Lunar Orbit Anders, Lovell, Borman (*Apollo 8* mission) Launched by USA	24 December 1968

RULE OF THUMB

This book uses the metric system, called SI units by scientists. One meter, the basic unit of measurement in SI, equals about 39 inches or 1.0936 yards. One meter per second (m/s) equals a velocity of about 2.236 miles per hour. One kilometer equals 1,000 meters, so one kilometer per second (km/s) equals 2,236 miles per hour.

from a philosophical point of view. In both the scientific and philosophical realm, spaceflight provides a chance for us to better understand and define who we are as a society, to see where we have been, and to catch a glimpse of where we are headed.

Sometimes, it is hard to believe that over a quarter century has passed since humans set foot on other worlds. However, the time will soon come again because humans have been explorers ever since the beginning of time. This book will help you take the first steps toward understanding the exploring of space. What follows is a plain English examination (without the use of equations) of the machines that make spaceflight possible, and the mechanics that determine how these seemingly complicated machines fly. You really do not have to be a rocket scientist to understand rocket science. The best way to begin is by answering the question, What is spaceflight? That answer involves knowing how orbits work.

1.1 What Is an Orbit?

Consider the following scenario. What happens when you throw a ball in the horizontal (forward) direction from the top of a tall building? The ball starts to fly forward, immediately begins to drop towards the ground after leaving your hand, and continues to fly forward until it hits the ground. What happens if you throw the ball forward with a faster velocity than before? This time, the ball lands on the ground further away from the building than in the first case. However, in both cases, the ball takes the same amount of time to fall to the ground. In fact, a ball dropped from rest (not thrown) from the top of the same building will fall to the ground in same amount of time as if it were thrown forward. Gravity provides the reason for the similar fall times because the Earth only pulls objects in the downward direction. This pull is independent of how fast or slow an object moves in the horizontal (forward) direction.

Circular Orbit

The harder you throw the ball, the farther it will travel in the forward direction before falling to the ground. Now, suppose that you throw the ball hard enough so that in the time gravity pulls the ball one meter closer to the ground, the ball travels forward far enough for the surface of the Earth to also curve downward by one meter. In this case, the ball never hits the ground, as it falls all the way around the Earth over and over again while remaining at the same altitude. Spaceflight engineers call this type of trajectory a circular orbit (see Figure 1 on page 9).

Many people often wonder, "what keeps an orbiting object up?" The truth of the matter is that nothing keeps the object up. In fact, gravity keeps the object down by deflecting the object's forward motion to follow the curvature of the Earth's surface. If gravity

Figure 1: *Free Fall Motion*

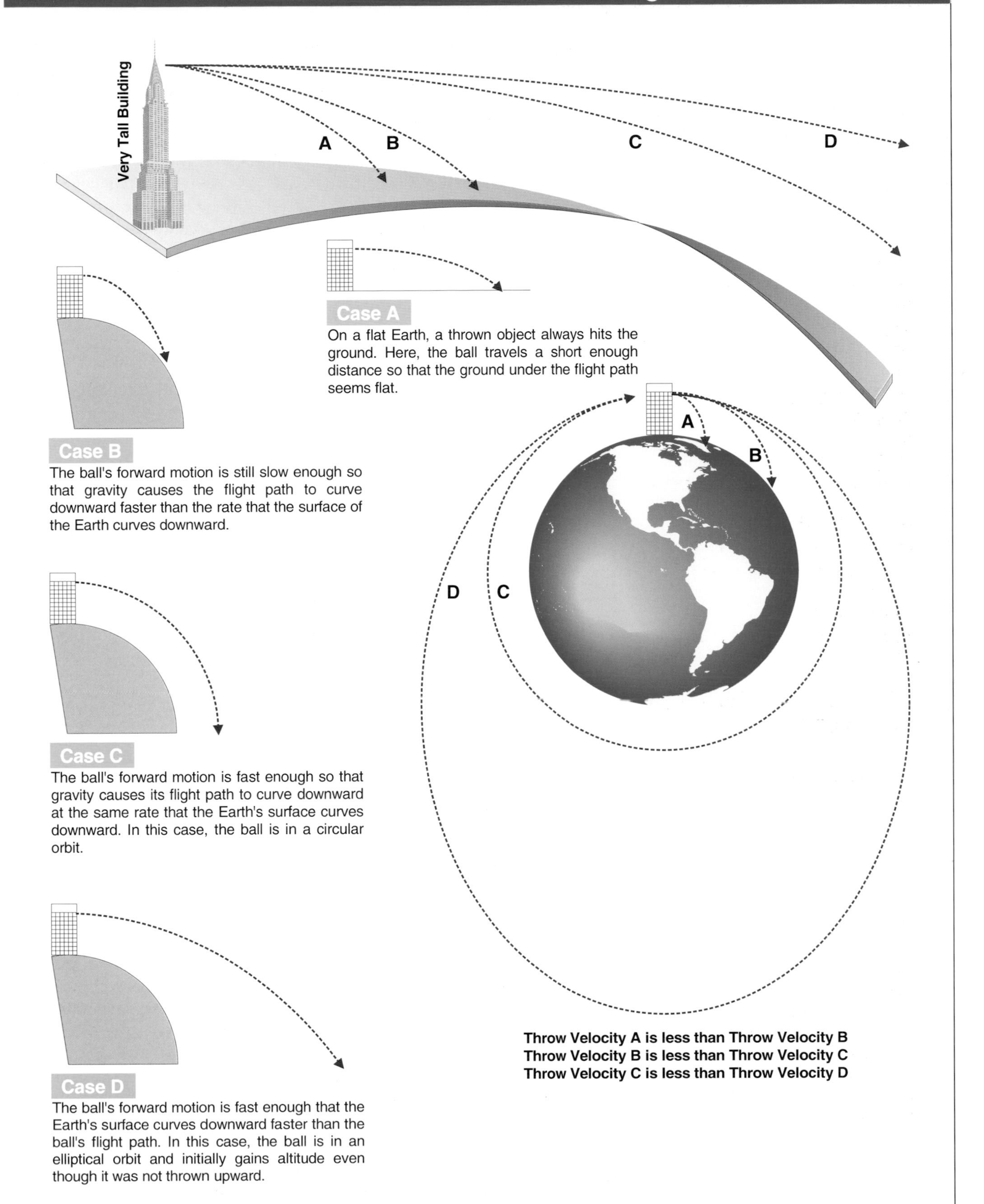

Orbital Velocities

Altitude	Orbital Velocity
Sea Level (impossible in reality)	7,905 m/s 28,459 km/hr 17,684 mi/hr
300 kilometers	7,726 m/s 28,239 km/hr 17,547 mi/hr
1,000 kilometers	7,350 m/s 26,460 km/hr 16,442 mi/hr
10,000 kilometers	4,933 m/s 17,760 km/hr 11035 mi/hr
100,000 kilometers	1,936 m/s 6,969 km/hr 4,330 mi/hr
384,400 kilometers (Moon's Orbit)	1,010 m/s 3,636 km/hr 2,259 mi/hr

Values assume circular Earth orbits
m/s = meters per second
km/hr = kilometers per hour
mi/hr = miles per hour

Scientific Fact!

Orbital velocity depends on the altitude of the orbit. Specifically, satellites orbiting at higher altitudes move slower than satellites orbiting at lower altitudes.

suddenly disappeared, an orbiting object would immediately start to fly away from the Earth in a straight line.

How fast do orbiting objects move? Consider the fact that the ability to sustain an orbit depends on fast enough forward motion along the curvature of the Earth's surface to keep the object from falling to the ground. Objects at higher altitudes experience a weaker gravitational pull than objects at lower altitudes. Therefore, an object at a higher altitude can travel forward at a rate slower than an object at a lower altitude and still maintain orbital motion. In other words, circular orbital velocity decreases with increasing altitude. For example, an object orbiting near the surface of the Earth must move at a velocity of 28,459 kilometers (17,684 miles) an hour, equivalent to 7,905 meters (25,936 feet) every second. Objects traveling at this velocity can circle the world in about 90 minutes. In contrast, our Moon orbits at an altitude of roughly 384,400 kilometers (238,885 miles), moves at a leisurely rate of about 3,636 kilometers (2,259 miles) per hour and takes about 28 days to circle the Earth. This relatively slow velocity still exceeds the standard cruising velocity of a passenger jet more than four times over.

Figure 1 on page 9 may give the mistaken impression that an object needs to start falling from the north pole to enter orbit. However, keep in mind that the top and bottom of a sphere come from arbitrary definitions. Map makers choose the north pole as the top of the Earth to create a reference to define north, south, east, and west. In reality, the Earth curves downward from every point along its surface. Therefore, any point can serve as the top, and an object can enter orbit from any point along the surface.

Elliptical Orbit

Objects in orbit around the Earth need not maintain a constant altitude. Consider the tall building example again. What happens if you now throw the ball in the forward direction with a velocity slightly faster than that required to maintain circular orbital velocity? In the time gravity pulls the ball downward by one meter, the ball travels forward far enough for the surface of the Earth to curve downward by one meter and then some. In other words, the surface of the Earth curves downward at a rate faster than the rate that the ball falls toward the ground. This phenomenon allows the ball to gain altitude even though it was not thrown upward.

As the ball falls around the world gaining altitude, gravity gradually slows the upward progress to the point where the ball reaches a maximum altitude halfway around the world from the throw point. Here, gravity wins out over the upward progress. The ball then begins to descend around the other half of the world while losing altitude in the process. Eventually, the ball returns to its original position with its original velocity and the entire process starts over again. This type of orbit takes the shape of an oval-like figure called an ellipse. The greater the increase in velocity over circular orbital velocity at the original throw point, the greater the difference between the original altitude and the maximum altitude halfway around the world on the elliptical orbit. Chapter 2 discusses elliptical motion in greater detail.

Generalized Concepts

Spaceflight engineers use the term *satellite* to describe any object in orbital motion. The dynamics that govern orbital motion remain the same no matter what body a satellite orbits (for example, the Sun, Earth, or Moon). In all cases, the satellite must move fast enough in the horizontal direction to avoid falling to the surface of the body being orbited. Some people refer to this body as the parent, or attracting, body. In general, for the same orbital altitude, a larger parent body requires a satellite to move with a faster velocity than a smaller parent body. This differential occurs because a larger parent body tends to pull satellites downward at a faster rate than a smaller parent body. As a result, satellites must move faster when in orbit around a larger parent body to avoid falling to the surface.

All satellites belong to two general categories. Natural satellites represent the class of objects not sent into space by humans. For example, the Moon is a natural satellite of the Earth, and the Earth is a natural satellite of the Sun. An artificial satellite or spacecraft, on the other hand, describes a machine sent into orbit by humans. Since 1957, humans have sent spacecraft to orbit the Earth, Sun, Moon, and other planets. This book uses the terms *spacecraft* and *artificial satellite* interchangeably.

Earth vs. Lunar Orbit

Altitude	Earth Orbit Velocity	Lunar Orbit Velocity
Surface	7,905 m/s	1,680 m/s
	28,459 km/hr	6,047 km/hr
	17,684 mi/hr	3,758 mi/hr
200 km	7,784 m/s	1,591 m/s
	28,023 km/hr	5,727 km/hr
	17,413 mi/hr	3,558 mi/hr
1,000 km	7,350 m/s	1,338 m/s
	26,460 km/hr	4,818 km/hr
	16,442 mi/hr	2,994 mi/hr

m/s = meters per second
km/hr = kilometers per hour
mi/hr = miles per hour

RULE OF THUMB

Although the Moon is in orbit around the Earth, it is possible to send a spacecraft to orbit the Moon. In general, it is possible to be in orbit around a body that is in orbit around another body. Another example is that the Moon orbits the Earth at the same time that the Earth orbits the Sun.

1.2 Boundary of Space

Major problems exist with the scenario of throwing a ball off a tall building to achieve Earth orbit. Obviously, a satellite orbiting at the altitude of a tall building stands a good chance of crashing into another tall building or mountain. In addition, a satellite also faces the problem of colliding with billions of tiny air molecules from the atmosphere that surrounds the surface of the Earth. The frequency and intensity of these collisions increases with increasing velocity. For example, compare the force of air on your face when you walk down the street with that when you stick your head out a car window while driving down the freeway. At orbital velocity, the energy from the collisions with air molecules produces more than enough heat to melt the strongest metals known to humans. Spaceflight engineers refer to this phenomenon as the friction heating effect.

Scientific Fact!

Any object that reaches orbital velocity within the Earth's atmosphere will be melted by the heat created as it collides with air molecules.

All Earth-orbiting satellites must orbit in space to avoid the frictional heating effect. Most spaceflight engineers consider space as the region above the Earth's atmosphere. However, no physical space-atmosphere boundary exists in reality because the atmosphere gradually thins with increasing altitude. In fact, traces of the gasses

In reality, there is no physical boundary where the atmosphere ends and space begins.

Above:

Hurricane Elena, with wind speeds exceeding 177 kilometers (110 miles) per hour, was photographed in the Gulf of Mexico on 1 September 1985 by astronauts on Space Shuttle *Discovery*. This oblique view from low Earth orbit gives an idea of the entirety of the storm and its features. Notice the spiral bands of thunderstorms leading to the center, or "eye," of the storm. Elena measured several hundred kilometers in diameter.

Historical Fact

The race to launch the world's first astronaut was won by the Soviet Union on 12 April 1961 when they launched Yuri Gagarin on a one-orbit mission. Alan Shepard followed Gagarin into space scarcely two weeks later on 5 May 1961 to become America's first astronaut. However, he did not reach orbit. Shepard rode a Redstone rocket to an altitude of 186 km and back during a trip that lasted for only 15 minutes. Incidently, Russians use the term *cosmonaut* instead of *astronaut*.

that we breathe exist beyond 161 kilometers (100 miles) above the surface of the Earth. At this altitude and above, the extreme thinness of the air allows satellites to move at orbital velocity without the danger of incineration from frictional heating. For comparison purposes, most passenger jets cruise at an average altitude of only 9 kilometers (about 30,000 feet), and the best military jets have a hard time climbing above 30 kilometers (about 100,000 feet).

Many different definitions exist as to where space really begins. For example, the United States awards astronaut status to any citizen who rides or flies in a plane or other vehicle above the altitude of 80.47 kilometers (50 miles). On the other hand, flight engineers who deal with American piloted space missions define space to begin at an altitude of 121.92 kilometers (400,000 feet or 75.76 miles). This altitude represents what some engineers call the *entry interface*. They consider objects above the interface altitude to be in "space," and objects below the interface altitude to be "in the atmosphere." In general, consider space as the region high enough above the Earth's surface that the atmospheric pressure drops to nearly zero. The range of altitudes that support human life from a breathing point of view, the range of altitudes at which planes fly, and the range of altitudes where weather patterns occur lie well below any semantic space-atmosphere boundary. For this reason, some people also refer to space as a vacuum. This term describes a region without air.

The lack of air prevents airplanes from flying in space for two major reasons. First and most important, planes remain airborne by using a phenomenon called *lift*. Planes fly at a speed fast enough so that air passing over the top of the specially shaped wings has a lower pressure than the air passing under the wings. This pressure differential lifts the plane into the air. The second reason is that both propeller and jet airplane engines require oxygen from the atmosphere to burn fuel.

Flight in the vacuum of space would not be possible without the existence of orbits. The reason is that an orbit allows a satellite to essentially circle the Earth without continuously expending fuel. Once a satellite achieves orbital velocity, gravity keeps it moving around the world. Airplanes, on the other hand, must continuously expend fuel to remain airborne. A plane with supply tanks to carry all of its fuel into space to sustain flight would weigh an outrageous amount, or would not possess the ability to remain aloft for very long. Engineers of the future may eventually devise methods to sustain flight in space without the use of orbits. However, this type of technology breakthrough is not likely to happen for quite some time to come.

1.3 Rocket Science

Another problem with throwing the ball off of a tall building to reach orbit is that nobody possesses the arm strength to accom-

plish this gargantuan feat. Placing a satellite in an orbit around the Earth requires lifting the satellite at least seventeen times higher than the altitude limit of the highest climbing plane, and accelerating the satellite to a velocity at least six times faster than the fastest flying jet. Currently, only rockets powered by rocket engines provide this amount of energy. Understanding how rockets work first requires understanding three basic principles postulated by a seventeenth-century British-mathematician and physicist named Isaac Newton. His three principles, known as Newton's Laws, provide the basis for understanding the motion and dynamics of a rocket.

Many consider Newton as one of the most brilliant scientists in history. Some of his greatest accomplishments include the invention of the mathematical field of calculus, theories of gravity and universal gravitation, and the laws of forces and motion. Although he conceived much of his work by the age of twenty-four in the year 1666, Newton did not publish the work until 1687. That publication, *Philosophiae Naturalis Principia Mathematica,* is known today simply as *Principia* and is regarded as the greatest science book ever written. Although Newton derived the theories appearing in *Principia* over 300 years before the beginning of the space age, they still serve as the foundation of modern spaceflight mechanics, more popularly known as rocket science.

Forces and Mass

A few important concepts merit discussion in order to facilitate the explanation of Newton's Laws. The first concept deals with invisible pulling and pushing actions called forces. Some common examples of forces include the force of gravity that pulls objects towards the surface of the Earth, the force of a person's arms and hands that push a grocery cart through a store, and the force that a rigid brick wall provides to stop a runaway car. In some instances, the source of the force (for example, hands pushing a cart) can be seen. However, the actual energy that causes the pushing or pulling action always remains invisible. Forces also tend to set objects into motion, change the direction of motion, or stop objects from moving. The degree to which a force affects an object's motion depends on a physical property of the object called the mass. This term describes the total quantity of matter (all of the molecules and atoms) that makes up an object.

Most people use the terms *weight* and *mass* interchangeably in everyday speech. In reality, weight describes the extent of the gravitational force on an object. The strength of Earth's gravity near the surface measures 32.2 pounds for every unit of mass in an object. This force gradually decreases with increasing altitude. For example, a young child with 3.115 American units of mass (called a slug) weighs 100 pounds at sea level, but only 99.8 pounds at the top of Mount Everest. In *Principia,* Newton showed that the strength of an object's gravity decreases as an inverse square to the distance away from the center of the body's mass. In plain English, that means

Above:
In the weightless condition experienced in orbit, astronauts have more choices in selecting areas to eat. Here, Dr. Sally Ride contemplates eating her dinner while upside down.

Strength of Earth's Gravity

Altitude Above Sea Level	Weight of 75 kg Object
0 km	100% of that at sea level (165.35 pounds)
10 km	99.69% of that at sea level (164.84 pounds)
100 km	96.94% of that at sea level (160.29 pounds)
1,000 km	74.73% of that at sea level (123.57 pounds)
10,000 km	15.17% of that at sea level (25.08 pounds)
343,400 km (*)	0% of that at sea level (0.0 pounds)

(*) Gravity is zero here because the object is at a point between Earth and the Moon where the gravity from the two pulls in equal strength, but in opposite directions. Since Earth is much larger, this point is much closer to the Moon than to the Earth.

Above:

Bloom County astronauts Opus and Steve Dallas illustrate the difficulties of working in the Space Shuttle while in a weightless environment.

Drawn by Berke Breathed

doubling an object's distance from the center of the Earth will decrease the gravitational force that the Earth pulls on it by four times, while tripling the distance will decrease the gravitational force by nine times.

In the metric system, the unit called the kilogram measures mass, and the unit Newton measures weight and force. One kilogram of mass weighs about 2.2 pounds and experiences a gravitational force of about 10 Newtons (in reality, 9.81 Newtons) near the surface of the Earth. Although technically incorrect, most people commonly associate the term *kilogram* with weight. In everyday speech, when an object is said to weigh 10 kilograms, the true implication is that the mass of the object is 10 kilograms. The actual weight will vary depending on the altitude at which the object is weighed.

Incidently, this decrease in weight with increase in altitude does not explain why astronauts appear to float weightless in space. A piloted space mission typically employs an orbit with an altitude of 296 kilometers (184 miles). At this altitude, the Earth pulls on astronauts with a force of only 5% less than that experienced on the ground. We (and other objects), on the ground, experience a sensation of weight only because the force of gravity continually pulls us onto the floor or ground. Spacecraft and astronauts in orbit remain in a continuous state of free fall as they fall all the way around the world. Because the astronauts and the floor of the spacecraft fall at the same rate, they do not feel as if something were pulling them onto the floor. As a result, they experience no weight. This feeling is the same sensation that people experience just as elevators begin to drop downward, or what divers feel when plunging off of diving boards and into pools. An important concept to remember is that objects in orbit may be weightless, but they certainly are not massless.

Scientific Fact!

It is a common misconception that astronauts in orbit are weightless because there is no gravity in space. In fact, if a person stood on top of a hypothetical mountain the height of an average Space Shuttle orbit (about 300 km), they would weigh close to what they would weigh at sea level. The famous "zero gravity" effect comes from the fact that objects in orbit are in a state of free fall.

RULE OF THUMB

Inertia is a property that describes an object's resistance to being set in motion, or to change in velocity once set in motion. Larger objects tend to have more inertia than smaller objects.

Newton's Laws and Rocket Thrust

Newton's three laws deal with forces and what happens to objects when acted upon by forces. The first of the three laws deals with a property called inertia. Simply stated, an object in motion tends to stay in motion in a straight line unless acted upon by an external force, and an object at rest tends to stay at rest unless acted upon by an external force. For example, a spacecraft at rest does not

begin to move spontaneously. Something must first provide a push or a pull. Conversely, a spacecraft in motion does not spontaneously stop, slow down, speed up, or change directions. Again, something must first provide a push or a pull.

The first law explains why a satellite in orbit needs no fuel to maintain orbital velocity while an airplane must expend fuel on a continuous basis to remain airborne. Fuel allows airplane engines to continuously provide a force to overcome the external force of the air resistance that continuously acts to slow a plane. Satellites in the vacuum of space experience very little slowdown from air resistance. As a result, they need not expend fuel on a continuous basis after achieving orbital velocity. The first law also explains why satellites expend no fuel to move in a circular orbital path. Gravity provides the force that causes a satellite to follow the curvature of the ground instead of flying off in a straight line away from the Earth.

Newton's second law states that forces tend to cause objects to accelerate (change velocity). Stronger forces cause objects to accelerate (speed up or slow down) at faster rates than weaker forces. In addition, the second law also states that a more massive object takes more force to accelerate than a less massive object. This law, when expressed in the form of an equation, allows spaceflight engineers to calculate the amount of force required to change the motion of satellites.

The third law provides the basic principle that governs rocket motion. This law states that for every action, there is always an equal and opposite reaction. In other words, an object cannot push something without being pushed in return. Rockets move by forcing or pushing extremely hot gasses at a high speed out the back of a nozzle. In turn, the gasses push back on the rocket in the opposite direction. Thus, a rocket moves by expelling hot gas in the direction opposite that of the desired motion. Spaceflight engineers use the term *thrust* to describe the forward force generated by the escaping gasses.

Contrary to a popular belief, rockets do not obtain forward motion by pushing against the outside air. In reality, air outside a

Scientific Fact!

For every action, there is an equal and opposite reaction. That means that you cannot push something without being pushed in return.

RULE OF THUMB

It is a common misconception that rockets work by pushing against the air. In fact, they work better in space where there is no air.

Below:

Bloom County astronauts Opus and Milo watch in shock as astronaut Steve Dallas foolishly demonstrates how Newton's third law works.

Drawn by Berke Breathed

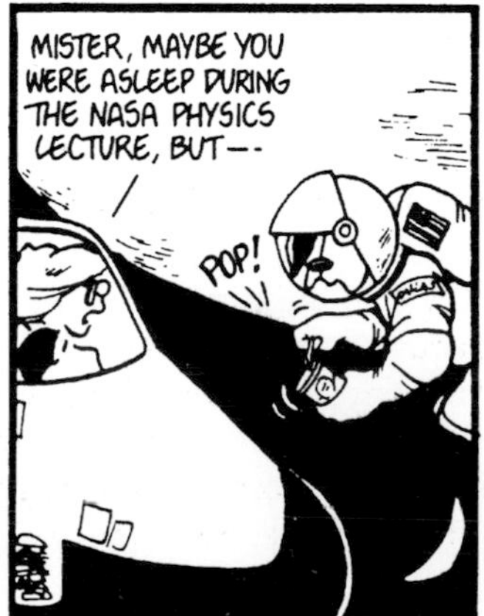

Above:
This view of Lake Poopo, Bolivia, was taken by astronauts on Space Shuttle *Discovery* on 13 September 1991. Extensive dry seasonal burning in the Amazon basin produces a thick haze as seen near the top of the photograph. The clarity difference in the picture is caused by the Andes Mountains (bottom half of picture) extending above the haze and into clean air. The view is looking northeast into Brazil.

Scientific Fact!

Thrust is a measure of how much force a rocket engine produces. Heavy rockets always take more thrust to accelerate from the launch pad than lighter rockets.

RULE OF THUMB

Propellant is a term that collectively describes a rocket's fuel and oxidizer. Almost all of a rocket's weight before launch can be attributed to the propellant. On the Space Shuttle, the propellant accounts for over 80% of the vehicle's weight at the time of launch.

rocket nozzle hinders the forward motion of the rocket because the escaping gasses tend to slow down after colliding with air molecules. This effect retards the amount of force that escaping gasses impart to the rocket, thereby slowing the rocket's forward progress. Although rockets work both in the atmosphere and in space, they operate more efficiently in space.

How a Simple Rocket Works

A simple rocket consists of a tall, skinny cylindrical body constructed out of relatively thin metal (see Figure 2 on page 17). The fuel and fuel supply tanks for the rocket engines sit inside the body of the cylinder while the engines that provide the thrust to propel the rocket sit at the bottom of the cylinder. Each engine looks like a bell-shaped nozzle and contains a device that injects raw rocket fuel into an area located at the top of the nozzle called the combustion chamber. Here, the fuel burns and turns into hot gas that tends to expand in all directions. The nozzle then channels the expansion of the hot gas in a direction opposite the desired direction of motion. Normally, the nozzles point directly along an imaginary line that runs through the middle of the rocket from the top to the bottom. However, most rockets have the ability to point their nozzles in a direction up to several degrees away from the imaginary line. This action, called gimbaling, provides the rocket with a limited amount of steering capability during flight.

A hollow streamlined cone attaches to the top of the cylindrical body with the base of the cone attached to the body and the tip of the cone facing upward. The tall cylindrical-shaped body configuration topped with a cone minimizes the cross-sectional area of the rocket that pushes against the air. This reduction in cross-sectional area reduces the amount of energy that the rocket must expend in pushing air out of the way. Typically, the satellite, or other payload destined for orbit, sits on top of the rocket inside the nose cone (some people use the terms *payload shroud* or *payload fairing* instead of *nose cone*). During the first few minutes of flight, the cone protects the payload from damage from the increasing wind that begins to pass by as the rocket accelerates upward through the lower parts of the Earth's atmosphere.

A rocket's ability to lift its payload into space depends on its ability to overcome the gravitational forces that tend to pull the rocket back toward the ground. Remember that gravitational forces take the form of weight. Thus, the rocket engines must produce enough force (thrust) to overcome the weight of the rocket body, engines, fuel, and payload. For example, suppose that a rocket with a total mass of 200,000 kilograms (440,925 pounds) sits on the ground. The engines must produce a total of 2,000,000 Newtons (440,925 pounds) of thrust just to negate the gravitational force. Any additional thrust that the rocket produces goes into accelerating (speeding up) the rocket towards orbital velocity. The more the excess thrust, the faster the rocket accelerates.

Figure 2: *A Look Inside a Simple Rocket*

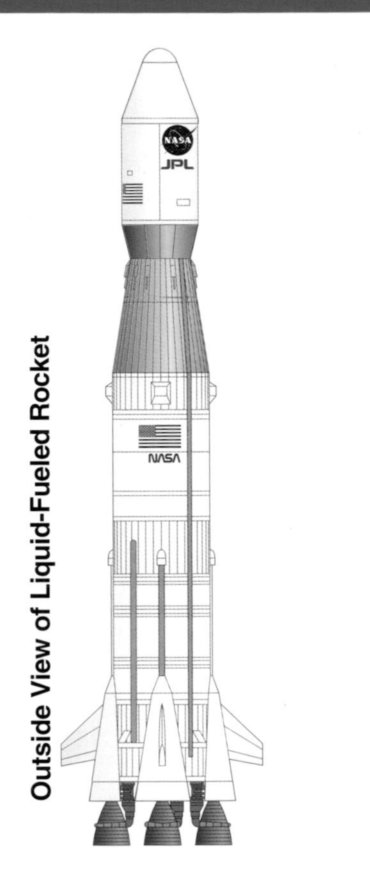

Inside a Simple Liquid-Fueled Rocket

Payload Shroud
Satellite to be launched into space sits in this canister (nose cone). It protects the satellite from aerodynamic forces during the first few minutes of the flight.

Oxidizer Tank
Contains an oxygen-rich liquid that supplies the oxygen so that the rocket engine can support a burning reaction. Remember that all chemical reactions involving burning require oxygen.

Fuel Tank
Contains the actual chemical liquid that the rocket engines burn to produce thrust.

Engine Combustion Chambers
The oxidizer and fuel mix here in a violent burning reaction. The net product is hot exhaust gas that tends to expand in all directions.

Exhaust Nozzles
Channels the exhaust gas in the direction opposite that of the desired direction of motion. At launch, they channel the gas downward to force the rocket upward and into the air.

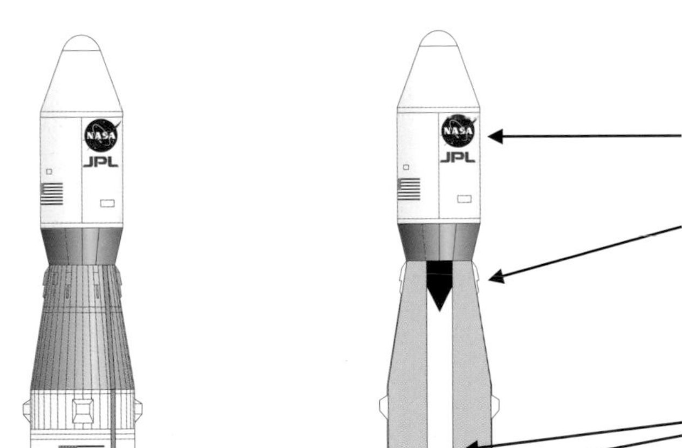

Inside a Simple Solid-Fueled Rocket

Payload Shroud
Same as for the liquid-fueled rocket.

Igniter
Shoots a long pillar of fire down the clearing in the middle of the rocket body at the time of lift-off. This ignites the solid propellant and starts the burning reaction. It is the black object in the middle of the rocket near the top.

Solid Propellant Mold
Flammable putty-like material that is both the oxidizer and fuel mixed into one substance. The mold runs down the entire length of the rocket body.

Combustion Chamber
A hollow clearing in the middle of the mold where the burning takes place. The mold burns from inside out. With increasing time, the burning increases the cross-sectional area of the clearing and brings more propellant into contact with the burning process. This increases the thrust that the rocket produces.

Cross section of propellant mold shown with increasing time from left to right

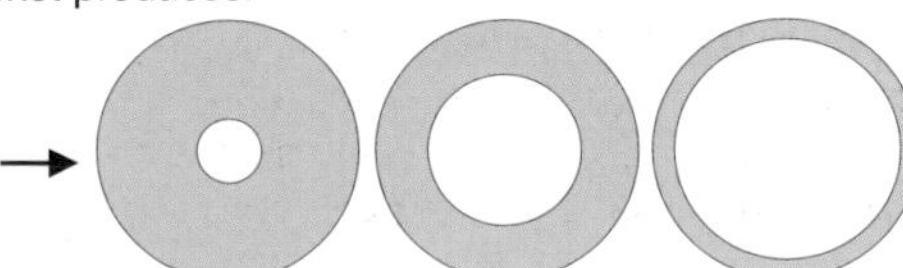

RULE OF THUMB

The term *payload* describes the contents that a rocket delivers into space, such as satellites, astronauts, or both. It does not include the rocket itself, or the rocket's propellant.

ISP stands for *specific impulse*. This number rates the relative efficiency of a rocket's propellant. It works analogously to the octane rating system of automobile gasoline – the higher the number, the more efficient the propellant.

Scientific Fact!

ISP measures how efficiently the propellant burns, not how much thrust it produces. For example, it is possible for a rocket to burn a propellant with a low ISP rating and still produce more thrust than a rocket that burns a propellant with a high ISP rating. In general, a lower ISP rating means that the rocket will need to carry more propellant. By today's standards, an ISP rating of over 450 is excellent, 400 to 450 is very good, 350 to 400 is good, and 250 to 350 is acceptable. In the future, exotic types of propulsion, such as nuclear electric, nuclear thermal, or ion, may come into use. These types have ISP ratings of over 1000.

In spaceflight jargon, many of the terms relating to rocket engines and their operation tend to center around the fact that rocket engines must burn fuel to provide thrust. For example, the term *ignition* refers to the moment when a rocket engine activates, and the verb *fire* describes the action of an engine in the process of thrusting. In addition, the terms *rocket burn* or an *engine burn* also refer to a rocket engine in the process of thrusting. The *burn time* measures how long a rocket engine thrusts before it shuts off.

Propellants and Specific Impulse

Rocket engines typically use two different types of chemicals at the same time to support the burning reaction that produces the hot exhaust gasses. Together, these chemicals form what rocket scientists refer to as the propellant. The first chemical, called the fuel, provides the substance that the engine burns to create the hot exhaust gas. The second chemical, called the oxidizer, supplies the oxygen for the burning reaction. Rocket engines require this dual-chemical scheme because all burning reactions require a substance to burn and oxygen to sustain the burn. Airplane and automobile engines get oxygen to support the combustion of fuel from the Earth's atmosphere. Rockets, on the other hand, cannot depend on the atmosphere to supply oxygen because they operate both in the atmosphere and in space. Consequently, they must carry their own oxygen supply to support the burning reaction in the combustion chamber of the engine.

Many different types of propellant combinations exist. A system called the specific impulse (ISP) rating system provides a method to gauge how much thrust a propellant produces when it burns. The rating number, measured in seconds, depends on various factors, such as the velocity with which the exhaust gasses from the combustion leave the rocket nozzle, and how much internal energy is stored in the oxidizer and fuel. Propellants with higher ISP ratings burn more efficiently than propellants with lower ratings. Think of the ISP number as the time that one pound of propellant will burn while producing one pound of thrust. For example, one pound of propellant with a relatively high ISP rating of 400 can produce one pound of thrust for 400 seconds. Remember that rocket engines operate less efficiently in the atmosphere than in a vacuum because air molecules hinder the escape of the exhaust gasses. As a result of reduced exhaust velocity, a propellant's ISP always rates lower in the atmosphere than in a vacuum.

Liquid Propellants

Liquid chemicals represent one of the two major classes of rocket propellants. One of the most potent combinations available for propulsion involves using a combination of liquid hydrogen (LH_2) as the fuel and liquid oxygen (LOX) as the oxidizer. This combination results in an ISP of 400 seconds at sea level and 453 seconds in a vac-

uum. Unfortunately, LH_2 and LOX fall into the cryogenic (super cold) category of liquids. They require special handling, super chilled tanks for storage, and temperatures near absolute zero (about 200 degrees below zero) to remain in liquid form. At temperatures above near absolute zero, these liquids tend to boil into gaseous form, which makes them useless as propellants. Other types of liquid propellants do not require storage at cryogenic temperatures. One common combination burns a fuel called Aerozine 50 (similar to kerosene) with a nitrogen tetroxide (N_2O_4) oxidizer. Although this combination allows for storage at normal room temperatures and normal handling procedures as compared to cryogenics, the ISP rates at a mere 254 seconds at sea level and 320 seconds in the vacuum of space.

Most liquid propellants require a spark to start the combustion process. Some types, called hypergolic propellants, burn spontaneously upon mixture of the fuel and oxidizer. Engines that employ hypergolics do not require an ignition system and function more reliably than engines with an igniter. Unfortunately, hypergolic propellants corrode almost everything they come into contact with and are extremely poisonous. Many satellites use an extremely reliable hypergolic combination of hydrazine fuel and nitrogen tetroxide oxidizer to produce thrust for minor space maneuvers.

Solid Propellants

Solid chemicals represent the second major class of rocket propellants. This type consists of a flammable putty or rubber-like material that contains a mixture of both the fuel and the oxidizer. Rockets that burn solid propellants, also called solid rockets or solid motors, look like the simple cylindrical-shaped rocket body presented earlier. However, instead of propellant tanks, solid propellant putty fills the inside of the rocket body from top to bottom. A narrow cylindrical-shaped clearing that lies in the middle of the putty mold runs from the top to the bottom of the rocket, acts as the combustion chamber, and allows the putty to burn evenly from top to bottom. A nozzle attached to the bottom of the rocket channels the gasses from the combustion in the proper direction.

An igniter located inside the rocket at the top shoots a pillar of flame down the entire cylindrical-shaped clearing to start the burning process. The flame only ignites the portions of the putty in contact with the cylindrical-shaped clearing. As the putty in contact with the clearing burns away, the diameter of the cylindrical clearing gradually enlarges in a process analogous to what happens when a person eats away at a doughnut from the inside out. This increase in the diameter of the clearing allows more putty to burn at any one time. The amount of thrust that a solid rocket produces at any given instant in time directly depends on the amount of putty burning at that time. Therefore, a simple solid rocket with a cylindrical-shaped clearing in the middle of the propellant mold produces relatively little thrust at first, gradually produces more thrust with increasing time as the

Comparison of Propellants

Propellant Type	ISP Rating
Liquid Hydrogen (fuel) **Liquid Oxygen** (oxidizer) Used on the Space Shuttle Expensive, difficult to store	453 seconds
Aerozine 50 (fuel) **Nitrogen Tetroxide** (oxidizer) Used on medium sized-rockets Moderate cost, easier to store	320 seconds
RP-1 (fuel) **Liquid Oxygen** (oxidizer) Often used on first stages Moderate cost, difficult to store	303 seconds
Hydrazine (fuel) **Nitrogen Tetroxide** (oxidizer) Hypergolic used on satellites Relatively cheap, very corrosive	300 seconds (average)
Gunpowder Included for comparison Not used in spaceflight	100 seconds

All propellants listed here are liquids, except for gunpowder. The ISP ratings for the liquids represent ratings in the vacuum of space. ISP ratings in the atmosphere will be slightly lower. Gunpowder rating is at sea level.

RULE OF THUMB

Hypergolic is a term that describes two chemicals that spontaneously ignite upon mixing. In other words, no spark is needed to begin the burning reaction.

Historical Fact

The ancient Chinese built the world's first solid rockets thousands of years ago. These rockets were similar to fireworks of today and used gunpowder for propellant.

Rockets From Scratch

The world changed forever on a cold day in Auburn, Massachusetts in March 1926. It was then that American rocket pioneer Dr. Robert Goddard unofficially inaugurated the space age by launching the world's first mechanical rocket. Although the ancient Chinese had launched simple gunpowder-fueled rockets resembling fireworks thousands of years before Goddard's time, his design represented the first liquid-propelled rocket in history. Unfortunately, nobody took notice of Goddard's accomplishment as the tiny rocket managed a meager 2.5-second, 56-meter (184-foot) climb over his Aunt Effie Ward's snow-covered farm. Little did Goddard know that the picture he posed for in front of his rocket before the flight would become one of the most famous in aviation history (see photograph to the right).

Goddard's triumph came only after years of frustration and much undeserved public humiliation. Seven years prior, he had published a scientific paper called *A Method of Reaching Extreme Altitudes* in which he described the results of his extensive scientific research on rockets. At the end of the article, he suggested the possibility of eventually landing a rocket on the Moon as a demonstration of the possibilities of spaceflight. Goddard was initially against publishing the paper and hoped that few would read it. He enjoyed no such luck as the paper's publishing sponsor, the Smithsonian, circulated a press release hoping to stimulate funding for future rocket research.

Immediately, headlines began to appear in newspapers across the country announcing Goddard's Moon suggestion. The majority of the articles indicated that the editors had ignored the scientific content of Goddard's paper and concentrated on lampooning him instead. For example, the *San Francisco Examiner* reported, "SAVANT INVENTS ROCKET WHICH WILL HIT THE MOON," and the *Boston American* printed, "MODERN JULES VERNE INVENTS ROCKET TO REACH MOON." Many adventurous souls who read the articles flooded Goddard with requests to take part on the lunar journey. One of them, Claude Collins from Pennsylvania, expressed an interest in flying to the Moon and then on to Mars, provided that Goddard supply a $10,000 life insurance policy.

The New York Times failed to share in Collins' enthusiasm. In perhaps one of the great journalistic blunders of the century, they printed a harsh editorial criticizing the scientific competence of Goddard. Parts of it read, "That Professor Goddard does not know the relation of action to reaction, and the need to have something better than a vacuum (of space) with which to react...Of course he only seems to lack the knowledge ladled out daily in high schools." Of course, it was the writer who failed to understand that rockets do not work by pushing against the air. Much to their credit, the *Times* finally retracted their editorial forty-three years later after NASA landed astronauts on the Moon.

The blistering treatment from the press devastated Goddard. Although he pressed on with his rocket research, he also avoided all publicity regarding the experiments. Most of the country soon forgot about Dr. Goddard and his pioneering rocket work. However, Goddard continued to build rockets after the 1926 first flight. In the 1930s, he launched rockets that reached altitudes of almost two kilometers and velocities close to the speed of sound.

In the ultimate tribute to Dr. Goddard and his accomplishments, NASA named one of their research centers after him in 1959. Today, the NASA Goddard Space Flight Center in Greenbelt, Maryland, is one of the world's leading institutions in using satellite data to study the changing global environment, and also NASA's lead center in tracking Earth orbiting satellites.

While Goddard continued his experiments in the 1930s, another rocket scientist who would eventually play a crucial role in the American space program was beginning his career across the Atlantic Ocean in Germany. His name was Wernher von Braun. During his childhood, von Braun developed a keen interest in rockets to

take people to the Moon, largely due to the influence of his mother, an accomplished amateur astronomer. As a young man of twenty years, von Braun lived in dilapidated accommodations on an abandoned munitions dump so he could live rent free and gather scrap parts to build rockets.

In 1932, General Walter Dornberger from the German Army visited the munitions dump, watched the youthful von Braun and his colleagues test their small rockets, and subsequently awarded them a contract to build demonstration rockets for the Army. However, the young scientist and his associates detested working for the military. Von Braun subsequently rationalized that building rockets for the Army provided a way for the team of scientists to proceed with serious space exploration efforts. With the Army paying for the research, the days of begging for scrap parts and funds were over.

Unfortunately, von Braun's worst fears materialized in the late 1930s when Hitler and the Nazis came to power. Hitler largely ignored the rocket research at first. Then, toward the end of World War II, with Germany losing badly, Hitler ordered von Braun to convert his experimental rockets into weapons capable of delivering bombs to England. The new weapon was called V-2 (Vengeance Weapon 2). Fortunately, the V-2 offensive arrived too late to influence the course of the war. One of Hitler's failed dreams was to send a rocket armed with an atomic bomb to destroy New York City.

Von Braun was no Nazi, despite the "black mark" on his résumé. He despised Hitler's dreaded SS police force and was even arrested several times during his "war career." SS commander in chief Heinrich Himmler accused von Braun of not being interested in creating weapons of war and having an objective of using Army money for space travel research. Only an impassioned plea from General Dornberger saved the rocket scientist's life.

In May 1945, shortly before Germany's surrender, von Braun and his team of over 100 engineers and rocket scientists escaped the persecution of the Nazis and surrendered to Allied forces. They took with them to America enough spare parts to build more than 100 V-2 rockets for testing. Eventually, the United States government recruited the brilliant team of German scientists to work for NASA, where their genius was instrumental in launching America's first satellite, *Explorer 1*.

In the picture to the left, von Braun (right) is shown triumphantly holding a full-scale model of *Explorer 1* with the two other individuals responsible for the satellite's success: scientist James Van Allen (middle) and William Pickering (left) of NASA's Jet Propulsion Laboratory. The photograph was taken shortly after the launch of *Explorer 1* on 31 January 1958, the day America officially entered the space age. Later, von Braun went on to develop the gigantic *Saturn 5* rocket that propelled the first astronauts to the Moon.

Many people, using ethical standards of today, questioned America's decision to let von Braun work for NASA because he had built V-2 rockets for Hitler using slave labor imported from Nazi prisoner of war and concentration camps. Von Braun probably never had a choice in the matter. The brilliant rocket scientist would have certainly been executed as a traitor to the Nazis if he had not cooperated. Years after the war, von Braun would tell listeners about the Nazi's appalling misuse of his "new technology." During the war, he and fellow rocket scientist Kurt Debus, eventual director of NASA's Kennedy Space Center, often spent hours wondering if the day would ever come when rockets would be put to "proper use." ❑

Profile of the Delta 2 Rocket

Characteristic	Value
Number of Stages	4 stages, no crew 0th stage (solid) 1st stage (liquid) 2nd stage (liquid) 3rd stage (solid)
Auxiliary Boosters	9 small solid rockets attach to side of 1st stage to form stage 0
Height at Liftoff	37.6 meters (123.4 feet)
Mass at Liftoff	231,879 kilograms (511,190 pounds)
Maximum Payload	
Low Earth Orbit	5,225 kilograms
Communications Orbit	1,800 kilograms
Moon	1,330 kilograms
Mars	1,070 kilograms
Manufacturer	McDonnell Douglas, Huntington Beach, CA

(*) Boeing has since bought the Delta manufacturing rights from McDonnell Douglas. (Photograph provided by McDonnell Douglas Space Systems Company in Huntington Beach, CA)

diameter of the clearing increases, and produces a lot of thrust at the end of the burn (see bottom of Figure 2 on page 17).

Some solid rockets use different shaped clearings in the middle of the propellant mold. One common variation utilizes a many-sided star-shaped clearing. In contrast to the cylindrical clearing configuration, the star configuration produces a lot of thrust at the beginning of the burn because more putty initially comes into contact with the combustion process. Some star configurations produce an increasing amount of thrust with increasing time, while others star shapes produce a constant thrust level throughout the burn. In all cases, star configurations tend to burn out quicker than the standard cylindrical types.

The biggest advantages of solid propellants over their liquid counterparts are that solids store indefinitely at room temperature, handle easily compared to hydrazine and nitrogen tetroxide, and the engines require almost no moving parts. In contrast, engines that burn liquids require many sophisticated pumps. However, a significant drawback exists. Once a solid rocket ignites, the rocket must thrust until all the putty burns out. All liquid engines can be shut off before the propellant runs out in case of an emergency, and some types of liquid engines possess the capability for multiple shut-offs and restarts during a flight.

Specific impulse ratings for solid rockets tend to lag behind their liquid burning counterparts. A typical solid propellant in use today utilizes a mixture of ammonium perchlorate fuel and aluminum polymer oxidizer, producing ISP values of about 266 seconds. Some solid propellants rate higher, up to 300 seconds. Despite these relatively low ratings, a large enough solid rocket can produce more thrust than many liquid burning engines.

Staging

Staging provides a method of increasing the efficiency of a rocket. A serially staged rocket consists of several (two or more) cylindrical-shaped rocket bodies stacked on top of each other. The payload typically sits in a protective nose cone at the top of the rocket stack. Each body, called a stage, contains its own propellant, propellant tanks, rocket engines, instrumentation, and structure to hold everything together (see Figure 3 on page 23). At the beginning of the flight, only the bottom (first) stage pushes the entire rocket stack. After the propellant in the first stage runs out, small explosive charges disintegrate the joints that hold the first stage to the next (second) stage in the stack. The first stage then drops off, allowing the second stage's rocket engines to take over the job of pushing what remains of the rocket stack. This process continues until the top stage's engines shut down and the satellite is placed in orbit. In a staged rocket, the first stage always contains much more propellant and uses more powerful engines than the other stages. The reason is that the first stage must lift itself, the other stages, and the payload

Figure 3: *What a Staged Rocket Looks Like*

Most rockets in use today employ a technique called staging. This design method breaks up the entire rocket into two or three smaller rockets stacked on top of each other. When a stage runs out of propellant, it drops off, and the stage above it takes over the job of pushing the rocket. Staging increases the efficiency of the entire rocket by throwing away parts of the rocket that contain empty propellant tanks (dead weight).

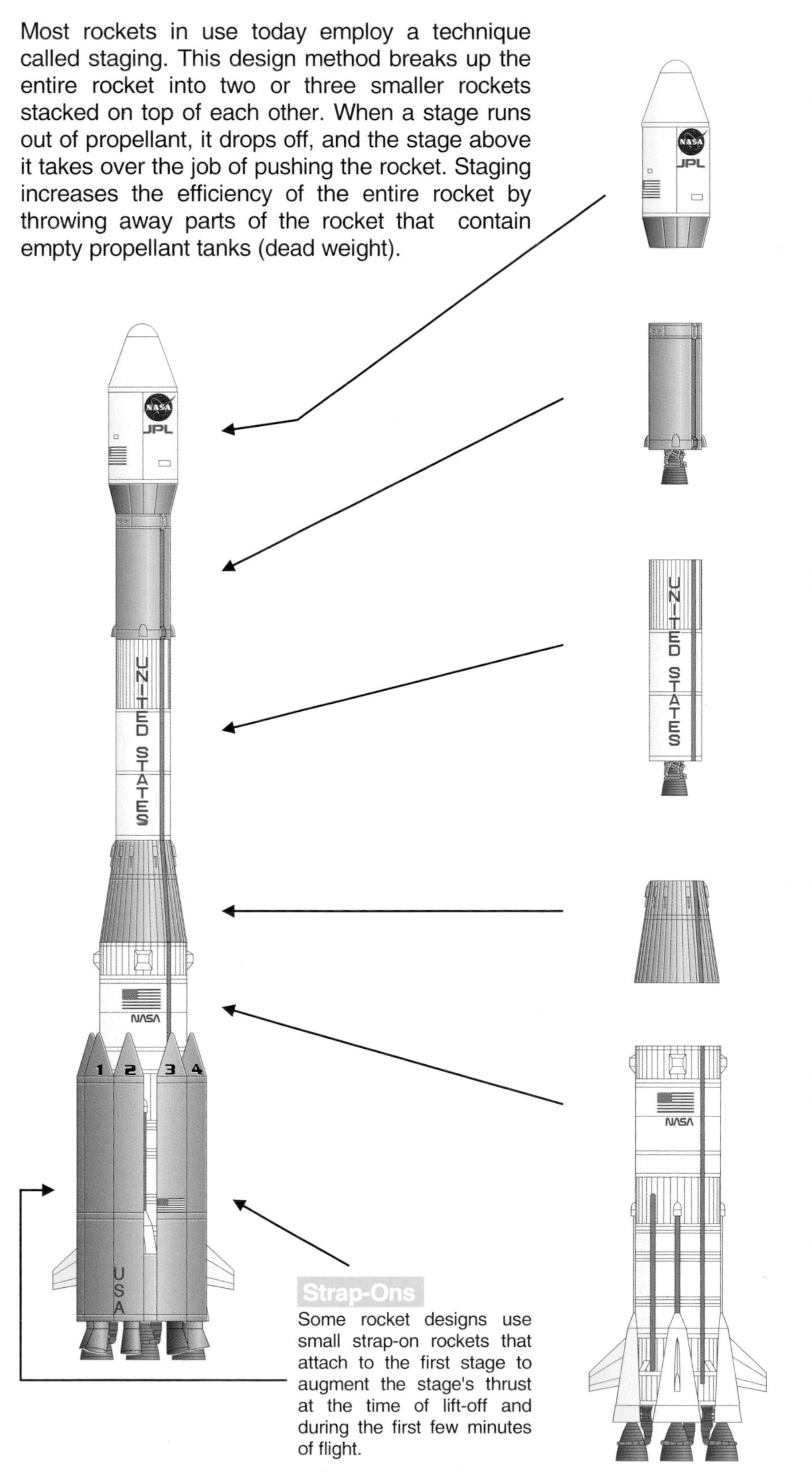

Payload Shroud

Satellite to be launched sits in this canister. This object protects the satellite from aerodynamic forces during the first few minutes of the flight. Some people also call it the payload fairing or the protective shroud or nose cone.

Third Stage

Usually has the job of placing the satellite into its final orbit around the Earth after the first two stages place the rocket into a preliminary orbit around the Earth.

Second Stage

Has the job of pushing the third stage and the satellite into low Earth orbit after the first stage's propellant runs out. It always has smaller engines than the first stage because it has less mass to push than the first stage.

Protective Skirt

This object acts as a cover for the second stage's engine when the first stage is still thrusting. Some rockets also have a skirt between the second and third stages.

First Stage

Always the largest, as well as the most powerful stage on the entire rocket. It has the job of pushing all of the stages above it from the ground and up past the densest parts of the lower atmosphere. This stage almost always contains more than half of the rocket's entire supply of propellant. On most rockets, all of the stages plus the payload shroud stacked together measure at least 31 meters (over 100 feet) tall.

Strap-Ons

Some rocket designs use small strap-on rockets that attach to the first stage to augment the stage's thrust at the time of lift-off and during the first few minutes of flight.

Profile of the Atlas 2AS Rocket

Characteristic	Value
Number of Stages	3 stages, no crew 0th stage (solid) 1st stage (liquid) 2nd stage (liquid)
Auxiliary Boosters	4 small solid rockets attach to side of 1st stage to form stage 0
Height at Liftoff	47.5 meters (155.8 feet)
Mass at Liftoff	234,000 kilograms (515,882 pounds)
Maximum Payload	
Low Earth Orbit	8,600 kilograms
Communications Orbit	3,830 kilograms
Moon	2,800 kilograms
Mars	2,270 kilograms
Manufacturer (*)	General Dynamics, San Diego, CA

(*) Lockheed Martin Corporation bought the Atlas manufacturing rights from General Dynamics in 1994 and moved production to Denver, CO.

(Photograph provided by General Dynamics Corporation's Space Systems Division in San Diego, CA. The photo is of the Atlas 2A. It is the same as the 2AS, but without the strap-on solid rockets)

into the air. In many cases, the first stage accounts for over 50% of a rocket's liftoff mass.

The advantage of staging is that once a portion of the rocket's total propellant supply runs out, the propellant tanks and structure that supported the expended propellant drop off. This technique reduces the overall propellant usage because staged rockets do not accelerate portions of the structure that do not serve a useful purpose. In theory, the performance of a rocket increases as the number of stages increases. However, increasing the number of stages past three or four decreases performance because each stage contains "overhead" that allows the stage to function as a separate entity capable of lifting the entire rocket stack. For example, each stage must possess its own engines and body structure, which complicates the operating procedures and the design of the rocket. Also, separating the spent stage and starting the engines in the stage above is not an easy process.

Many rockets today employ a technique called parallel staging in addition to serial staging. This method works by attaching one or more smaller auxiliary rockets to the bottom (first) stage. These smaller rockets, sometimes called strap-ons, or stage zero, typically thrust in unison with the engines from the first stage to produce additional thrust that helps the rocket accelerate during the first few minutes of flight. Once these auxiliary rockets expend all of their fuel, explosive bolts separate them from the first stage. A popular combination employs two or three serial stages that run off of liquid propellants combined with several strap-on solid rockets that attach to the bottom stage. Most rockets today measure more than 30.5 meters (100 feet) tall.

The existence of rocket stages sometimes leads to confusion regarding the use of the word *booster*. Some people use the term to describe the strap-on auxiliary rockets that attach to the first stage. Some call the first stage the booster stage because the first stage does the majority of the boosting of the payload to orbit. However, most people refer to the entire rocket stack without the payload as the booster.

Reaching Orbit

Most rockets begin their trip from the ground to orbit from a cement slab called the launchpad. A metal tower on the launchpad, called the launch tower or gantry, allows technicians on the pad access to parts of the rocket not accessible from the ground. This tower also provides tubes to aid in the transfer of propellant from ground storage tanks into the rocket. On the launchpad, the rocket stands in a vertical orientation with the engines pointing toward the ground and the nose cone pointing toward the sky.

The beginning of the flight, called the launch or liftoff, occurs when the rocket engines on the first stage activate to ignite the pro-

pellant stream flowing into their combustion chambers. Then, the rocket heads straight up into the air and takes a few seconds to climb past the tower. During the first minute or two after a launch, a rocket spends most of its time flying almost directly upward to clear the densest parts of the atmosphere that lie below 30.48 kilometers (100,000 feet) in altitude. Acceleration occurs slowly during this phase of flight because the rocket still weighs a tremendous amount from all the propellant in the first stage.

Remember that orbital velocity measures about 27,400 kilometers per hour in the horizontal direction. As a result, rockets must accelerate in the vertical direction to reach space, and in the horizontal direction to reach orbital velocity. Immediately after the rocket clears the launch tower, the engine nozzles gimbal to rotate the rocket slightly off vertical. In this orientation, gravity will slowly rotate the rocket from a slightly off vertical orientation to a horizontal orientation by the time the propellant in the upper stage runs out. This maneuver, called the gravity turn, helps the rocket to gradually direct its energy from the upward climb into space to the forward push needed to reach orbital velocity.

Explosive bolts disconnect the first stage from the rest of the rocket after the propellants in the stage run out. At this time, the rocket has shed a lot of structural and propellant mass. Engines on the second stage typically produce less thrust then those on the first stage. However, the second stage and above always contain significantly less propellant mass than the first stage. As a result, the rocket accelerates much faster than before. Rockets also typically jettison the nose cone sometime during the second stage burn. The rationale behind the nose cone jettison is that the payload no longer requires protection from wind damage once the rocket has climbed past the densest parts of the atmosphere. Jettisoning the cone reduces the amount of mass that the rocket engines accelerate.

The propellant in the second stage in some rockets runs out with the rocket just short of orbital velocity. Explosive bolts separate the second stage from the satellite payload. Then, a small rocket on the payload called a thruster gives the satellite the extra horizontal push needed to reach orbital velocity. In the meantime, thrusters on the second stage cause it to tumble back into the upper atmosphere. Air friction will eventually incinerate the stage into harmless ashes. Disposing of the second stage in the atmosphere avoids cluttering space with useless junk such as empty rocket parts. Orbiting junk that collides with functioning satellites creates even more junk by obliterating the satellite.

Satellites bound for Earth orbits higher than 1,000 kilometers (621 miles) or other planets typically ride on three-stage rockets. In this type of configuration, the job of the second stage involves placing the third stage and payload into a low Earth orbit, typically 300 kilometers (186 miles) in altitude or lower. At a later time, the third stage

Profile of the Titan 4 Rocket

Characteristic	Value
Number of Stages	3 stages, no crew 0th stage (solid) 1st stage (liquid) 2nd stage (liquid)
Auxiliary Boosters	2 large solid rockets attach to side of 1st stage to form stage 0
Height at Liftoff	62.2 meters (204.0 feet)
Mass at Liftoff	860,000 kilograms (1,895,975 pounds)
Maximum Payload	
Low Earth Orbit	22,200 kilograms
Communications Orbit	5,760 kilograms
Moon	10,000 kilograms
Mars	8,290 kilograms
Manufacturer	Lockheed-Martin, Denver, CO

(Photograph provided by Lockheed-Martin Corporation in Denver, CO)

▶ Figure 4: *Size Comparison of American Historical Rockets*

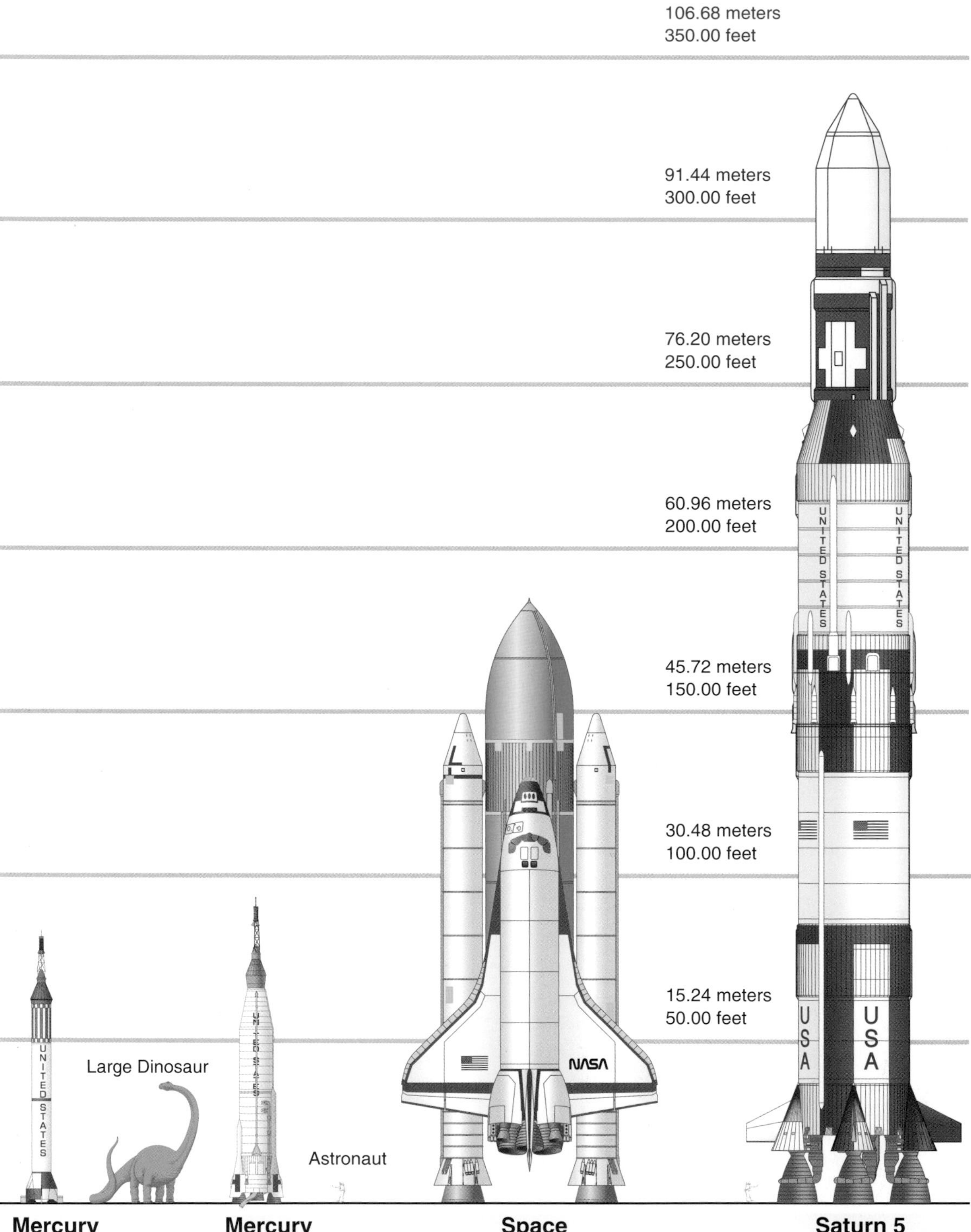

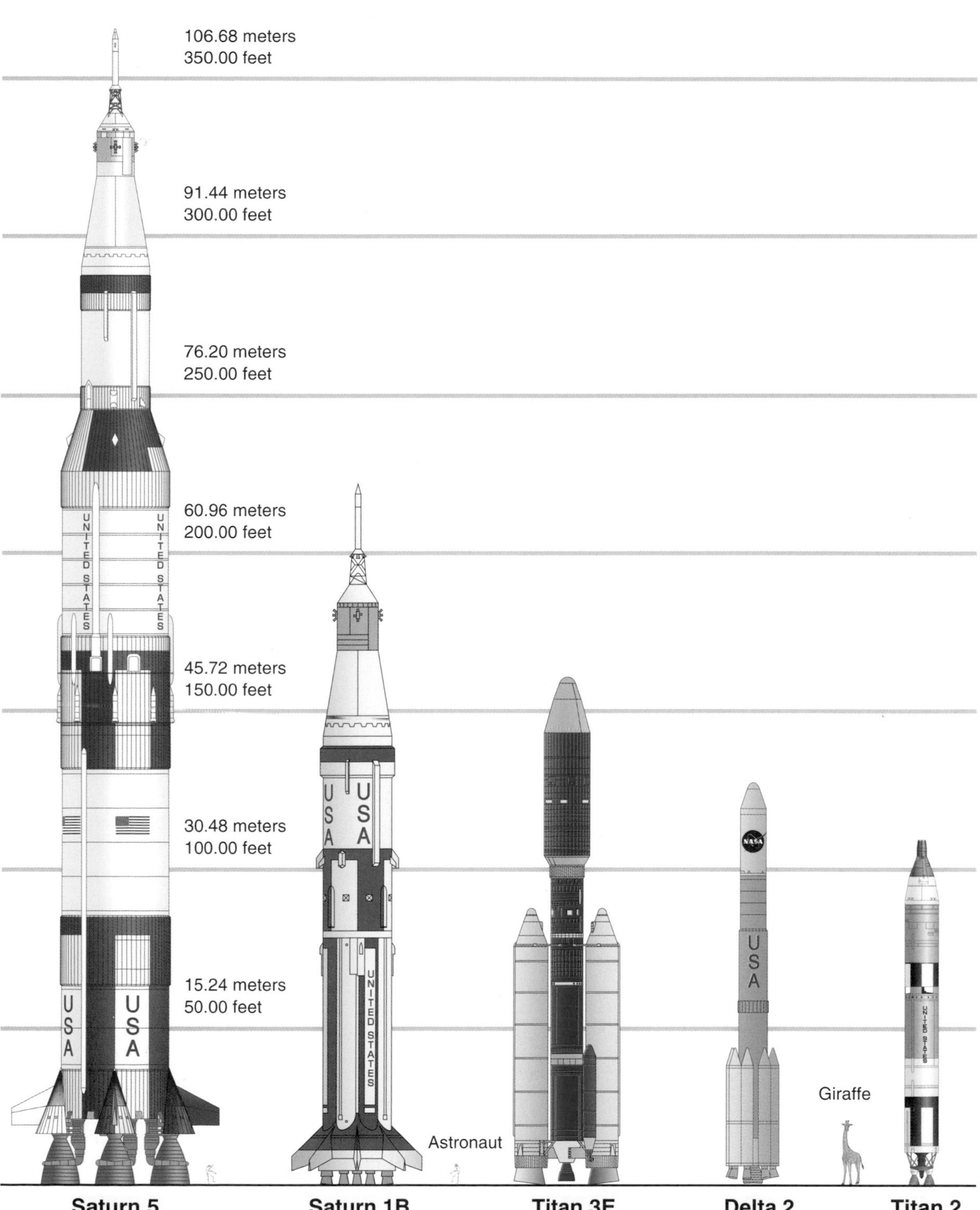

Saturn 5 Apollo
110.6 meters tall
2,912,925 kg
Used 1967–1972

Saturn 1B Apollo
68.3 meters tall
587,300 kg
Used 1966–1975

Titan 3E Centaur
48.8 meters tall
640,000 kg
Used 1974–1977

Delta 2 7925
37.6 meters tall
231,870 kg
Still in use today

Titan 2 Gemini
33.2 meters tall
185,000 kg
Used 1964–1966

Profile of the Space Shuttle

Characteristic	Value
Number of Stages	2 stages, up to 7 crew 0th stage (solid) 1st stage (liquid)
Auxiliary Boosters	2 large solid rockets attach to the shuttle to form stage number 0
Height at Liftoff	56.1 meters (184.1 feet)
Mass at Liftoff	2,041,000 kilograms (4,500,000 pounds)
Maximum Payload	
Low Earth Orbit	23,500 kilograms
Communications Orbit	4,500 kilograms
Moon	4,000 kilograms
Mars	3,200 kilograms
Manufacturer	Rockwell International, Palmdale, CA

The Space Shuttle is limited to low Earth orbit. In order to send satellites higher, it carries the satellite attached to another rocket, and then launches them from orbit. Note that the payload capacity listed for low Earth orbit only includes what is inside the shuttle. The shuttle itself also reaches orbit and weighs about 75,000 kilograms.

fires to place the satellite into its final, higher orbit, or onto a trajectory bound for another planet.

Spaceflight engineers use the term *coast period* to describe the time between the second stage achieving low Earth orbit and the third-stage burn. These coasts last for about 30 to 60 minutes on average (between 33% and 66% of an orbit), and are timed so that after the third-stage burn, the satellite will arrive at the right place relative to the Earth at the right time. The location of the "right place" and the occurrence of the "right time" depend on the specific mission parameters. For example, the "right place" for a communications satellite that provides television links between Asia and America would be somewhere over the mid–Pacific Ocean.

All rockets in use today, with the exception of the Space Shuttle, fall into the expendable launch vehicle (ELV) category. This fancy terminology means that the stress of flight, heat created by the engines burning propellant, the fall back to Earth after jettison, and incineration in the upper atmosphere render the various parts of an ELV useless after one flight. Although building a new rocket for every launch sounds wasteful, building a reusable rocket would make the cost of launch prohibitively expensive, though plans for affordable designs are in development.

1.4 What Satellites Look Like

Contrary to popular belief, satellites do not look like the flying saucers portrayed by cartoons or science fiction literature of the past. Satellites take on a variety of shapes and sizes depending on the task that they perform. While airplanes must take on smooth, streamlined, aerodynamic shapes to remain airborne, satellites orbit in the vacuum of space and face no such restriction. Remember, a satellite in orbit remains in a perpetual state of free fall. Whether a satellite looks like a sleek airplane or a twisted piece of metallic junk will not affect how well it "falls." Gravity works equally well on both. In general, satellite designers employ two general shapes depending on the design philosophy of a concept called *attitude control*. This term refers to how a satellite keeps its instruments and communications antennas pointed in the proper direction.

3-Axis Stabilized Satellites

Satellites designed using the first of the two general satellite shapes look like a box with wing-like projections. The box portion, called the central bus by spaceflight engineers, contains all of the electronics needed to operate the satellite, while the wing-like projections function as solar panels that provide electricity. Exact dimensions of the central bus vary from satellite to satellite. Some appear cubical, while others look tall and skinny, or short and wide. In addition, since most orbits contain a segment where the satellite moves behind

Figure 5: *What Satellites Look Like* ◀

Satellites come in all sorts of shapes and sizes. Unlike airplanes, satellites do not need to take on a streamlined, aerodynamic shape because they fly in the vacuum of space. In general, the shape of a satellite is dictated by the type of job it performs. Essentially, satellites take on whatever shape is necessary for their mission. This figure illustrates six different types of satellites used by NASA over the years. Keep in mind that the satellites are not drawn to scale relative to each other.

Galileo Orbiter

Mission to study Jupiter
Launched in 1989

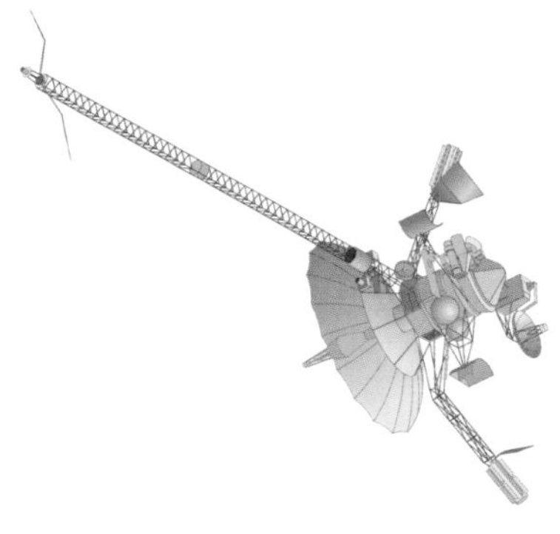

Hubble Telescope

Orbiting astronomical observatory
Launched in 1990

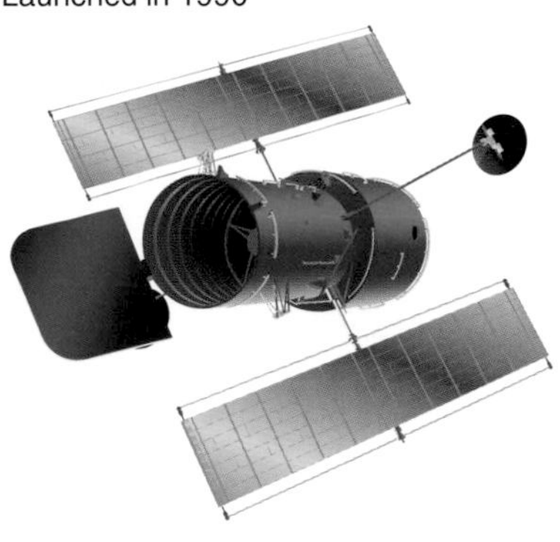

TDRSS Series

Communications relay from orbit
First one launched in 1983

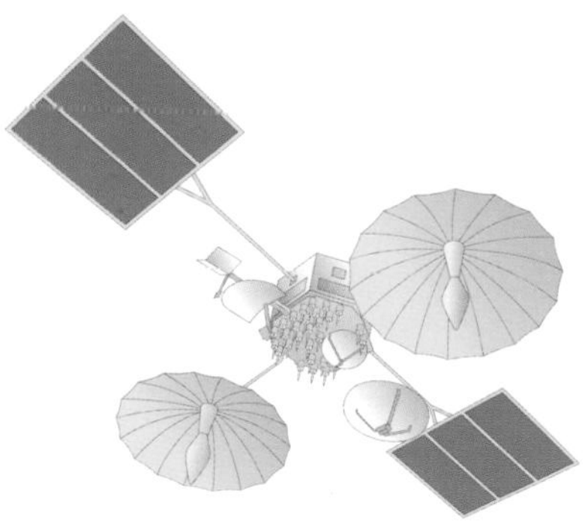

Lunar Module

Landed astronauts on Moon
Used 1969 to 1972

Space Shuttle

Transports astronauts to orbit
First used in 1981

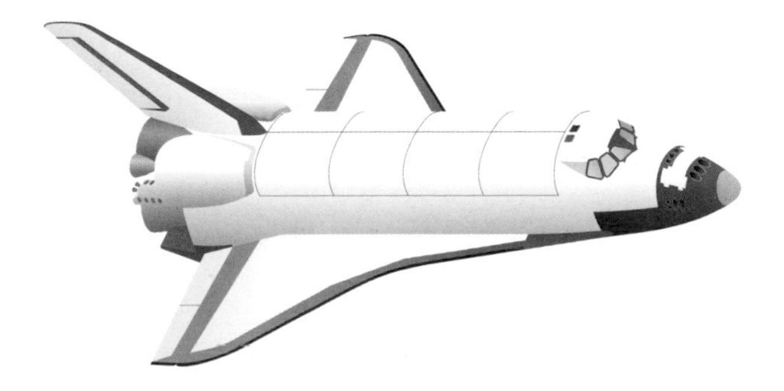

Solar Maximum

Study of the Sun from Earth orbit
Launched in 1980

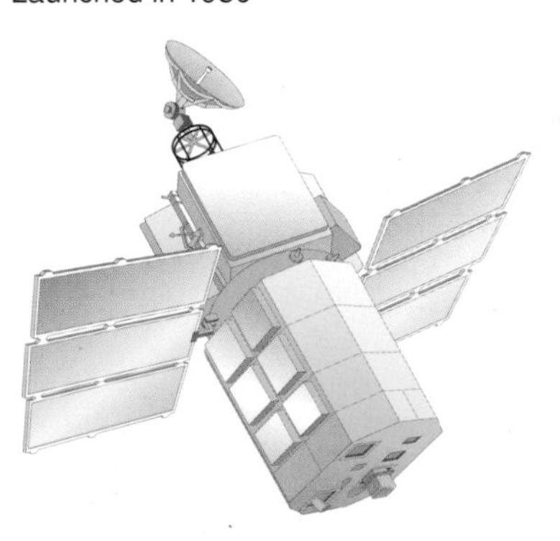

Recycling Rockets

No airline C.E.O. would ever contemplate a business strategy that both mandated the construction of a new aircraft for each flight, and required each passenger to secure a reservation several years in advance. Yet, as ludicrous as this type of plan sounds, companies that provide rocket launch services have been operating in an analogous fashion for the last four decades. Every customer with a satellite to be launched must find a way to secure between $50 to $100 million in order to pay companies, such as Boeing or Lockheed Martin, for the costs of building and launching a new rocket. In turn, these costs are ultimately passed to the consumer in the form of higher services fees for phone calls, internet usage, or even soft drinks and other consumer products. Yes, the cost of new rockets indirectly affects soft drink prices because television stations that use satellites employ advertising fees to pay for their operational costs.

Believe it or not, rocket companies and their stock holders are not committing "highway robbery." The reason is that the best technology today is hard pressed to produce a rocket capable of multiple flights. In fact, all rockets in use today, with the exception of the Space Shuttle, are what spaceflight engineers call *expendable launch vehicles* (ELVs). The stress of flight, the heat created by the rocket engines, and the fall back to Earth after jettison of the stages, and incineration in the upper atmosphere renders all ELVs useless after only one flight.

In 1993, NASA reviewed its space transportation needs for the 21st century and recognized the importance of significantly reducing the cost of delivering satellites and other payloads into space. Costs currently range from between $5000 per kilogram for a low-cost rocket, such as the Delta 2, to over $20,000 per kilogram for the Space Shuttle. Many experts in the aerospace industry now speculate that if economical solutions to the technological hurdles can be found, then single-stage-to-orbit (SSTO) type of fully reusable launch vehicles (RLVs) will offer greater potential savings than today's ELVs and Space Shuttle.

The term *single-stage-to-orbit* means that the rocket will not discard any components, such as empty propellant tanks or expended rocket boosters, while on its ascent to orbit. The term *fully reusable* means just that. Between flights, the vehicle will simply undergo inspection, refueling, and cargo loading. Much like a commercial airliner, there will be no major components to be completely assembled or refurbished after each flight.

As if history were repeating itself, this is not the first time NASA has touted these concepts. In the late 1960s, a project was started with the goals of creating a fully reusable rocket that was "economical" and "easy to operate." The end result was today's partially reusable and somewhat complicated Space Shuttle system. Although only a few question the immense value of the shuttle to America's space program, the original goals were clearly not met. With today's technology being 30 years more advanced than the shuttle's, the chances of success are much higher. NASA's latest efforts involve two vehicles, the X-34 and X-33. The prime objective of these two projects is to develop and demonstrate new technologies for the next generation of launch vehicles that will eventually replace the Space Shuttle.

Experiment vehicle X-34 is a winged-body spacecraft that will provide a basis for the realistic assessment of the development and operational costs of an RLV. This launcher is being designed as a lightweight payload carrier capable of delivering 400 to 900 kilograms (880 to 1,980 pounds) to low-Earth orbit. In 1995, NASA signed an agreement with Orbital Sciences Corporation to jointly develop this small, reusable space booster.

The other rocket, called X-33, is designed to prove the concept of an eventual, larger RLV system that will employ SSTO characteristics. In July 1996, NASA Administrator Daniel Goldin and Vice President Al Gore announced that Lockheed Martin had won the competition to develop the X-33. Lockheed's design features a winged-body configuration and a revolutionary liquid-hydrogen/oxygen-fueled engine called an *aerospike*. Unlike a conventional rocket engine that pushes hot gas out of a bell-shaped nozzle, the aerospike resembles an upside-down bell with the thrusters on the outside. These thrusters push the hot gas down "ramps" along the outside surface of the upside-down bell.

This vehicle is not intended to reach orbit, but will instead launch from Edwards Air Force Base in Southern California, climb on an arcing trajectory over the western United States to an

Below:

Artist rendition shows the external view of the *Venture Star*, a reusable launch vehicle. NASA is hoping to use this type of technology to design the eventual replacement to the Space Shuttle.

(Photo provided courtesy of Lockheed Martin Skunk Works Division)

altitude of 50.3 kilometers (165,000 feet) and a velocity of nine times the speed of sound, then land 724 kilometers (450 miles) away at the Dugway Proving Ground in Utah. Later flights will achieve a top altitude of 91.4 kilometers (300,000 feet) and a velocity more than 14 times the speed of sound, with a landing 1,530 kilometers (950 miles) away at the Malmstrom Air Force Base in northern Montana. Nearly 15 flights, starting in late 2000 or early 2001, are planned so that the engineers can validate critical technology needed to develop the *Venture Star*, an larger version of the X-33 capable of reaching low-Earth orbit. If all goes according to plan, NASA officials hope that *Venture Star* operations will commence before 2010, perhaps as early as 2005. Once operational, this new vehicle will be able to lift about 22,680 kilograms (50,000 pounds) into low-Earth orbit. As important, it may eventually replace the aging Space Shuttle fleet. In 2010, the first space shuttle to be launched, the venerable *Columbia*, will be celebrating its 29th year in operation for NASA. ❑

	Space Shuttle	**Venture Star (*)**	**X-33 (*)**
Length (m)	56.1	38.7	20.4
Width (m)	23.8	39.0	20.7
Empty Mass (kg)	269,000	89,360	28,580
Liftoff Mass (kg)	2,041,000	991,570	123,800
Propellant Type	Liquid H_2, Liquid O_2 Solid Rocket Boosters	Liquid H_2, Liquid O_2	Liquid H_2, Liquid O_2
Propellant Mass (kg)	1,725,000	875,000	95,710
Main Propulsion Scheme	3 Main Engines, 2 Solid Rockets	7 Aerospikes	2 Aerospikes
Liftoff Thrust	28,467,200	13,388,500	1,823,700
Maximum Velocity (m/s)	7,725 (orbital speed)	7,725 (orbital speed)	Mach 15
Payload Capacity (kg)	23,500	22,680	n/a
Payload Bay Size (m)	4.57 x 18.29	4.57 x 13.72	1.52 x 3.05

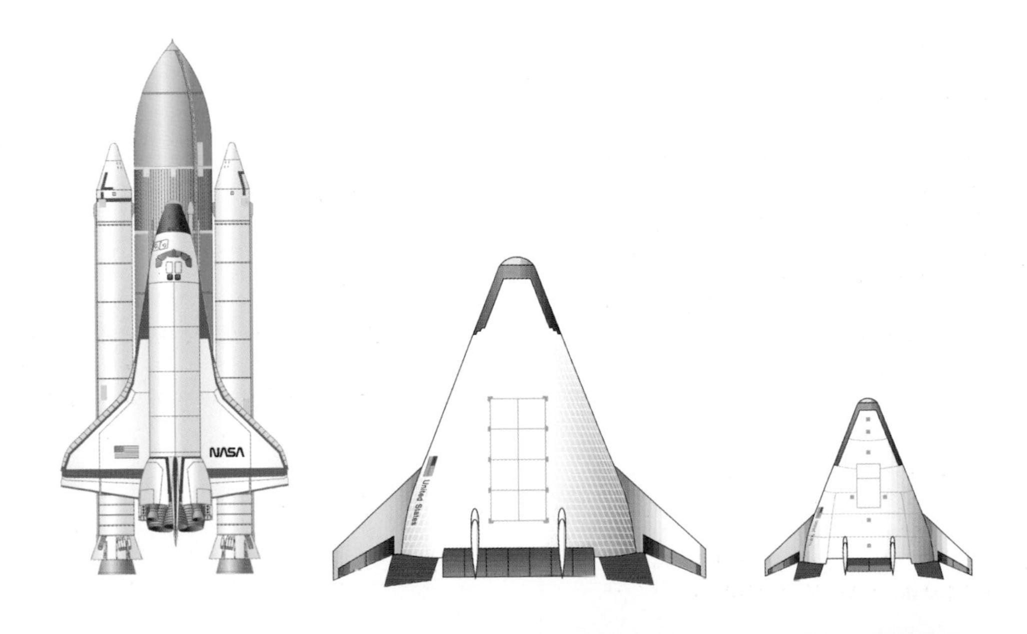

RULE OF THUMB

Remember that a satellite must fit in the cylindrical or cone-shaped payload canister located at the top of a rocket. Unfortunately, satellites often possess appendages that stick out, such as solar panels and antennas. As a consequence, satellites must ride into space with various parts folded up so that they fit inside the rocket. The appendages deploy only after the satellite reaches orbit.

Scientific Fact!

While in space, many cylindrical-shaped satellites spin like a toy top to keep from wobbling as they orbit. Spaceflight engineers refer to this theory as "conservation of angular momentum" and call the technique "spin stabilization."

the Earth relative to the Sun, the central bus must also contain batteries to power the electronics during those times. Recharging of the batteries takes place during the "daylight" segment of the orbit.

Box-shaped satellites employ an attitude control technique that spaceflight engineers call *3-axis stabilization*. This method involves the use of tiny rocket engines, called attitude control thrusters, to point their communications antennas and scientific instruments. These thrusters are mounted on the sides of the central bus. When needed, they squirt out tiny amounts of propellant to nudge the satellite into its proper orientation. Once the attitude control propellant runs out, the satellite becomes useless even if the electronics remain functional. Typical propellant supplies last for five years or more.

Spin-Stabilized Satellites

The other general type of satellite design employs a cylindrical central bus that looks like a large soup can. On these satellites, the solar cells that provide power to the electronics are mounted on a thin sheet that surrounds the cylinder, similar to a label surrounding a can of soup. Attitude control is maintained by spinning the entire bus around an imaginary line that runs through the center of the satellite, from the top of the cylinder to the bottom. Spaceflight engineers call this imaginary line the *longitudinal axis*. Once the satellite "spins up," its orientation tends to remain fixed in the direction pointed to by the longitudinal axis.

This technique works in a fashion similar to a toy top spinning on a table. As long as the top spins, it remains pointed straight up. However, as the rate of spin slows down, the top begins to wobble. Another way to think about this technique involves throwing a toy frisbee. Correctly thrown, a frisbee spins as it moves forward. A frisbee thrown without a spin will wobble as it flies. Satellites that spin in space work analogously to the examples of the two toys. The spin keeps the satellite from wobbling.

Use of spin stabilization comes with an inherent disadvantage. Once the satellite spins up, reorienting it so that its longitudinal axis points in a different direction becomes extremely difficult and requires a lot of energy. For this reason, the primary users of spin stabilization are communications satellites because they tend to remain pointed at a fixed location on the ground for long periods of time. Scientific and weather satellites must constantly change their orientation to point their cameras and other instruments at locations of interest. Consequently, they tend to use propellant as their primary means of attitude control.

Not all satellites look cylindrical or box shape. Some take on spherical shapes. Others combine aspects of the box and cylinder designs. For example, the *Hubble Space Telescope* appears cylindrical, but possesses wing-like solar panels similar to box-shaped satellites.

To complicate matters, quite a few satellites look box shaped with wing-like solar panels, but use spin stabilization for attitude control.

1.5 What Is NASA?

Spaceflight remains a relatively new form of transportation compared to the automobile or airplane. During the early days of spaceflight, between the late 1950s and early 1960s, the only two groups in the world that launched rockets were the Russian military and the American military. The reason behind military domination of early spaceflight was that the cold war generated most of the research money for spaceflight, and the only rockets available were missiles modified to carry satellites instead of nuclear weapons. Today, the Russian military and the United States Air Force remain two of the largest users of rockets in the world. Most military satellites carry no weapons. Instead, they take high-resolution pictures of the trouble spots in the world to keep their governments informed of rapidly changing global situations.

In 1958, American president Dwight Eisenhower created a civilian space agency because he feared the consequences of a military domination of space. Ironically, he was a former five-star general in the Army. This new governmental agency, called the National Aeronautics and Space Administration (NASA), was charged with advancing American civilian interests in space. Over the years, NASA became the most famous rocket-launching group in the world as it sent many weather, communications, and scientific research satellites into Earth orbit; sent robotic explorers to distant planets; and landed humans on the Moon. Today, NASA leads the world in all spaceflight and human space-exploration efforts.

NASA operates many facilities scattered across the United States. Kennedy Space Center, located on 140,000 acres of swamp on the central Florida coast, is the most famous of all NASA's sites. Its launchpads serve as America's "spaceport." Another well-known facility, the Johnson Space Center, sits in the plains southeast of Houston, Texas. Johnson functions as the training ground of America's astronaut corps and acts as the mission control center for all astronaut spaceflights. Missions involving robotic satellites fall into the realm of the Jet Propulsion Laboratory in the foothills near Pasadena, California, and the Goddard Space Flight Center in Greenbelt, Maryland. The Jet Propulsion Laboratory plans and implements scientific missions to other planets, while Goddard keeps watch on the home planet and claims the title of the world's leader in studying the Earth from space. NASA also operates facilities dedicated to scientific research. The Marshall Space Flight Center in Alabama and the John H. Glenn Research Center in Ohio conduct rocketry research, while the Ames Research Center in California and the Langley Research Center in Virginia look after the aeronautics portion of NASA by studying aircraft flight dynamics within the Earth's atmosphere. ❑

Space Launch Totals

Country	Number of Space Rocket Launches
USSR / Russia	2,465 total First launch in 1957
United States	1,027 total First launch in 1958
ESA	77 total First launch in 1979
Japan	49 total First launch in 1970
China	38 total First launch in 1970
India	6 total First launch in 1980
Israel	2 total First launch in 1988

Data provided by NASA Division of Special Studies (Code Z) and is current as of 1 January 1995. United States numbers represent sum of Department of Defense, NASA, and private commercial launches.

Above:

The two images above represent the official logos of the National Aeronautics and Space Administration. NASA was created by Congress in 1958 to promote American civilian interests in space. Employees of the space agency refer to the top logo as the "meatball" and the bottom one as the "worm."

Above the Clouds

Chapter 2: Orbital Mechanics Without Math

Photographs of the Earth taken by astronauts in orbit constantly remind us of how lucky we are to live on Earth. As Apollo astronaut Michael Collins said about the Earth, “I know how lucky we are to have this planet because I’ve seen another one, namely the Moon.” Many scientists and environmentalists have credited space photography and the Apollo program for providing a catalyst for the modern environmental movement. For the first time in history, people literally began to view the Earth as a fragile, little planet. From space, political boundaries suddenly disappear, and the Earth appears as a majestic blue-and-white oasis amidst the black void of the endless universe.

This new and unique perspective on viewing the Earth critically depends on a branch of engineering science called orbital mechanics. The reason is that the ability to take a picture of a specific portion of the Earth from orbit depends on the ability to determine a satellite’s present and future locations on the orbit. Without this knowledge, satellites turn into expensive pieces of orbiting junk whether or not the mission involves picture taking.

Unfortunately, when most people hear the term *orbital mechanics*, they imagine old professors filling chalkboards with nasty equations from the depths of the most cryptic math and physics books ever published. This chapter removes the equations and confusion from the field of orbital mechanics and replaces them with simple geometric explanations. By the end of this chapter, you will know the details of orbital motion, how to characterize and differentiate between orbits, and how spaceflight engineers predict the ground locations that satellites fly over during the course of an orbit. In addition, by reading this chapter, you will also have a chance to view spectacular photographs of the Earth taken by Space Shuttle astronauts.

2.1 How Elliptical Orbital Motion Works

Although the introduction of spaceflight into human civilization occurred less than 40 years ago, theoretical mathematicians and other scholars with nothing better to do with their time have been studying orbital motion for over 400 years. In 1609, an astronomer named Johannes Kepler published three fundamental laws regarding how the geometry of ellipses governs the motion of orbiting bodies. His results came after years of frustrating research studying data that Tycho Brahe, a Danish astronomer of the late 1500s, gathered while observing the orbit of the planet Mars around the Sun. Although Kepler’s three laws supply the basic concepts that

Opposite Page:

This photograph shot from Space Shuttle *Discovery* in May 1991 shows the Strait of Gibraltar and the western end of the Mediterranean Sea. In the image, Spain occupies the top of the photograph and Morocco lies at the bottom.

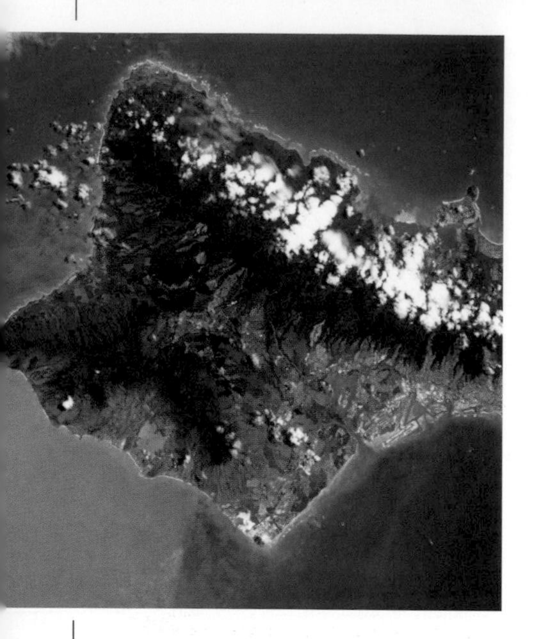

Above:

The island wake emerging from the southwestern side of Oahu, Hawaii (light-colored area, lower left) is caused by wind from the northeast obstructed by the cloud-covered Koolau Mountain Range. The lighter-colored water indicates a smoother surface with a slower water current than the darker, rougher, faster-moving water on either side. Oahu, at 1,536 square kilometers, is the chief island of Hawaii. Pearl Harbor, Honolulu, and the International Airport are clearly visible on the south side of the island. This photograph was taken on Space Shuttle *Columbia* in July 1994.

form the foundation of modern orbital mechanics, they were based strictly on empirical observations. Verification of the truth did not occur until 1666 when scientific genius Isaac Newton used calculus to prove Kepler's laws.

Elliptical Geometry

Kepler's first law deals with the shape of orbits. Essentially, this law uses dry, scientific terms only a mathematician would be proud of to state what you learned by reading the beginning of the previous chapter. Specifically, satellites in orbit around the Earth move in elliptical paths. If you read the last chapter, you might be thinking, "What happened to the circular orbit?" As you will see later, circles represent special cases of ellipses.

Ellipses look like ovals, and an easy way to visualize ellipses certainly involves thinking of them as ovals. However, while an oval is just a colloquial way to describe a squashed or elongated circle, ellipses have precise mathematical definitions. The first important part of the definition deals with an imaginary line called the major axis (see Figure 6 on page 37). Every ellipse contains an infinite number of imaginary lines that run through the center, from one edge to the opposite edge. The major axis, analogous to a circle's diameter, is the longest of these lines and defines the size of the ellipse. A longer major axis results in a larger ellipse. The best way to understand this concept is to look at the ellipse and imagine folding it exactly in half the long way. The fold line represents the major axis.

The second important part of the ellipse deals with two imaginary points called foci (plural for focus). Each one of the two foci sits on the major axis, with both equidistant from the center. When a satellite orbits the Earth, the center of the Earth *always* coincides with one of the two foci. This concept represents the fundamental concept of Kepler's first law. Incidentally, the choice between the two is not important at this time.

These foci play a major role in defining the exact shape of an ellipse in that the sum of the distances from any one point on the ellipse to each of the two foci always equals the length of the major axis (see Figure 6 on page 37). However, for a given major axis length, an infinite number of ellipses exist because the two foci can lie anywhere on the major axis as long as both remain equidistant from the center of the ellipse. Each one of these many ellipses varies in how squashed it looks (see Figure 7 on page 39).

Spaceflight engineers use the term *eccentricity* to measure the amount of squash in an ellipse. The act of bringing the foci together so that both lie at the center of the ellipse sets the eccentricity equal to 0.0 and creates a perfect circle. Spreading the foci apart increases the amount of squash in the ellipse and increases the eccentricity value towards a maximum value of 1.0. Thus, ellipses with eccentricities such as 0.0, 0.001, 0.01, or 0.1 look circular, while ellipses with eccen-

Historical Fact

Kepler derived his three laws using data collected about Mars' orbit around the Sun. Consequently, when he published the laws between the years 1609 and 1619, Kepler thought that they applied only to planets in orbit around the Sun. Nearly 70 years later, Isaac Newton used calculus to prove that Kepler's laws apply to any satellite in orbit around any celestial body, including the Earth. Strangely enough, Newton misplaced the papers containing the mathematical proof for over five years. He searched for and found them only to win a bet from Edmond Halley, discoverer of Halley's comet.

An early 17th-century astronomer named Kepler postulated three laws regarding orbital motion. The consequence of the first law is that all Earth-orbiting satellites move on an elliptical-shaped path with the center of the Earth at one focus and the other focus empty. This law is valid for orbits around any body in the universe. For example, all of the planets in the solar system orbit the Sun in an elliptical path with the center of the Sun at one focus of the ellipse.

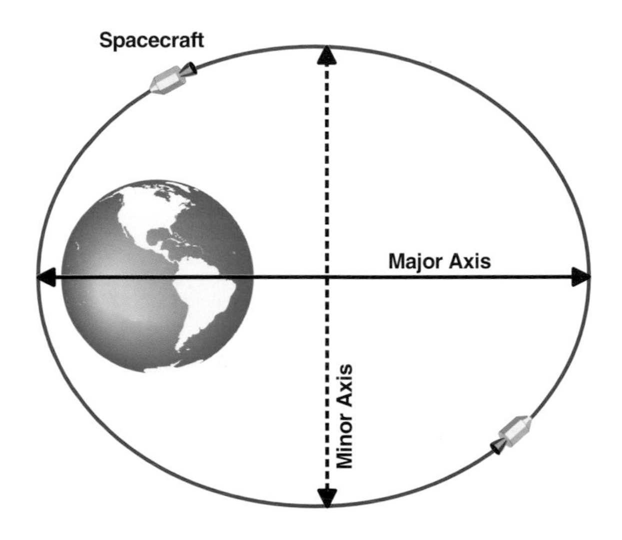

What Does an Ellipse Look Like?

It looks like an oval or a squashed circle. The drawing to the left shows an elliptical orbit (solid red line) around the Earth.

The dark solid line is the major axis. It is the longest possible line that can be drawn from one end of the ellipse to the other through the center.

The dark dotted line is the minor axis. It is the shortest possible line that can be drawn from one end of the ellipse to the other through the center.

The major and minor axes always meet at the center of the ellipse and are perpendicular to each other. The semi-major and semi-minor axes are one half of the major and minor axes, respectively.

What Are the Foci of an Ellipse?

Every ellipse has two foci. Each focus lies on the major axis and the distances between each focus and the center of the ellipse are equal. In the drawing to the right, the round dots mark the position of each focus, and the solid square marks the position of the center of the ellipse.

Line A connects the left focus with point C on the ellipse, and line B connects the right focus with point C. Notice below that the lengths of lines A and B added together equal the length of the major axis (twice the semi-major axis).

Line D connects the left focus with point F on the ellipse, and line E connects the right focus with point F. The drawing below also shows that the lengths of lines D and E added together equal the length of the major axis. In fact, the sum of the distances between the two foci and any point on the ellipse always equals the length of the major axis.

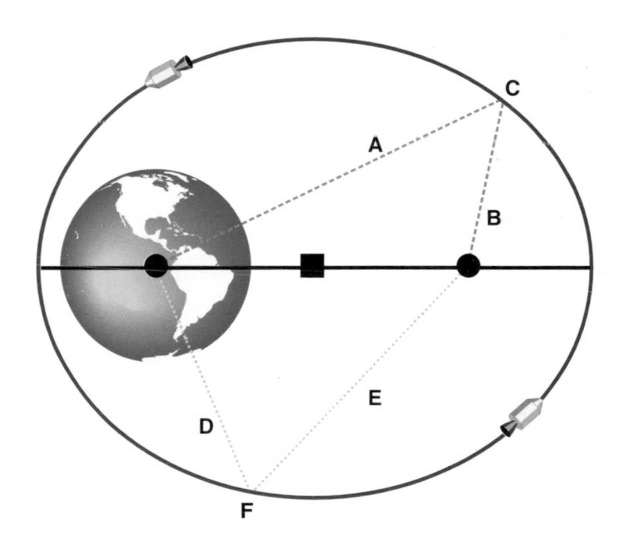

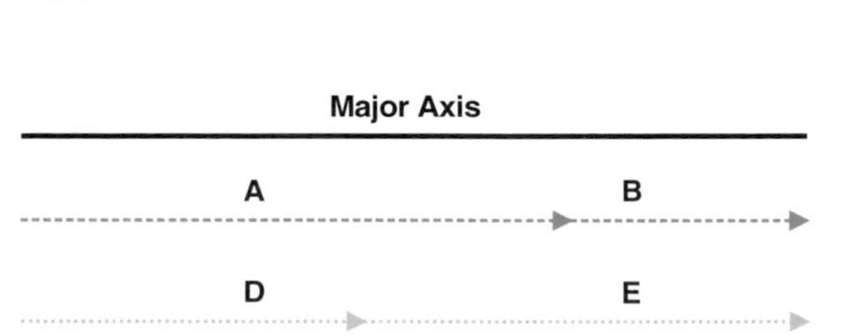

Important Concept

In the drawing above, and in the figures to come later on, it may appear that the focus is in the Pacific Ocean, the North Pole, or near the African continent, depending on the view of the Earth used in the figure. Remember that this is a consequence of the three-dimensional picture. The focus is always at the center (inside of) the Earth.

Above:

This magnificent view was taken in July 1985 from Space Shuttle *Challenger*. The photograph is a view above the eastern Himalayas looking southward across the great plain of the Ganges into the Bay of Bengal. A large, white cluster of mature monsoonal thunderstorms over the Khasi Hills is in the center of the picture. Within the cluster, squall lines and areas of downdrafts (microbursts) can be seen. This area averages 1,200 cm (472 inches) of rain every year.

Orbital Terminology

Terms for Closest and Farthest Point	Proper Usage
Perigee Apogee	Orbit Around Earth
Perihelion Aphelion	Orbit Around Sun
Pericynthion Apocynthion	Orbits Around Moon
Periapsis Apoapsis	Any Orbit

tricities like 0.8, 0.9, or 0.99 look extremely squashed. One of the many methods of calculating an ellipse's eccentricity value involves dividing the distance between the foci by the major axis length.

Figure 7 on page 39 shows various orbital paths around the Earth with the same major axis length, but different eccentricities. The drawings in the figure also illustrate the major property of Kepler's first law in that the center of the Earth always lies on one of the foci. On nearly circular orbits, the second foci often lies inside (below the surface of) the Earth. Conversely, increasing the eccentricity value high enough may push the focus close enough to the edge of the ellipse and cause portions of the orbital path to intersect the surface of the Earth. Satellites on these types of orbits face the danger of smashing into the ground.

A satellite's distance from the center of the Earth varies as it orbits between two points on opposite ends of the ellipse called the perigee and apogee. These two points always lie on the major axis and mark the point of closest approach, and farthest separation, respectively, to the center of the Earth (see Figure 8 on page 40). Circular orbits lack unique perigee and apogee points because the distance from the Earth always remains constant. The orbits shown in Figure 8 on page 40 also illustrate the fact that for two orbits of the same size (major axis length), a larger eccentricity results in both a shorter distance from the Earth's center to perigee and a longer distance from the Earth's center to apogee.

This line of thinking requires some caution. In the mid-1960s, American astronauts flew on a mission that utilized an orbit with a perigee altitude of 322 kilometers (200 miles) and an apogee altitude of 1,287 kilometers (800 miles). A television news reporter covering that mission commented, "Wow! What an eccentric orbit!" Upon first examination, that orbit seemed very eccentric, as the apogee altitude was four times farther out than the perigee altitude. However, keep in mind that eccentricity calculations depend primarily on the distance to the center of the Earth, not altitude from the ground. He forgot to factor in the distance from the center of the Earth in his thinking. The Earth's radius measures approximately 6,400 kilometers (4,000 miles), which when factored in, results in a perigee distance of 6,732 kilometers and an apogee distance of 8,019 kilometers. In reality, the apogee was only 1.2 times farther out than the perigee.

Angular Measurements

Spaceflight engineers always need to keep track of a satellite's position as it orbits the Earth. One way to solve this problem involves using X, Y, and Z coordinates. However, this scheme is not very useful because the mathematical equations that describe an ellipse in three dimensions are difficult to visualize without a computer. Use of angular measurements provides a simple but powerful solution to this visualization problem. During the course of one orbit or revolution around the Earth, a satellite moves through 360° of

Figure 7: *How Eccentricity Affects the Shape of an Orbit*

Eccentricity measures how squashed an ellipse looks. Values close to zero indicate a circular looking ellipse. The closer the eccentricity value gets to one, the more squashed the ellipse looks. Spreading the ellipse's two foci apart while keeping the size of the orbit (major axis) constant squashes the ellipse and increases the eccentricity value.

Notice that one of the two foci always lies at the center of the Earth. In the drawings, the solid red line represents the orbit, the solid dark line shows the location of the major axis, the dashed dark line is the minor axis, and the two round dots are the two foci. All four orbits shown in the diagram below have the same major axis length, but different eccentricity values.

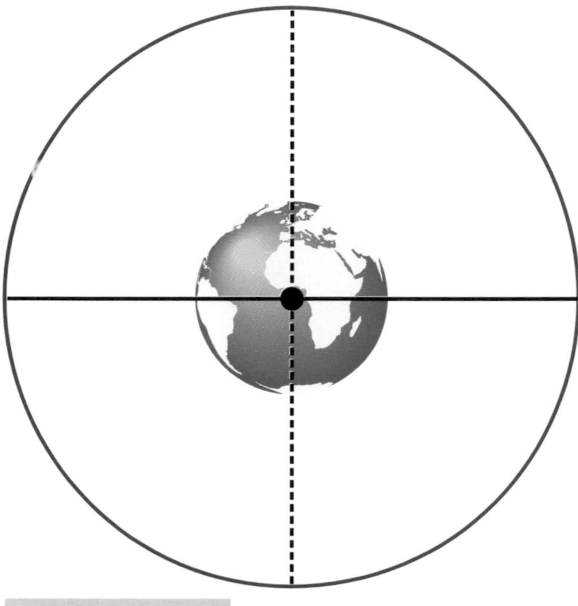

Eccentricity = 0.0

This orbit is a perfect circle. Both foci lie at the center of the circle. Remember, a circle is a special case of an ellipse.

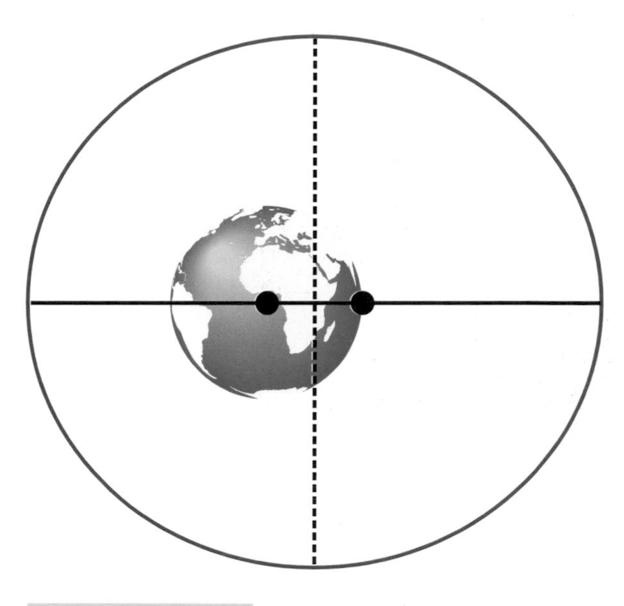

Eccentricity = 0.2

This orbit is a little more squashed than the circular orbit to the left, but is still pretty circular. Notice that the second focus almost lies inside the surface of the Earth.

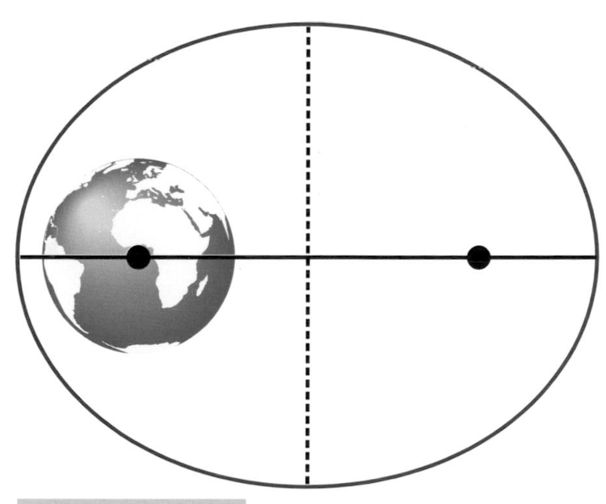

Eccentricity = 0.6

Pushing the two foci apart (toward the edges of the ellipse) squashes the ellipse. Since the center of the Earth must stay at one focus, increasing the eccentricity also moves the surface of the Earth closer to the edge of the ellipse.

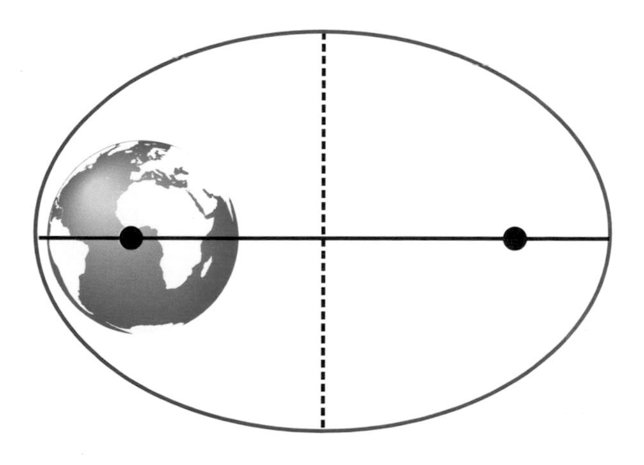

Eccentricity = 0.7

For this size orbit, this is about as far as the foci can be put apart. Theoretically, the eccentricity value can go higher (up to 1.0), but squashing the ellipse any more will push the two foci so close to the edge that the orbit will run into the Earth.

► Figure 8: *Common Parts of an Orbit*

Perigee: Point on orbit where satellite is closest to the Earth. Marked by the symbol ▼

Apogee: Point on orbit where satellite is farthest from the Earth. Marked by the symbol ▲

True Anomaly: Angle on orbit in between perigee and the satellite's position, as measured from the focus of the ellipse

Altitude and Radius: Distance between satellite and the surface of the Earth or center of Earth, respectively

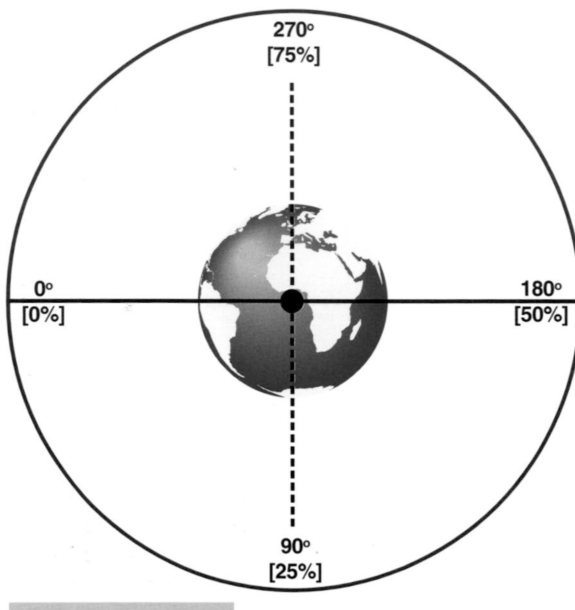

Eccentricity = 0.0

This orbit is a perfect circle. Since perigee and apogee do not exist, true anomaly can be measured from anywhere.

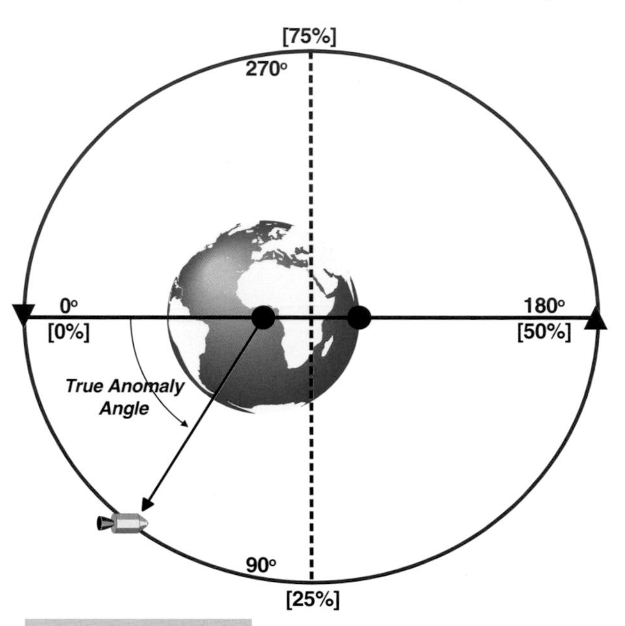

Eccentricity = 0.2

This orbit is a little more squashed than the circular orbit to the left, but is still pretty circular. Notice that on an elliptical orbit, true anomaly is usually 0° at perigee.

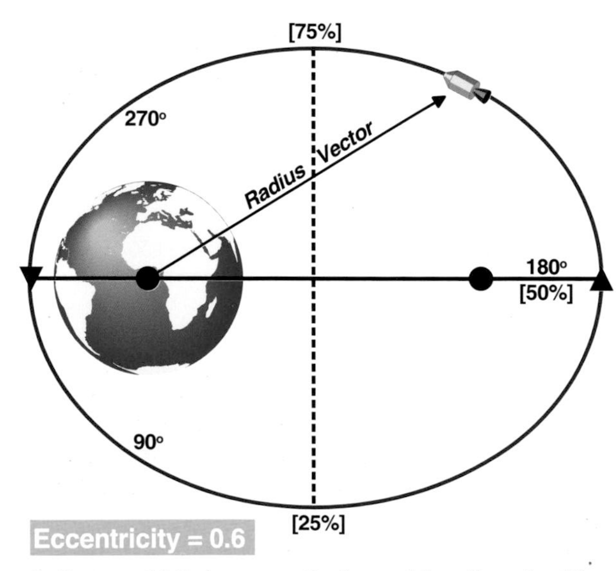

Eccentricity = 0.6

As the eccentricity increases, the focus of the ellipse (and the center of the Earth) moves closer to perigee. Therefore, notice that the 90° and 270° true anomaly points move closer to perigee as well.

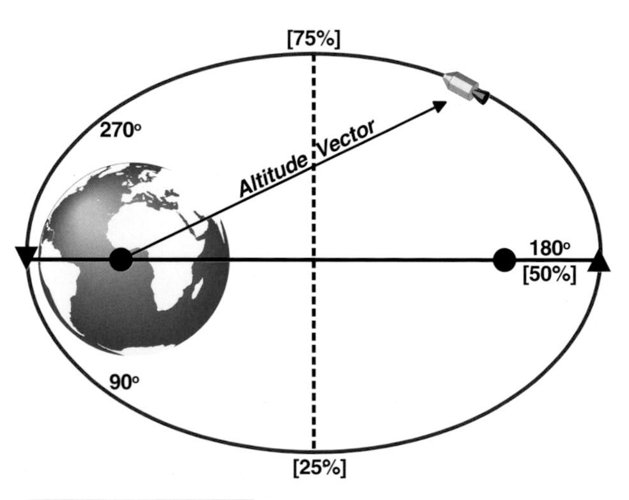

Eccentricity = 0.7

Another consequence of increasing eccentricity is that the 25% and 75% completion points slide closer to apogee relative to the 90° and 270° marks. These percentages represent the *distance traveled*, not travel time along the orbit.

angle. A parameter called true anomaly measures the amount of angle between perigee and a satellite's current position, with positive angles measured in the direction of the satellite's motion. When a satellite passes through perigee, spaceflight engineers say that the true anomaly measures 0°. Since apogee always lies at the opposite end of an orbit from perigee, the true anomaly at apogee is 180°, halfway between 0° and 360° (see Figure 8 on page 40). On a perfectly circular orbit, any arbitrary point can serve as the 0° point because unique perigee and apogee points do not exist.

Over the course of one orbit, satellites always spend half of their time moving from perigee to apogee. They spend the other half returning to perigee from apogee. Therefore, true anomaly values between 0° and 180° mean that the satellite's location lies somewhere after perigee, but before apogee. Conversely, values between 180° and 360° mean that the satellite has passed apogee and is heading back to perigee.

At this time, it might be tempting to think that since 180° always marks the halfway point of an orbit, then 90° and 270° must represent the 25% (one-quarter) and 75% (three-quarter) completion points, respectively. That statement only holds true for circular orbits. In reality, increasing an orbit's eccentricity slides the 25% and 75% points toward apogee, and the 90° and 270° points toward perigee (see Figure 8 on page 40).

Equal Areas and Equal Times

Kepler's second law of orbital motion deals with the velocity of a satellite despite the fact that the explanation sounds like geometry. The best way to understand Kepler's thinking involves pretending that an imaginary line from the center of the Earth to the satellite is a big stick of chalk, and that the inside of the ellipse is a chalkboard. As the satellite orbits the Earth, the chalk will gradually color an imaginary pie-shaped wedge inside the ellipse (see Figure 9 on page 42). The size of this imaginary wedge will depend on time. Specifically, wedges of greater area take longer to color in than smaller wedges. Kepler postulated that the size of any two of these imaginary wedges will be equal as long as the times it took to color in the two wedges are also equal. This fact holds true no matter the location on the ellipse.

How do imaginary wedges of equal areas relate to a satellite's orbital velocity? Remember that a satellite makes its closest approach to the center of the Earth at perigee and farthest separation at apogee. This fact dictates that the imaginary wedges colored in at perigee will look short, while the wedges at apogee will look long. In order for both to contain equal areas, the short one must also be fat, and the long one must also be skinny (see Figure 9 on page 42). Essentially, satellites must move a greater distance along the orbit at perigee than at apogee in order to color in the same size imaginary

Above:
Astronauts on Space Shuttle *Discovery* shot this photograph of the Florida peninsula in January 1985. Their Kennedy Space Center launch site is clearly visible on the Atlantic Coast (right side, middle of photograph) as a chunk of land extending outward into the ocean. This photograph also shows Lake Okeechobee (middle of peninsula), Florida Keys (at the bottom), and part of the Bahamas (lower right edge). The lighter areas in the Gulf of Mexico (left edge) indicate relatively shallow water compared to the deep Atlantic.

Scientific Fact!

Kepler's second law states that a satellite's velocity reaches a maximum at perigee and falls to a minimum at apogee. He arrived at this conclusion by studying the orbit of Mars around the Sun. Fortunately for Kepler, the orbit of Mars is more elliptical than most of the other planets'. If he had chosen to study a planet with a more circular orbit (Venus, Jupiter, or Saturn), the limitations of his crude measurements would probably have prevented him from arriving at the same conclusion.

► Figure 9: *A Look at Kepler's Second Law and Orbital Motion*

Kepler's second law deals with the speed of a satellite on an elliptical orbit. To understand this law, imagine a line connecting the center of the Earth to a satellite. Imagine that line sweeping out or filling in the interior of the ellipse (in a process similar to dragging the fat end of a stick of chalk across a chalkboard) as the satellite moves. The law states that the satellite always takes an equal amount of time to sweep out equal amounts of area on the ellipse.

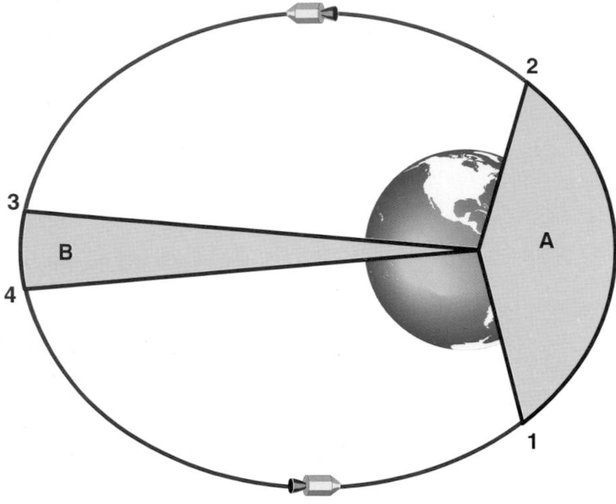

A Graphical Illustration of the Second Law

The law says that a line connecting the center of the Earth to a satellite sweeps out equal areas in the interior of the ellipse in equal times. Therefore, if the zone marked A and the zone marked B have equal areas, then a satellite takes the same amount of time to move from point 1 to point 2 and from point 3 to point 4. This means that a satellite moves much faster near the perigee than it moves near the apogee.

How the Second Law Affects Orbital Motion

The orbit on the right has a semi-major axis length of 9,800 km (major axis length of 19,600 km) and an eccentricity equal to 0.2. A satellite on this orbit takes 2 hours and 40 minutes to complete a single orbit. The tick marks on the orbit show the satellite's position at 10-minute intervals. The farther the marks are spread apart, the faster the satellite moves. Notice that the satellite speeds up going from apogee to perigee, and slows down moving from perigee to apogee.

Perigee

Apogee

A Graphical Illustration of How Eccentricity Affects Orbital Motion

The orbit to the right is exactly the same size as the orbit immediately above it, and has exactly the same period. However, this orbit is more squashed and has an eccentricity of 0.6. The tick marks on the orbit show the satellite's position at 10-minute intervals. Notice that the satellite moves much faster at perigee and much slower at apogee as compared to the orbit above this one.

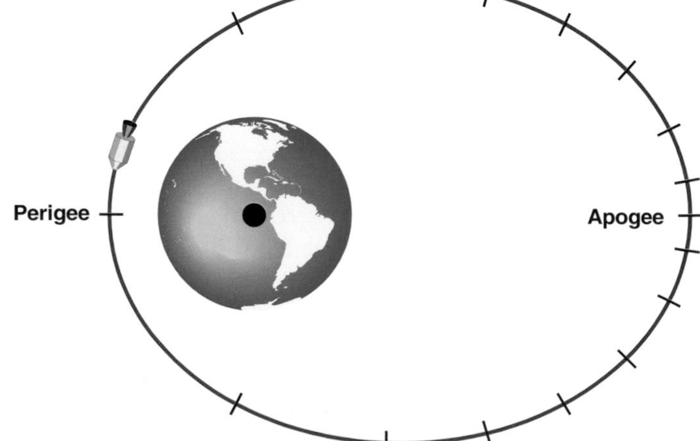

wedge in both cases. Consequently, satellites must move faster at perigee than at apogee.

In fact, a satellite moves fastest during perigee passage, slows down while ascending toward apogee, moves slowest at apogee, and then speeds up as it falls back toward perigee. How much faster does a satellite move at perigee than at apogee? The answer depends on the eccentricity of the orbit. Increasing the eccentricity (squashing the orbit) while keeping the size the same will increase the differential between the perigee and apogee velocities. For example, if the eccentricity equals 0.1, the orbit will look nearly circular, and the perigee and apogee velocities will not differ by much. However, a satellite on an extremely squashed orbit with a 0.9 eccentricity will move much faster at perigee than at apogee.

Another important fact to keep in mind is that a satellite on an elliptical orbit always moves faster at perigee and slower at apogee than a satellite on a corresponding circular orbit. For example, suppose that a satellite moves around the Earth on an orbit with a perigee altitude of 300 kilometers and an apogee altitude of 1,000 kilometers. That satellite will move faster at perigee than a satellite on a corresponding circular orbit of 300 kilometers. In addition, the satellite will also move slower at apogee than a satellite on circular orbit of 1,000 kilometers.

The consequence of the second law should make intuitive sense. What happens when you throw a ball into the air? The ball slows down as it ascends into the air, reaches its slowest velocity at the height of the climb, then speeds back up as it plunges back toward the ground. This sequence of events matches what happens to a satellite on an elliptical orbit.

Orbital Periods

Kepler's third law deals with the length of time a satellite takes to make one complete revolution around the Earth. Spaceflight engineers refer to this time as the *orbital period*, or just *period* for short. The period always depends on the altitude of the orbit. Specifically, higher altitude orbits take a longer time to complete than lower altitude orbits. The reason is that satellites in high orbits move slower, and must travel a longer distance through space to complete one orbit than satellites on low orbits. For example, at a low Earth orbit altitude of 300 kilometers (186 miles), the Space Shuttle takes about 90 minutes to complete an orbit. In contrast, communications satellites at a high altitude of 35,786 kilometers (22,236 miles) move much slower. They take 24 hours to circle the Earth.

In mathematical terms, Kepler's third law states that the square of the orbital period is directly proportional to the cube of half the major axis length. When this law is deciphered into plain English, it states what you just learned by reading the previous paragraph. However, the law also contains a hidden message that is not immedi-

Elliptical Orbit Velocities

Eccentricity Value	Perigee Velocity
0.0 (circular)	Velocity same on all parts of orbit, no unique perigee
0.1	122% of apogee velocity 105% of circular velocity
0.3	186% of apogee velocity 114% of circular velocity
0.5	300% of apogee velocity 122% of circular velocity
0.7	567% of apogee velocity 130% of circular velocity
0.9	1900% of apogee velocity 138% of circular velocity

Although the terms *perigee* and *apogee* apply specifically to Earth orbits, the comparisons in this table apply for satellites in orbit around any body (Sun, Earth, Moon, Mars, etc.) at any altitude. Circular velocity assumes an orbit altitude equal to the perigee altitude. Divide by 100 to find "how much faster" corresponds to a given percentage. For example, 300% means three times faster.

Earth Orbit Periods

Altitude of Orbit	Period
Sea Level	1 hr, 24 min, 29 s
300 kilometers	1 hr, 30 min, 31 s
1,000 kilometers	1 hr, 45 min, 07 s
10,000 kilometers	5 hr, 47 min, 40 s
100,000 kilometers	3 days, 23 hr, 54 min
384,400 kilometers	28 days, 3 hr, 18 min

hr = hours, min = minutes, s = seconds
An orbit at sea level is not possible in real life.
384,400 kilometers is the altitude of the Moon's orbit.

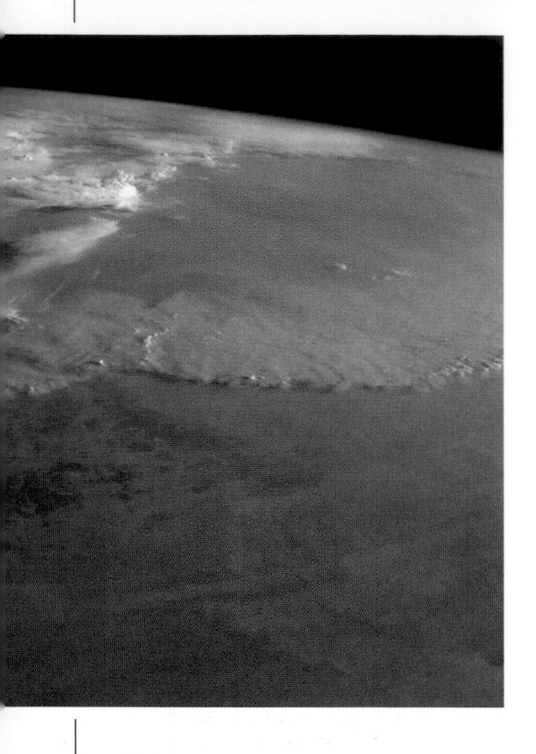

Above:

A late afternoon dust storm crosses the Sahara Desert in May 1992. The dust cloud is the light gray feature in the middle of the photograph. The front of the advancing storm can be clearly seen as a semi-horizontal line running across the middle of the picture. Notice that small cumulus clouds have formed over the most vigorously ascending parts of the dust front (left side of the image). The direction of movement parallels streaks of sand (lower part of the photograph) on the desert surface created by previous dust storms. This photograph was taken from Space Shuttle *Endeavour.*

RULE OF THUMB

Key Facts of the J2000 Coordinate System

- Z axis points through the North Pole
- XY plane contains the equator
- Axes do not rotate with the Earth
- Coordinates on ground constantly change
- Coordinates in space always stay the same

ately obvious. Notice that Kepler said the time it takes a satellite to complete one orbit depends on the size of the orbit, not the shape. In other words, any two orbits of different eccentricity take the same amount of time to complete as long as their major axis lengths equal.

2.2 How to Describe an Orbit

Orbits exist in many different shapes and sizes. However, the task of describing an orbit in a way that differentiates it from any other orbit requires more than specifying the major axis length and eccentricity. This task also requires knowing the location and orientation of the orbit in space because all orbits exist on imaginary flat surfaces called geometrical planes. These surfaces can be tilted, rotated, and oriented in many different ways because space contains three dimensions and is certainly not flat. A standard reference frame called the geocentric J2000 coordinate system provides a way to measure the tilt, rotation, and orientation of these imaginary surfaces that contain orbits.

Geocentric J2000 Coordinate System

Use of this coordinate system in orbital mechanics and spaceflight engineering ensures that different people describing the same orbit will describe the orbit in exactly the same way. Geocentric J2000 coordinates describe the location of any point on the Earth, or any point in space near the vicinity of the Earth. Like any standard three-dimensional Cartesian coordinate system, this one contains X, Y, and Z axes. *Geocentric* means that the center of the Earth serves as the origin (center) of the coordinate system. Furthermore, the XY plane contains the Earth's equator, the Z axis points through the North Pole, and the X axis points toward deep space in the direction of an imaginary line called the vernal equinox (see Figure 10 on page 45). This line connects the center of the Earth and the center of the Sun on the first day of spring, usually near March 20. The vernal equinox also happens to point toward a group of stars (constellation) called Aries.

J2000 denotes that the definitions for where the X, Y, and Z axes point are exactly valid only in the year 2000. In reality, the spin of the Earth causes the North Pole to gradually point to different locations in deep space over a 23,000-year cycle. The Moon also disrupts this pattern in another cycle that repeats every 18 years. Consequently, the location of the X axis with respect to the vernal equinox, the Z axis with respect to the north pole, and the XY plane with respect to the equator of the Earth differ slightly. These changes only amount to minuscule variations almost impossible to detect. However, these changes do affect most orbital and navigation calculations because they require an extremely high degree of precision. Consequently, spaceflight engineers often use slightly modified coordinate systems that account for these changes.

Figure 10: *Understanding the J2000 Coordinate System*

What the J2000 Coordinate System Is Used For

Spaceflight engineers use this coordinate system to describe points on the surface of the Earth, or in space near the vicinity of the Earth. It is a standard three-dimensional system. The cut-out drawing of the Earth to the right illustrates the J2000 system's properties.

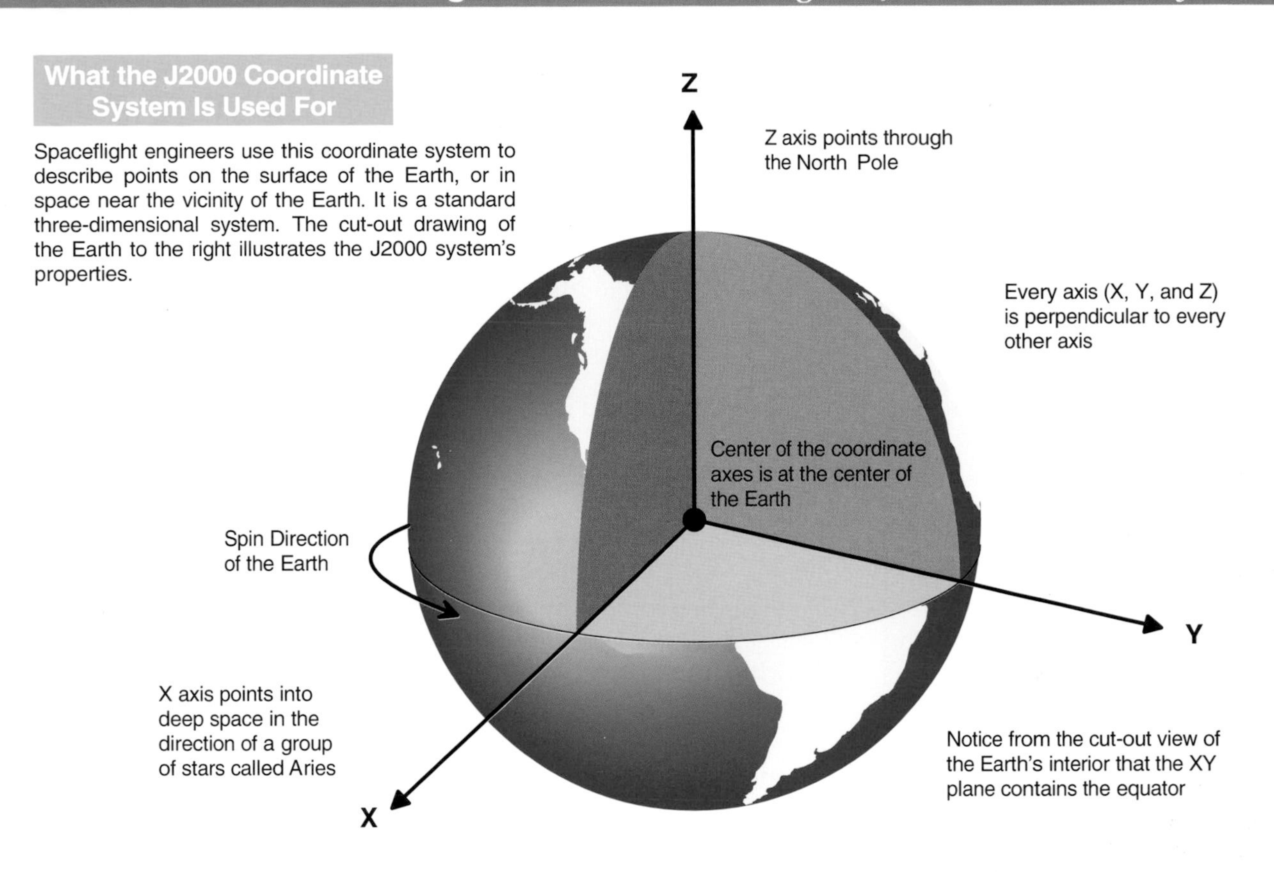

Important Concept

The coordinate axes stay fixed and always point to the same points in deep space. On the other hand, the Earth rotates about its axis counterclockwise within the J2000 coordinate system as viewed from the North Pole looking down. As a result, coordinates of points on the Earth constantly change while coordinates of points in space remain constant. In this example, the Earth is shown as seen from the North Pole at three different intervals six hours apart. Also, in these three drawings shown below, the radius of the Earth equals 1.0 for simplicity.

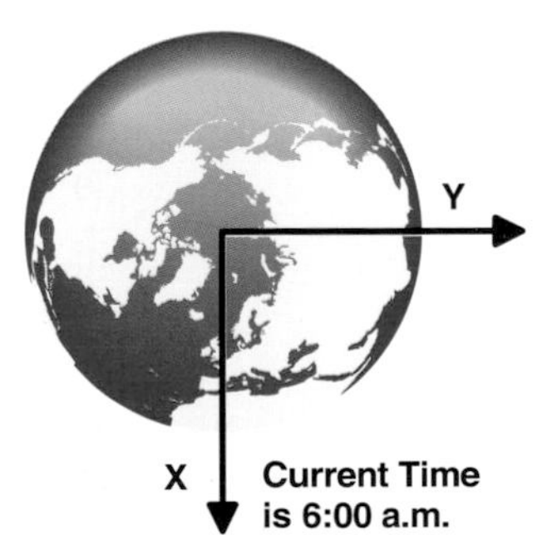

Here, the west coast of Africa on the equator starts at the coordinates X=1, Y=0, Z=0

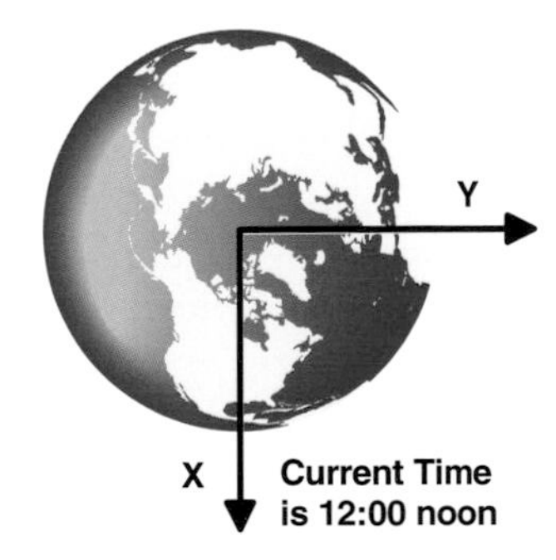

Six hours later, the Earth has rotated one quarter of a turn, and new coordinates of the west coast of Africa on the equator are X=0, Y=1, Z=0

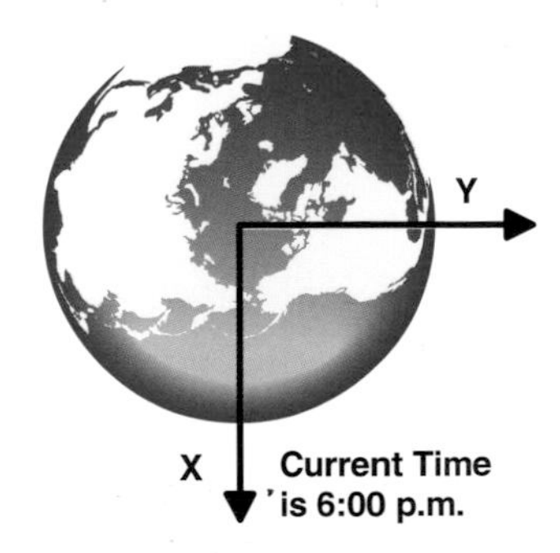

Another six hours passes and the Earth has rotated another one quarter of a turn. The new coordinates are X=-1, Y=0, Z=0

Above:

In March 1992, astronauts on Space Shuttle *Atlantis* shot this spectacular photograph of the southern tip of Greenland. The numerous "cracks" in the ice near the ocean are fjords carved by glaciers during the most recent Ice Age. Many of these fjords contain small settlements along the waterfront. The ice covering the surface of Greenland is estimated at nearly three kilometers thick. Often, giant chunks of this ice break off and fall into the sea as icebergs. Some icebergs exceed the size of small islands. During the winter, the fjords freeze over, limiting transportation to either dog sleds or helicopters.

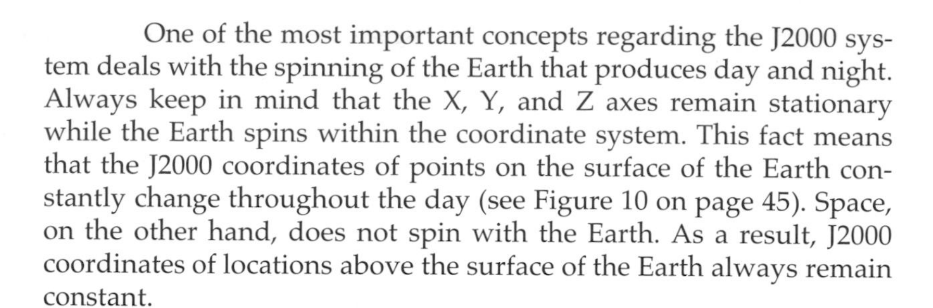

One of the most important concepts regarding the J2000 system deals with the spinning of the Earth that produces day and night. Always keep in mind that the X, Y, and Z axes remain stationary while the Earth spins within the coordinate system. This fact means that the J2000 coordinates of points on the surface of the Earth constantly change throughout the day (see Figure 10 on page 45). Space, on the other hand, does not spin with the Earth. As a result, J2000 coordinates of locations above the surface of the Earth always remain constant.

Classical Orbital Element Set

Six orbital parameters, measured in terms of the J2000 coordinate system, collectively form the classical orbital element set. Together, this set of numbers provides a complete set of parameters that uniquely specifies an orbit and differentiates it from any other orbit. In addition, the set also provides a method for visualizing the size, position, and orientation of any orbit in three-dimensional space without the use of complicated mathematical equations.

The first two parameters in the element set, major axis length and eccentricity, define the size and shape of an orbit, respectively. In reality, spaceflight engineers use a parameter called the semi-major axis to describe size. The semi-major axis length is exactly half the length of the major axis. The reason for this seemingly unintuitive system comes from the fact all the orbital equations that spaceflight engineers deal with use the semi-major axis as the major axis as the parameter that defines size.

All books and publications that deal with orbital mechanics use the symbol a for semi-major axis, and e for eccentricity. Remember that e takes on values between 0 and 1 depending on how circular or squashed the ellipse looks. The units for a vary depending on who took the measurement. Some people prefer to use kilometers or miles. Another popular method uses a unit of measurement where the distance equal to the radius of the Earth (denoted by the symbol R_e) equals 1. Use of this unit provides an easy method of visualizing the size of an orbit because most people do not remember the exact radius of the Earth. Always keep in mind that semi-major axis is not the same as orbital altitude.

RULE OF THUMB

Semi-major axis length describes the size of an orbit. Eccentricity describes the shape of the orbit (how squashed the ellipse looks). Remember, a circular orbit represents a special case of an elliptical orbit where the eccentricity value measures zero.

The next orbital element measures the tilt of the orbit plane with respect to the equator (see Figure 11 on page 47). Orbits contained entirely in the XY (equatorial) are tilted by 0°. Spaceflight engineers refer to these as equatorial orbits because all points along the orbit lie directly above locations along the equator. Tilting the orbit toward the North or South Pole increases the tilt value toward 90°. An orbit with a 90° tilt lies in a plane perpendicular (at a right angle) to the equator. Satellites on this type of orbit will fly directly over the North and South Poles. Consequently, spaceflight engineers call this type of orbit a *polar orbit*.

Figure 11: *What Tilted Orbits Look Like* ◄

The inclination of an orbit describes the degree of tilt of the orbit relative to the Earth's equator. It also provides a way to specify whether a satellite is moving in the posigrade (counter-clockwise) or retrograde (clockwise) direction as seen looking down at the Earth from the North Pole. Posigrade orbits always take on inclination values equal to the tilt angle, while retrograde orbits take on inclination values equal to 180° minus the tilt angle.

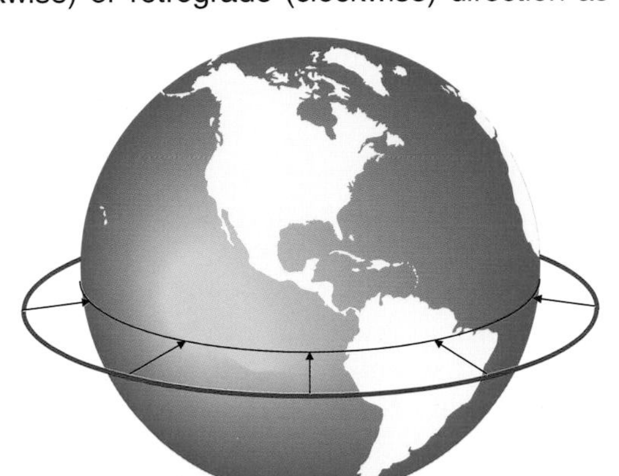

Orbital Tilt Angle = 0 Degrees

This type of orbit is called an equatorial orbit because all the points on the orbit lie above points on the equator. This is shown by the arrows pointing down from the orbit to the Earth. The inclination measures 0° for a posigrade orbit, and measures 180° for a retrograde orbit.

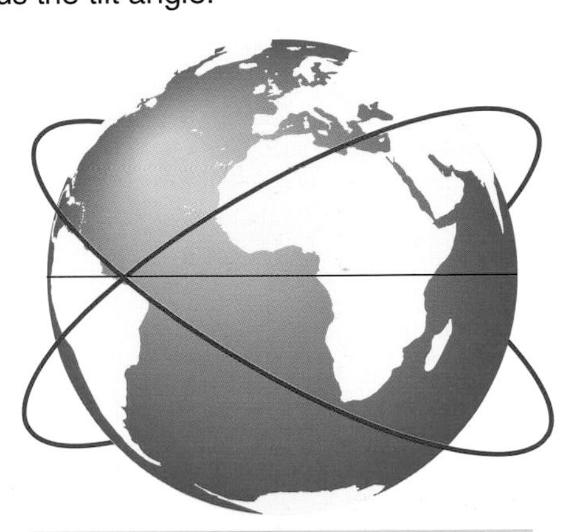

Orbital Tilt Angle = 30 Degrees

This type of orbit takes on an inclination value of 30° for motion in the posigrade direction, and takes on a value of 150° for retrograde motion. A satellite launched from NASA's Kennedy Space Center in Florida cannot reach orbits with a lower inclination than approximately 30° due to geometrical constraints. This picture shows just two of the many possible orientations of a 30° or 150° orbit.

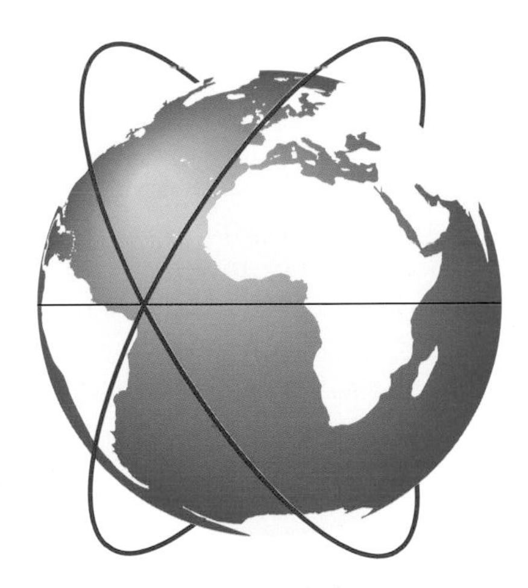

Orbital Tilt Angle = 60 Degrees

This type of orbit takes on an inclination value of 60° for motion in the posigrade direction, and values of 120° for retrograde motion. A satellite launched from NASA's Kennedy Space Center in Florida cannot reach orbits with a higher inclination than approximately 60°. This picture shows two of the many possible orientations of a 60° or 120° orbit.

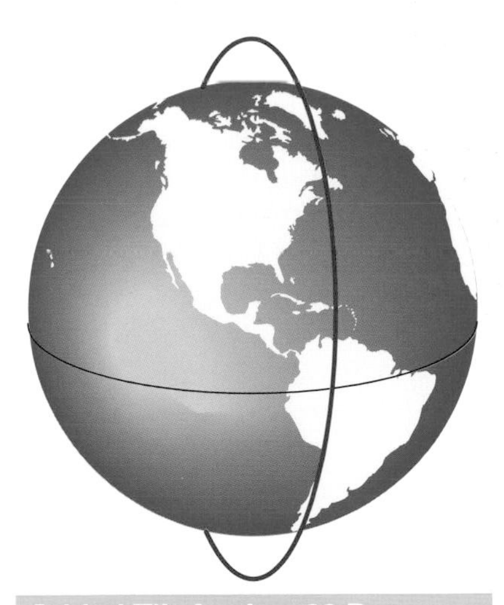

Orbital Tilt Angle = 90 Degrees

This type of orbit is called polar because a satellite on an orbit with this inclination passes over both poles. Posigrade and retrograde directions are not defined for polar orbits. American launches into polar orbits take place at Vandenberg Air Force Base in Southern California.

RULE OF THUMB

Inclination describes the tilt of an orbit as measured from the equator and also provides a way to tell which way a satellite moves along the orbital path. The posigrade direction (inclination values between 0° and 90°) is counterclockwise as seen looking down at the North Pole. On the other hand, retrograde (inclination values between 90° and 180°) is clockwise. Most satellites, including the Moon, move around the Earth in the posigrade direction. In addition, all of the planets in the solar system move around the Sun in the posigrade direction.

Visualizing Inclination

Value of Inclination	Implication
0°	Equatorial orbit west to east motion
0° to 45°	Orbit tilt close to equator west to east motion
45° to 90°	Orbit tilt close to pole west to east motion
90°	Polar orbit northerly or southerly motion
90° to 135°	Orbit tilt close to pole east to west motion
135° to 180°	Orbit tilt close to equator east to west motion
180°	Equatorial orbit east to west motion

Technically, a parameter called inclination measures the tilt of an orbit. The reason is that orbiting satellites can move either posigrade (counterclockwise as seen looking down from the North Pole) or retrograde (clockwise). Posigrade satellites orbit from west to east, while retrograde satellites move the other way. Inclination provides a way to describe the tilt of an orbit as well as a satellite's direction of motion on the orbit. Spaceflight engineers use the symbol i to represent inclination.

Inclination values between 0° and 90° denote posigrade orbits, while values between 90° and 180° represent retrograde orbits. The laws of trigonometry dictate that the angle at which a retrograde orbit is tilted to the equator takes on a value equal to 180° minus the inclination angle. For example, an i value of 150° corresponds to a retrograde orbit with a 30° tilt, while an i value of 180° describes a retrograde equatorial orbit (no tilt). Notice that an increase in tilt angle for a retrograde orbit results in a lower i value.

Remember that the center of the Earth always remains at the focus of the ellipse and in the plane of the orbit. This constraint dictates that a non-equatorial orbit intersect the equatorial plane in exactly two locations called nodes. The ascending node occurs at the point where a satellite moves from below (south of) the equatorial plane to above (north of) the equatorial plane. On the other hand, the descending node marks the location where a satellite moves from above (north of) the equatorial plane to below (south of) the equatorial plane. An imaginary line of nodes connects the two node points and runs through the center of the Earth. Equatorial orbits ($i = 0°$ or 180°) do not possess node points or a line of nodes because the entire orbit stays in the equatorial plane.

Do not confuse the concepts of below (south of) the equatorial plane and above (north of) the equatorial plane with up and down in terms of altitude. Upward motion, or a gain in altitude, occurs when a satellite moves from perigee to apogee. In contrast, downward motion, or a loss in altitude, occurs when a satellite moves from apogee to perigee. A satellite can move upward or downward at either the descending or ascending node depending on the location of perigee and apogee. For example, if perigee on a posigrade orbit occurs above the equator and apogee occurs below, then a satellite moves upward and south of the equator at the descending node, and downward and north of the equator at the ascending node.

The fourth parameter in the set depends on the location of the line of nodes. This element, called the right ascension of the ascending node or the longitude of the ascending node, measures the counterclockwise angle between the X axis and the ascending node point (see Figure 12 on page 49). The right ascension takes on values between 0° and 360°. An important concept to remember is that this orbital element locates the position of the ascending node within the J2000 coordinate system. It does not reveal the true anomaly of the satellite at the ascending node point. Most books use the symbol Ω,

Figure 12: *How Orbital Elements Describe an Orbit's Orientation*

Ascending Node: Point where the satellite moves from below to above the equatorial plane

Line of Nodes: Line through center of the Earth connecting both node points

Argument of Perigee: Angle along the orbit between the ascending node and perigee

Line of Apsides: Line through center of the Earth connecting apogee and perigee

Descending Node: Point where the satellite moves from above to below the equatorial plane

Longitude of the Ascending Node: Angle in the equatorial plane between the X-axis and the ascending node

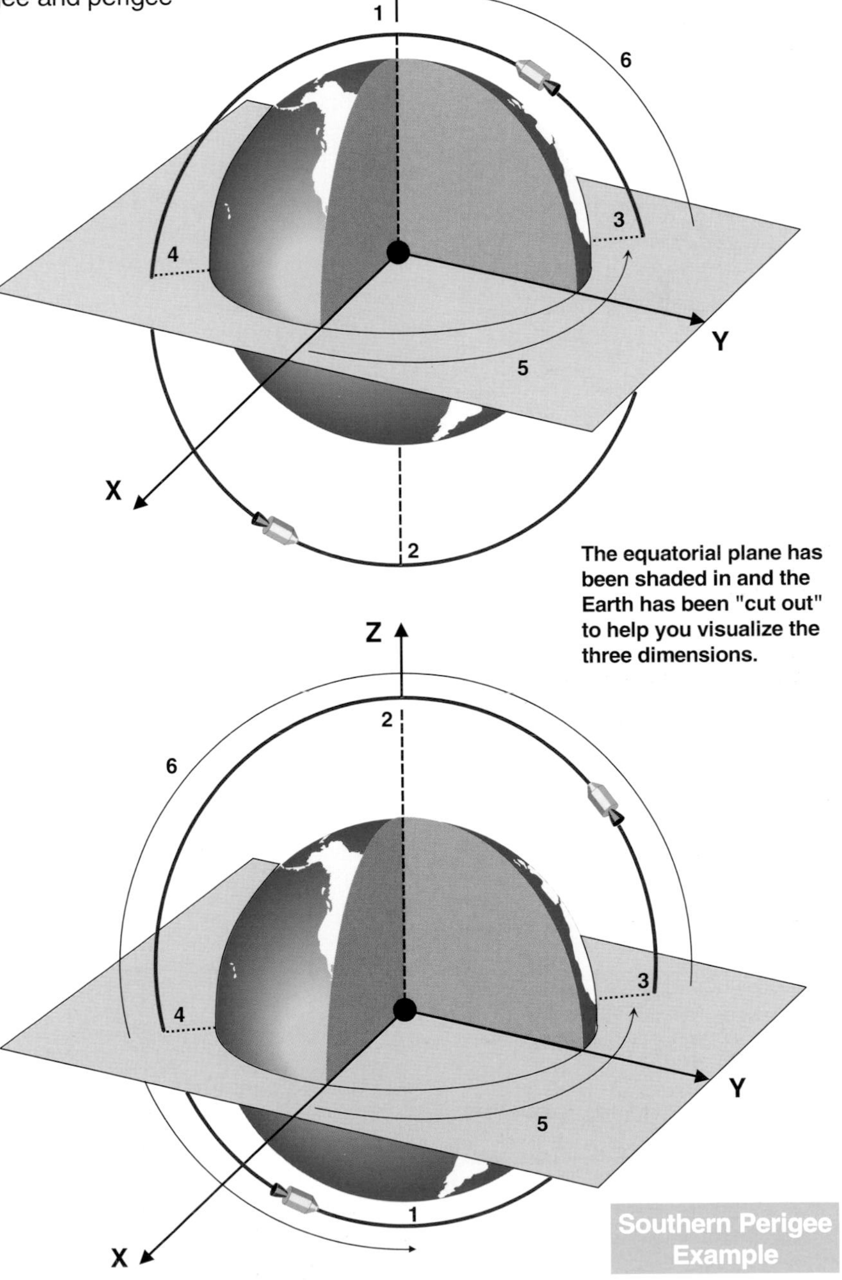

1 Perigee
2 Apogee
3 Ascending Node
4 Descending Node
5 Longitude of Ascn. Node Equals 135°
6 Argument of Perigee Equals 90°

Notice that an argument of perigee between 0° and 180° means that the satellite reaches perigee sometime after passing through the ascending node, but before reaching the descending node.

LEGEND	
Orbit	————
Line of Nodes	········
Line of Apsides	-------
Equatorial Plane	(shaded box)

Note, both upper and lower figures show counterclockwise orbits, 90° (polar) inclination

1 Perigee
2 Apogee
3 Ascending Node
4 Descending Node
5 Longitude of Ascn. Node Equals 135°
6 Argument of Perigee Equals 270°

Notice that an argument of perigee between 180° and 360° means that the satellite reaches perigee sometime after passing through the descending node, but before reaching the ascending node.

Above:
This photograph of the Sahara in Algeria was taken by astronauts on the Shuttle *Endeavour* after a rare rainstorm in October 1994.

Visualizing Location of Perigee

Argument of Perigee Value	Location of Perigee
0°	Over the equator, with satellite moving from south to north
0° to 90°	After passing equator, before arrival at most northern point
90°	Over most northern point
90° to 180°	After passing most northern point, before arrival at equator
180°	Over the equator, with satellite moving from north to south
180° to 270°	After passing equator, before arrival at most southern point
270°	Over most southern point
270° to 360°	After passing most southern point, before arrival at equator

Most northern, southern point is a latitude on Earth in the Northern and Southern Hemispheres, equal to the orbit's tilt angle

an uppercase Greek letter omega, to denote the right ascension of the ascending node. Ω is undefined for an equatorial orbit because this type of orbit lacks node points.

Spaceflight engineers refer to the fifth orbital element as the argument of perigee. This parameter measures the true anomaly between the ascending node and perigee (see Figure 12 on page 49). Most books use the symbol ω, the lowercase letter omega in the Greek alphabet, to represent this orbital element. Values for ω vary between 0° and 360°. A value between 0° and 180° means that perigee occurs north of (above) the equator and somewhere after a satellite passes the ascending node. On the other hand, an ω between 180° and 360° places the perigee south of (below) the equator and somewhere after the descending node. ω is undefined for a circular orbit because this type of orbit lacks unique perigee and apogee points.

The first five orbital elements provide a method to visualize the shape, size, and orientation of an orbital path. The sixth and last orbital element, called the time of perigee passage, provides crucial data toward pinpointing the location of a satellite on the orbit. Knowledge of a time in the past when a satellite passed through perigee, combined with knowledge of the current time, semi-major axis length, and eccentricity allow spaceflight engineers to use orbital equations to calculate a satellite's current true anomaly. In turn, knowledge of the true anomaly combined with the a, e, i, Ω, and ω allows the use of geometrical formulas to calculate a satellite's current J2000 coordinates.

Many people often wonder why some common and easy-to-understand orbital parameters, like the period or altitude at perigee, fail to appear in the classical set. Remember that orbital elements provide the minimum amount of information required to uniquely specify an orbit and to distinguish it from any other orbit. The addition of further parameters to the classical set results in redundant information because equations exist to determine every other orbital parameter as a function of the classical six. Unfortunately, use of the classical six parameters to visualize an orbit takes a lot of practice. This book will often use other, simpler parameters, like the period, to help you visualize orbits with a minimal amount of confusion.

2.3 Ground Track Drawings

The classical orbital element set provides a way to determine a satellite's location in space. However, a satellite's usefulness often depends on knowing the location of the satellite with respect to the surface of the Earth. For example, what if meteorologists need a satellite to monitor a developing hurricane in the Atlantic Ocean, or if the Department of the Interior needs to scan for natural resources in the Rocky Mountains, or if astronauts need to know whether their spacecraft will fly over a safe landing site? In all of these cases, a satellite will turn into a piece of expensive orbiting junk without the

knowledge of what the path on the surface of the Earth directly under the orbiting satellite looks like.

A drawing called a ground track map provides spaceflight engineers with extremely useful information by revealing the locations on the Earth that a satellite directly flies over during the course of one or more transits around its orbital path. The first step toward understanding what ground tracks look like involves learning about the standard system in use today for describing locations on the surface of the Earth.

Latitude and Longitude

The J2000 coordinate system provides a good method to describe locations in space. However, as seen in Figure 10 on page 45, this system creates confusion in the description of locations on the surface of the Earth because the coordinates change as the Earth spins. A different coordinate system, called the latitude-longitude system, rectifies the situation by providing a reference frame that rotates with the Earth. All ground track drawings use this coordinate system.

In the latitude-longitude system, two values describe the location of any point on the surface of the Earth (see Figure 13 on page 52). Latitude measures how far north or south a point lies from the equator. These values range from 90° North at the North Pole to 0° at the equator to 90° South at the South Pole. A latitude measurement of 45°, either north or south, represents locations halfway between the equator and one of the two poles. From any location on the Earth, any other location that lies due east or due west takes on the same latitude value. Thus, the latitude lines on a globe run horizontally in the east-west direction.

Longitude measures how far east or west a point lies from an imaginary line that runs from the North to South Pole through a portion of London called Greenwich. Most people refer to this line as the Prime Meridian, or Greenwich for short. Locations with a 0° longitude lie exactly on the Prime Meridian. Values for longitude range from 0° to 180° West for locations west of the Prime Meridian (Western Hemisphere), and from 0° to 180° East for locations east of the Prime Meridian (Eastern Hemisphere). From any location on the Earth, any other location that lies due north or south takes on the same longitude value. As a result, longitude lines on a globe run vertically in the north-south direction.

Remember that the Earth's round shape allows for eastward or westward travel to reach any location on the surface. This fact, combined with the fact that a full circle contains 360°, means that the values 180° East and 180° West both describe the same longitude halfway around the world from the Prime Meridian. Some people use values between 0° and 360° East to measure longitude. In this case, longitude values greater than 180° East describe locations in the

Classical Orbital Element Set

Symbol	Orbital Element and Purpose
a	*Semi-Major Axis* Measures the orbit's size
e	*Eccentricity* Describes the orbit's shape
i	*Inclination* Measures the orbit's tilt
Ω	*Longitude of the Ascending Node* Locates the equatorial crossing
ω	*Argument of Perigee* Locates the closest approach point
t	*Time of Perigee Passage* Determines position on the orbit

Where in the World?

City	Location	
Atlanta, GA	84° W	33° N
Cape Town, S. Africa	19° E	34° S
Chicago, IL	88° W	42° N
Fairbanks, AK	149° W	65° N
Houston, TX	95° W	29° N
London, England	0° E	52° N
Moscow, Russia	38° E	56° N
Nairobi, Kenya	37° E	1° S
Quito, Equador	78° W	0° N
San Francisco, CA	122° W	38° N
Sydney, Australia	151° E	34° S
Tokyo, Japan	140° E	36° N
Washington, D.C.	77° E	39° N

All values are approximate and have been rounded to the nearest degree

► Figure 13: *A Look at How the Latitude and Longitude System Works*

Latitude: Measures how far north or south a point on the Earth lies from the equator

Longitude: Measures how far east or west a point on the Earth lies from the Prime Meridian

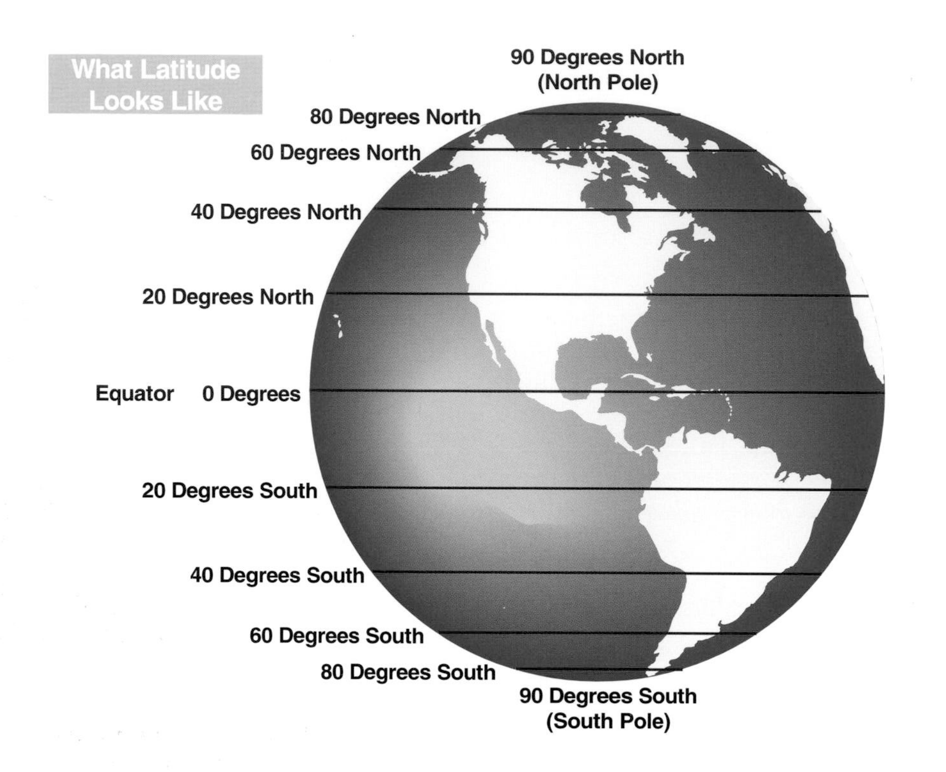

Lines of equal latitude run east-west along the globe. They are also called parallels.

Lines of equal longitude run north-south along the globe from the North to South Pole. They are also called meridians.

If you walk across one of the poles, you will wind up on the other side of the world, 180° in longitude away from where you started.

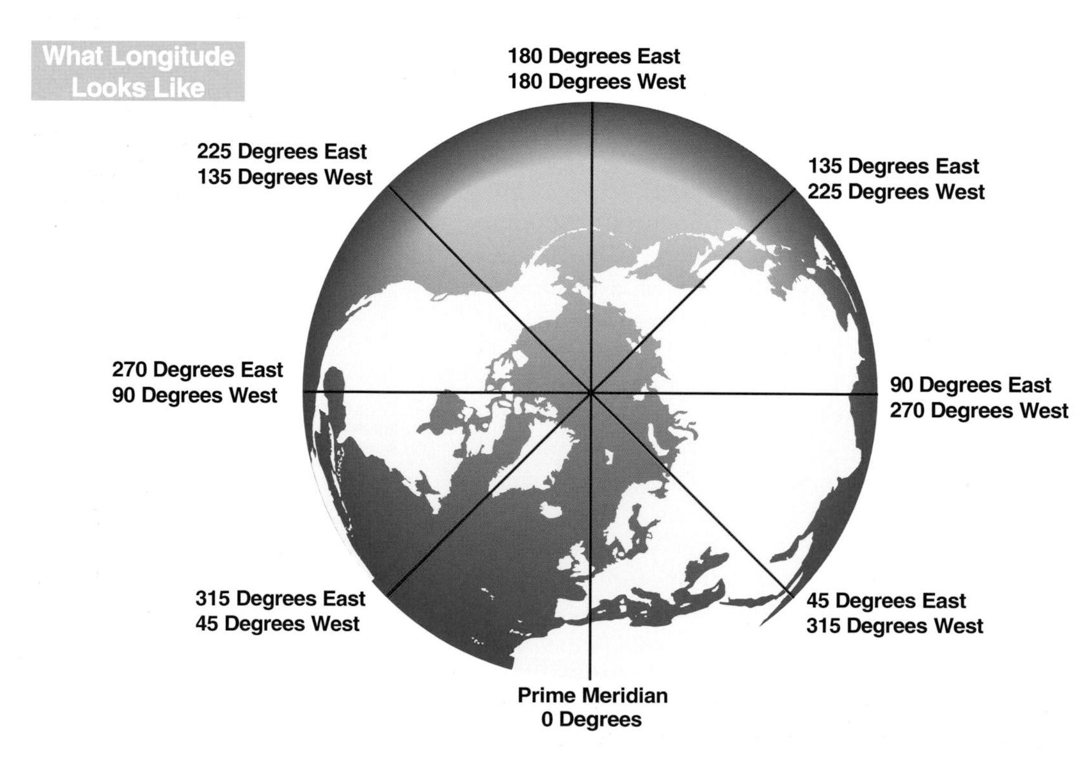

Western Hemisphere. Another method specifies longitude values between -180° and +180° where negative values indicate locations in the Western Hemisphere and positive values indicate locations in the Eastern Hemisphere. The bottom part of Figure 13 on page 52 illustrates this concept.

Ground track drawings almost always appear on Mercator projections of the Earth's surface. This type of map allows the entire surface of the round world to appear on a rectangle. Latitude lines run from the left to the right edges of the map, while longitude lines run from the North Pole at the top of the map to the South Pole at the bottom. The left and right edges of the map represent the same locations. Thus, anything that moves past the right edge will appear at the same latitude on the left edge, and vice versa. This effect allows a flat map to simulate the roundness of the Earth.

Ground Tracks of Low Circular Orbits

A ground track is the path on the surface of the Earth that falls directly under a satellite's orbit. The best way to understand how ground tracks look on Mercator projections involves thinking about the equator. When drawn on a globe, the equator appears as a closed, circular path that loops all the way around the world. Mercator projections of the Earth work just like a globe printed on the label of a soup can. Peeling the label off of the can yields the map. When printed on a Mercator map, the Earth's equator no longer looks like a closed, circular path. Instead, it turns into a horizontal line across the middle of the map.

The top of Figure 14 on page 54 shows a typical inclined circular orbit. Like the equator, the ground track under the orbit looks like a closed, circular path that loops all the way around the world when on a globe. One way to understand the shape of the ground track involves thinking about it as another circular orbit, but with an orbital altitude of sea level.

What does this ground track look like on a Mercator projection of the world? Remember that a satellite on an inclined orbit spends part of its time above (north of) the equator and the other part of its time below (south of) the equator. An inclined orbit's ground track follows the same property. Instead of looking like a horizontal line like the equator, the ground track looks like a line that has been twisted so that half of it lies above the equator and half lies below. In essence, the ground track looks like an S-shaped curve centered on the equator (see Figure 14 on page 54). Use your imagination to convince yourself that "putting the label back on the can" turns the S-shaped curve back into the inclined circular path that loops all the way around the world.

One extremely important property about a ground track deals with how far north and south the track extends from the equator. Geometrical constraints dictate that both the maximum northern

Above:
This east-looking photograph shows the great Ganges River delta region of Bangladesh and India. The cloud-covered Arakan Hills can be seen toward the top, center of the picture. Near the center, the Ganges and the Brahmaputra Rivers join into one as they flow into the Bay of Bengal. Astronauts in Space Shuttle *Atlantis* shot this photograph in December 1988. During the summer, monsoons often form in this region and cause heavy damage to crops and human life.

RULE OF THUMB

A Mercator projection is the "flat Earth" type of world map printed on a rectangle. On this type of map, locations near the poles, such as Greenland and Antarctica, appear larger than areas of similar size near the equator.

► Figure 14: *What the Basic Shape of a Ground Track Looks Like*

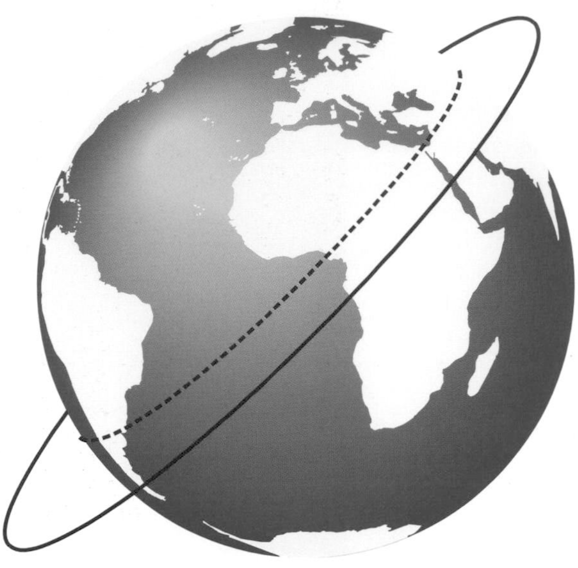

LEGEND	
Orbital Path in Space	———
Ground Track on Earth	- - - - - -

The ground track is the set of points on the surface of the Earth that lie directly below the orbital path. These diagrams show what the ground track would look like if the Earth did not rotate. On the globe map, the ground track (shown as a dotted line) forms an inclined circular path that extends all the way around the world. It looks like this no matter the orbit's size or eccentricity.

On the Mercator (flat Earth) type map projection shown below, the inclined circular path around the world turns into an S-shaped curve. Notice that since the Earth does not rotate, the ground track corresponding to one trip around the orbital path spans 360° worth of longitude along the Earth's surface. Also notice that the ground track extends as far north as it does south. The maximum latitude reached always equals the tilt angle of the orbit. This rule holds no matter the orbit's size or eccentricity.

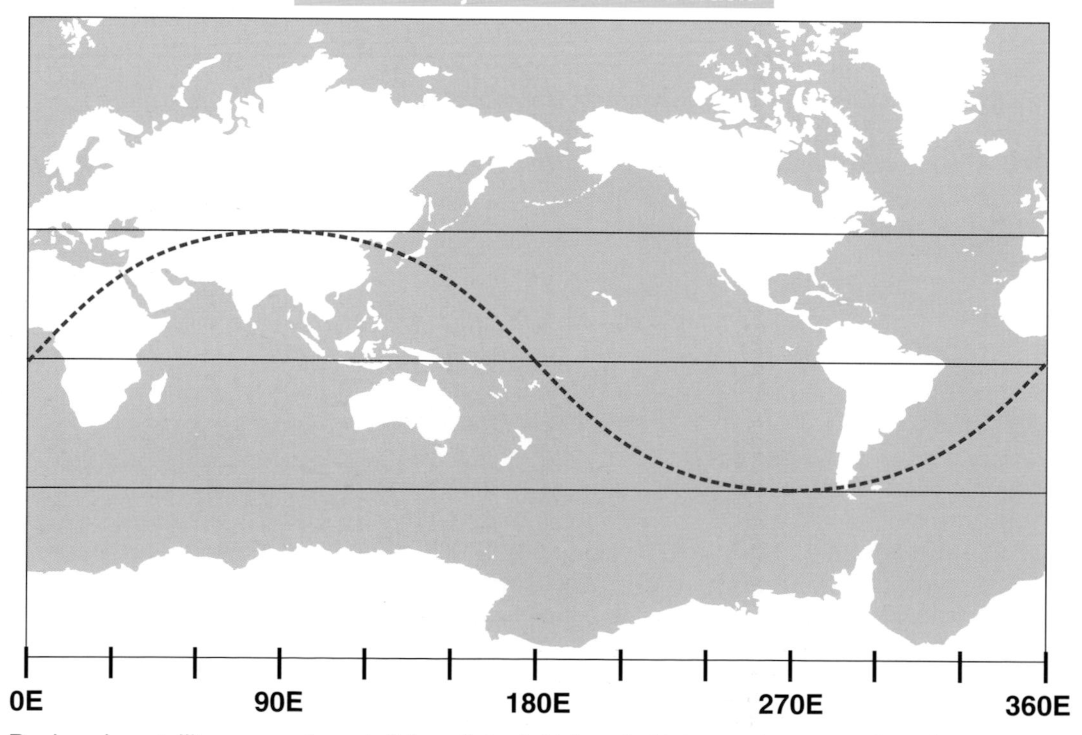

Posigrade satellites move from left (west) to right (east). Retrogrades move the other way.

and southern latitudes on the ground track must equal the tilt angle of the orbit. A consequence of this rule is that for a satellite to fly directly over a location on the Earth, the orbit's tilt angle must equal or exceed the latitude value of the ground site. In this example, the maximum latitude (either north or south) on the ground track measures 50°. This fact indicates that the inclination of the circular orbit in Figure 14 on page 54 measures 50° if the orbit is posigrade, or 130° if the orbit is retrograde. Keep in mind that the maximum latitude rule also applies to noncircular orbits.

The next important property to understand deals with which way the satellite moves on a Mercator projection of a ground track. Do they move left to right, or right to left? Look at the view of the Earth from the North Pole in Figure 13 on page 52. Notice that Europe lies to the east of, and counterclockwise from, North America. This example shows that eastward motion is counterclockwise and westward motion is clockwise. Consequently, satellites on posigrade (counterclockwise) orbits tend to move from left (west) to right (east) along the ground track. Conversely, a retrograde satellite tends to move from right to left.

Knowledge of the direction of motion along the ground track helps determine the location of the ascending node (south to north equatorial crossing) and the descending node (north to south equatorial crossing). In Figure 14 on page 54, the ascending node for a posigrade (west to east) orbit occurs at the left edge of the map, the descending node in the middle, and the next ascending node at the right edge. Conversely, the descending node for a retrograde (east to west) orbit occurs at the right edge of the map, the ascending node in the middle, and the next descending node at the left edge of the map.

Over the course of one orbit around the Earth, satellites pass through the ascending and descending nodes only once no matter the orbit's shape or size. Therefore, if a posigrade (west to east) satellite begins an orbit at the ascending node at the left edge of the map, then the ascending node at the right edge of the map must mark the beginning of the next orbit. This fact simplifies the task of looking at a ground track to determine the portion of the track that corresponds to a single revolution around the Earth. Essentially, just find the portion of the ground track that falls between two ascending or two descending nodes.

At this point, it might be tempting to think that 360° of longitude on the Earth always separates the two ascending nodes on the ground track, just as in Figure 14 on page 54. After all, common sense seems to dictate that a satellite must fly all the way around the world during the course of one orbital revolution in space. Unfortunately, this assumption only holds true if the Earth does not rotate. In fact, the ground tracks in Figure 14 on page 54 are drawn with the Earth's rotation frozen in place.

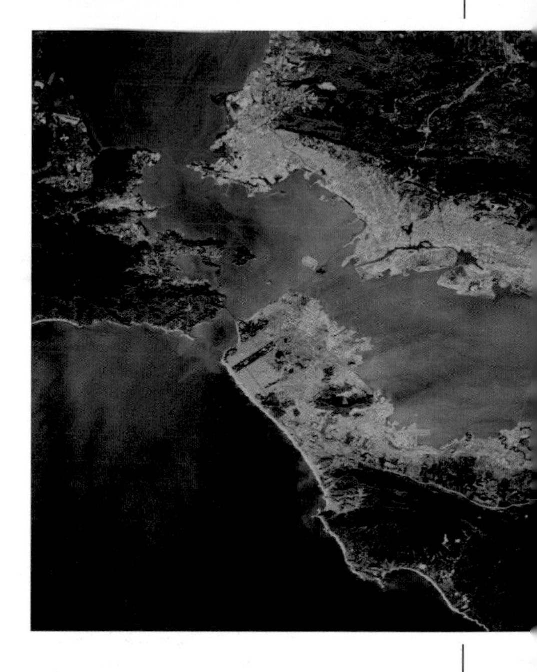

Above:

This photograph of the San Francisco Bay area was taken by the astronauts on *Discovery* in May 1991. In the image, San Francisco lies at the tip of the peninsula extending from the bottom to the upper middle of the picture. Golden Gate Park is clearly visible as a long, rectangular dark patch in the middle of the city. Also visible are the Golden Gate Bridge (line connecting San Francisco with the Marin peninsula extending down from the top), and the Bay Bridge (to the right of San Francisco).

RULE OF THUMB

On a low Earth orbit ground track, a posigrade satellite moves from left (west) to right (east). On the other hand, a retrograde satellite moves in the opposite direction. If the Earth did not rotate, the ground track corresponding to one orbit would cover 360° of longitude along the Earth's surface.

► Figure 15: *How the Earth's Rotation Warps the Ground Track*

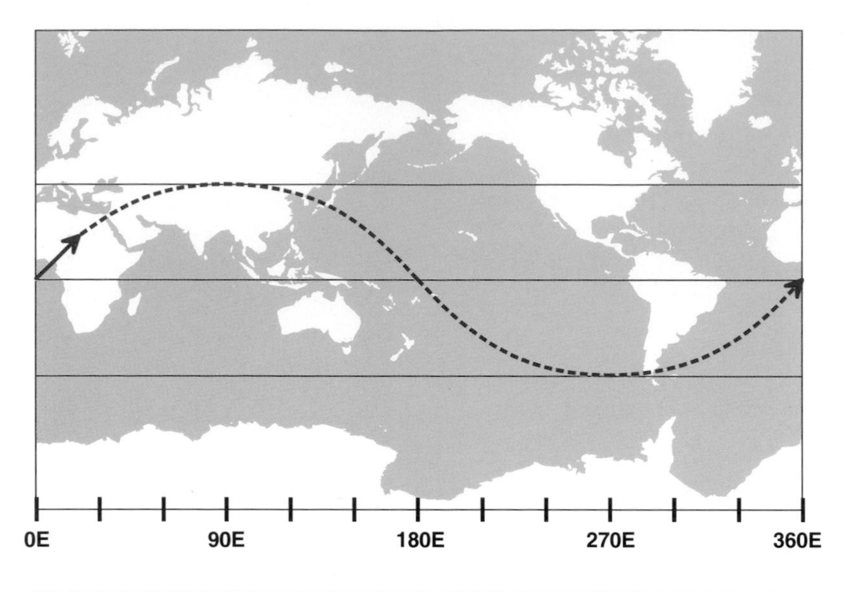

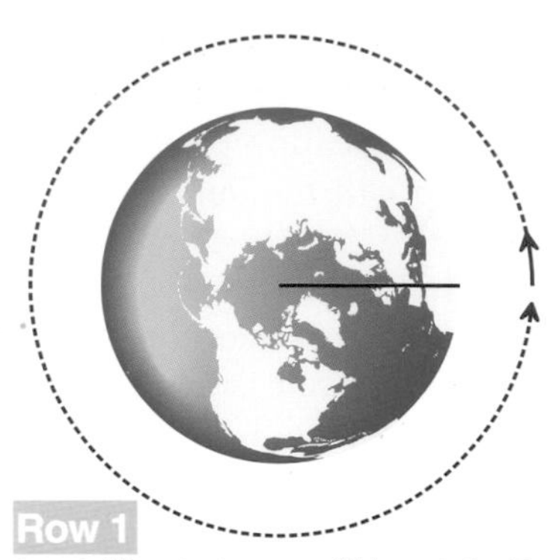

Row 1

A satellite starts over Africa at 0° East longitude on a two-hour posigrade orbit around the Earth. Dotted lines show the orbital path that the satellite needs to travel on to complete the orbit. If the Earth does not spin, the orbit will finish at 360° (0°) East.

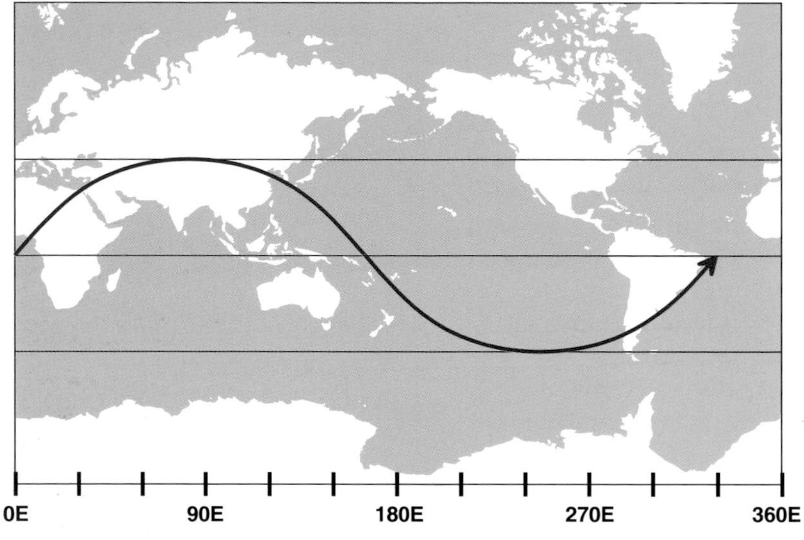

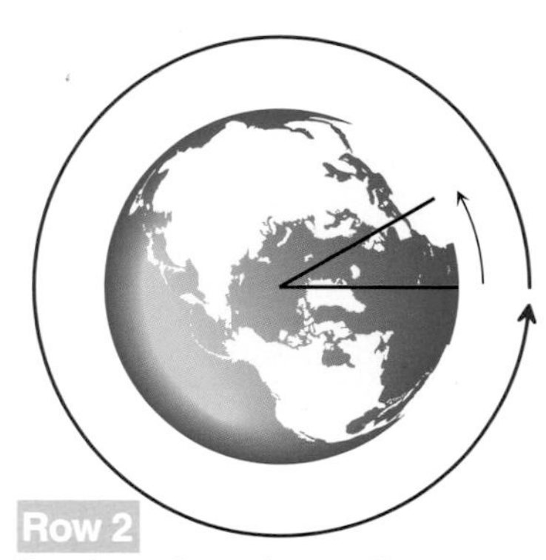

Row 2

Two hours later, the satellite returns to the same point in space from where it started, and one orbit is done. But, now the satellite finds itself over 330° East, a location 30° to the west of 360° (0°) East.

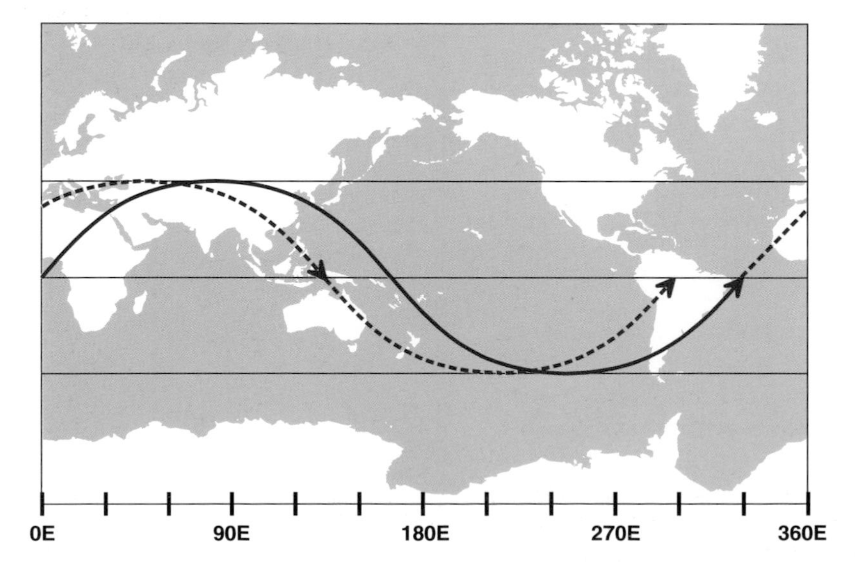

Row 3

The finish point of every orbit in this example will be 30° to the west of the starting point because of the Earth's spin. Since the satellite moves to the east on a posigrade orbit, the westward shift of the finish point means that every orbit only covers 330° worth of longitude along the Earth's surface instead of 360°.

Orbit number two (dashed line shown to the left) starts where orbit number one finished - over the Atlantic at 330° East. The orbit finishes 30° to the west at 300° East longitude over South America. Eventually, after several more orbits, a crisscrossed pattern will form on this map. Keep in mind that these numbers are specific to this example.

The Earth's rotation changes the entire picture because the orbital path and the ground track do not rotate with the Earth. All of the dynamics that govern orbital motion and the orientation of an orbit in space do not depend on the Earth's rotation. As a result, orbital paths remain constant within the J2000 coordinate system while the Earth rotates counterclockwise under the orbit. If you could hover motionless in space and look down at the Earth, you would see the ground moving eastward. This rate of motion amounts to 15° every hour because the Earth rotates through 360° in 24 hours. Therefore, if you looked down at the ground once every hour, you would see a location 15° in longitude to the west of the location that you saw an hour earlier.

One way to think about how this rotating Earth effect works involves thinking about running on an oval track, like the ones used in the Olympics. Normally, a runner starting at the starting line must run 400 meters to travel once around the track to arrive at the finish line, located at the same point as the starting line. Suppose that sometime during the race, somebody moves the finish line 40 meters in the same direction as the runner's motion. In other words, the finish line is moved 40 meters further up the track. Runners who run 400 meters expecting to finish the race will still be 40 meters short of the finish line. However, regardless of the finish line's location, they will have completed one lap around the track because they will have returned to the same physical point in space from where they started.

Now, consider a satellite with a two-hour posigrade orbit that starts at the ascending node at 0° longitude (see Figure 15 on page 56). During the two hours that the satellite takes to circle the world moving from west to east, the Earth also rotates by 30° from west to east. Think of this scenario as the ground track staying fixed in space while the Earth's surface slides to the east (right) under the track. In other words, the so-called finish line relative to the ground moves by 30° to the east, and away from the satellite during the course of the orbit.

When the satellite returns to the same physical point in space from where it started, it will still need to travel 30° more to the east to reach the longitude from where it started. In addition, the ground location under the satellite will be a point with a longitude of 330°, 30° to the west of the original starting point. However, regardless of the location of the finish line, the satellite will still have completed one orbit because it will have returned to the same location in space from where it started.

The net result is that a satellite with a two-hour period only flies over 330° of longitude along the surface of the Earth to complete one orbit instead of the full 360°. Therefore, the first orbit starts at 0° longitude and ends at 330°, the second orbit starts at 330° and ends at 300°, the third orbit starts at 300° and ends at 270°, and so on. This continuous westward shift causes a ground track drawing that displays several consecutive orbits to take on a "crisscrossed"

RULE OF THUMB

The rotation of the Earth causes a satellite on a posigrade orbit to fly over less than 360° of longitude during the course of one complete orbit. In general, when a satellite completes an orbit in the posigrade direction, it will find itself to the west (relative to the ground) of its position one orbit earlier.

Scientific Fact!

In general, a satellite on a posigrade orbit moves eastward. However, at high enough altitudes, the eastward orbital velocity of a satellite will be less than the eastward velocity of the ground (moving due to the Earth's spin). In this case, the satellite will appear to move westward along its ground track even though the orbit is in the posigrade direction!

► Figure 16: *What a Retrograde Ground Track Looks Like*

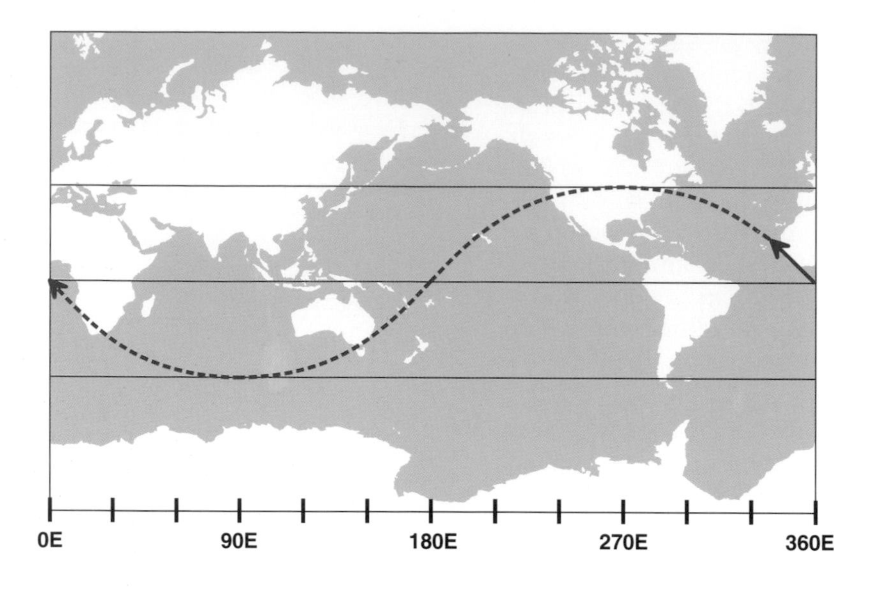

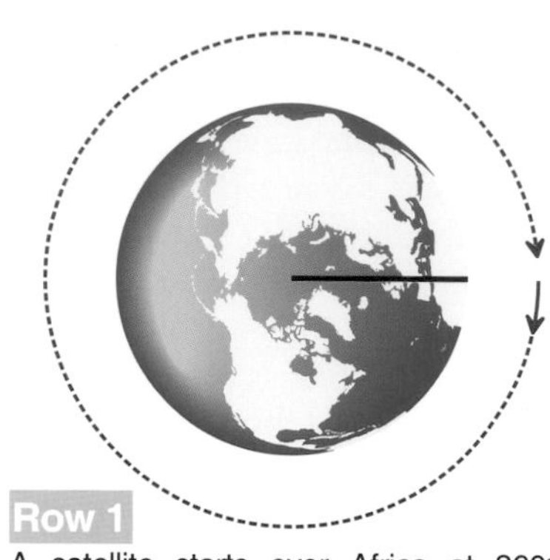

Row 1

A satellite starts over Africa at 360° East longitude on a retrograde two-hour orbit around the Earth. Dotted lines show the orbital path that the satellite needs to travel on to complete the orbit. If the Earth does not spin, the orbit will finish at 0° (360°) East.

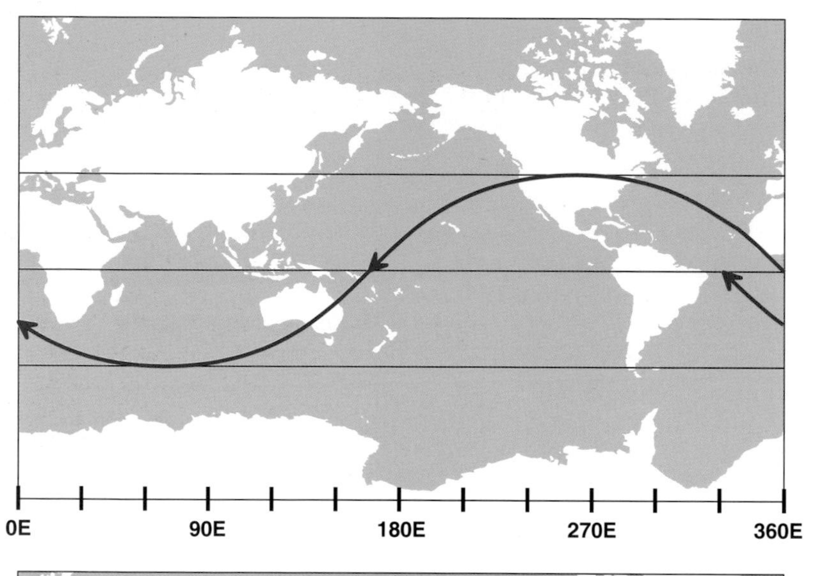

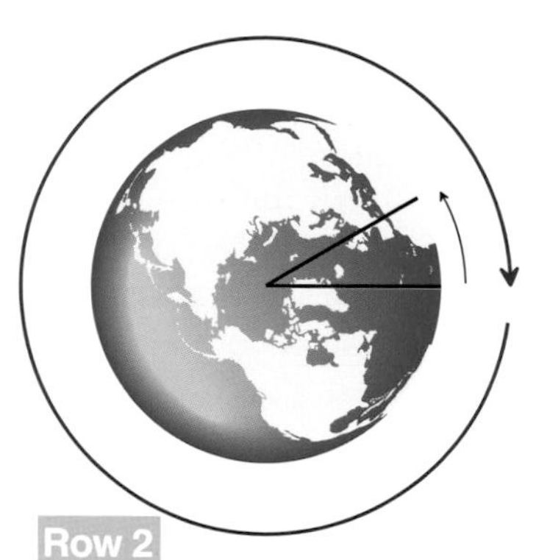

Row 2

Two hours later, the satellite returns to the same point in space from where it started, and one orbit is complete. It has gone around the world relative to the ground slightly more than once.

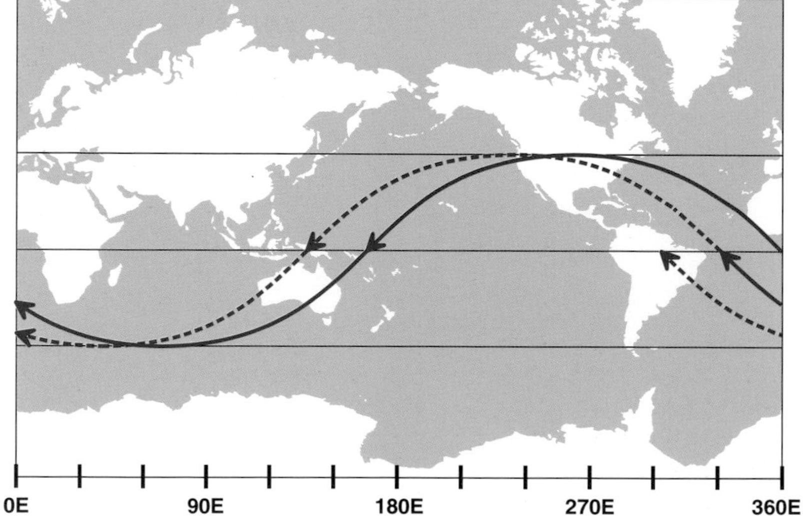

Row 3

The reason is that the Earth spins by 30° in two hours from west to east. The finish point will always be 30° to the west of the start point. Since the Earth spins in the opposite direction that the satellite moves in, each orbit covers 390° worth of longitude on the world map instead of 360°.

Orbit number one (solid line) begins at 360° (0°) and ends at 330° East. Orbit number two begins at 330° East and ends at 300° East. This may seem the same as the posigrade example, but keep in mind that the satellite moves around the Earth relative to the ground more than once during the course of one orbit.

appearance (see bottom of Figure 15 on page 56). However, after 12 complete orbits, the ascending node returns back to its original ground-relative location at 0° longitude, and the pattern repeats for orbits number 13 through 24.

The number of orbits that a posigrade satellite must complete before its ground track pattern repeats depends on the westward displacement of the ascending node, relative to the ground, on each successive orbit. If the westward displacement is an even multiple of 360°, then the pattern repeats relatively quickly. For example, the ground track pattern repeats after eight orbits for a three-hour orbit (45° displacement per orbit), and after six orbits for a four-hour orbit (60° displacement per orbit). However, if the westward displacement does not divide evenly into 360°, a long time will pass between repeat cycles. For example, the ground track pattern takes 240 orbits to repeat for a 2.5-hour orbit. Many orbits have ground track patterns that never repeat. Eventually, a satellite on a *non-repeating ground track* orbit will fly directly over every single location between its maximum northern and southern latitude range.

In general, if distance traveled is measured relative to the moving ground of the rotating Earth, then posigrade satellites fly around the world less than once during the course of one orbit. Increasing a satellite's period decreases the amount of ground it covers. However, do not confuse the concept of longitude traveled with true anomaly. Remember, longitude is used to measure distance traveled with respect to the moving ground, while true anomaly measures distance traveled along the orbit in space. Keep in mind that a satellite always completes 360° of true anomaly along the orbit in space during the course of one revolution around the Earth. This fact holds true regardless of how much, or how little longitude a satellite flies over during the course of one orbit. If the Earth did not rotate, then a satellite would fly over 360° of longitude and true anomaly during one orbit.

Now, what do the ground tracks of retrograde (west to east) orbiting satellites look like? Retrograde satellites orbit in the opposite direction of the Earth's spin. Consider another race on the 400-meter oval track. Just like last time, completing the race involves running all the way around the track and reaching the finish line, located at the same place as the starting line. In this case, suppose that somebody moves the finish line by 40 meters in the opposite direction of the runners' motion during the race. The effect is that the runners will reach the finish line 40 meters early, but will not have completed a lap. They will still need to run 40 more meters past the finish line to return to the same physical point in space from where they started.

This example translated into orbital mechanics means that a retrograde-orbiting satellite flies around the world slightly more than once relative to the ground during the course of one orbit. In the case of the two-hour orbit, the ground under the starting point moves 30° to the east during the course of the orbit. Since a retrograde satellite

Above:
Mount Everest, the highest mountain in the world at 8.8 kilometers (29,028 feet), appears almost in the center of this cloud-free photograph. The challenging "North Face" of the mountain is in shadow and faces the bottom edge of the picture. NASA analysts consider this image, taken by astronauts on Space Shuttle *Columbia* in October 1993, as one of the best Mount Everest photos ever taken.

RULE OF THUMB

The rotation of the Earth causes a satellite on a retrograde orbit to fly over more than 360° worth of longitude during the course of one revolution. Retrograde-orbiting satellites represent only a small fraction of the total number of satellites sent into space since the dawn of the space age in 1957.

RULE OF THUMB

The ground track of a polar orbit looks like straight lines running in the north-south direction. Strictly speaking, a polar orbit measures 90° in inclination. However, from a qualitative perspective, most spaceflight engineers consider orbits anywhere in the range of 80° to 110° as polar.

Scientific Fact!

Earth-observing satellites that transmit data to the ground to help with environmental research or weather prediction usually orbit in high-inclination, near-polar orbits. Many of them orbit in a special type of polar orbit that spaceflight engineers refer to as *Sun-synchronous.* In this type of orbit, the angle between the orbit and the Sun always remains constant. This special orientation causes the local time under the satellite's equatorial crossing to remain fixed no matter what part of the world is under the satellite as it crosses the equator.

One advantage of Sun-synchronous orbits is that the shadows cast by objects on the ground, as seen from the satellite, will always be at the same lengths and orientations. Such a characteristic provides a common reference point for scientists to compare satellite pictures of different parts of the world.

moves westward, it will encounter the longitude of the starting point before returning to the same point in space from where the orbit started, just like the runners on the racetrack reaching the finish line earlier than expected. In order to complete the orbit, the satellite must fly past the ground-relative finish line. In total, the satellite will fly more than 390° of longitude during the two hours instead of 360°. Figure 16 on page 58 illustrates this concept.

Ground Tracks of Circular Low Polar Orbits

Satellites on polar orbits circle the Earth with an orbital inclination of 90°. In other words, the plane of the orbit lies tilted at a right angle to the equator. Satellites in this type of orbit move exactly in the north-south direction and fly directly over the North and South Poles. These aspects cause the ground track to appear completely different from the two previous examples (see Figure 17 on page 61). Instead of looking like an S-shaped curve, the ground track looks like two straight lines running exactly north-south, each separated from the other by 180° in longitude (half the world).

A satellite that moves north along one segment of the ground track will eventually reach the North Pole at the top edge of the map. The satellite will then immediately appear halfway around the world on the other segment of the ground track and will head south from the North Pole toward the South Pole, located at the bottom edge of the map. Likewise, after the satellite reaches the South Pole, it will reappear halfway around the world on the original segment of the ground track and will head north from the South Pole.

This discontinuity in the ground track at the poles occurs because a Mercator projection represents an attempt to display a round world without edges on a flat map with edges. The seemingly instantaneous reversal in the direction of motion at the poles occurs because the only valid direction of travel from the North Pole is south. Conversely, the only valid direction of travel from the South Pole is north. If you have a hard time picturing the fact that moving across one of the poles changes the longitude by 180°, look at the bottom of Figure 13 on page 52.

The shape of the polar ground track described above assumes that the Earth's rotation has been frozen. How does adding the Earth's rotation back into the picture change the shape? Consider a satellite on a circular polar orbit with a two-hour period. Each one of the two segments of the ground track in Figure 17 on page 61 represents half of the circular orbit. Consequently, a satellite takes one hour to move from one pole to the other along one of the segments. During that hour, the Earth rotates by 15° from east to west. If a satellite starts at the South Pole and moves north, its longitude position at the North Pole will be 15° to the west of the longitude position from where it started. A similar 15° westward displacement occurs when the satellite moves from the North to the South Pole along the other segment. The middle of Figure 17 on page 61 illustrates this process.

Figure 17: *What a Polar Ground Track Looks Like* ◀

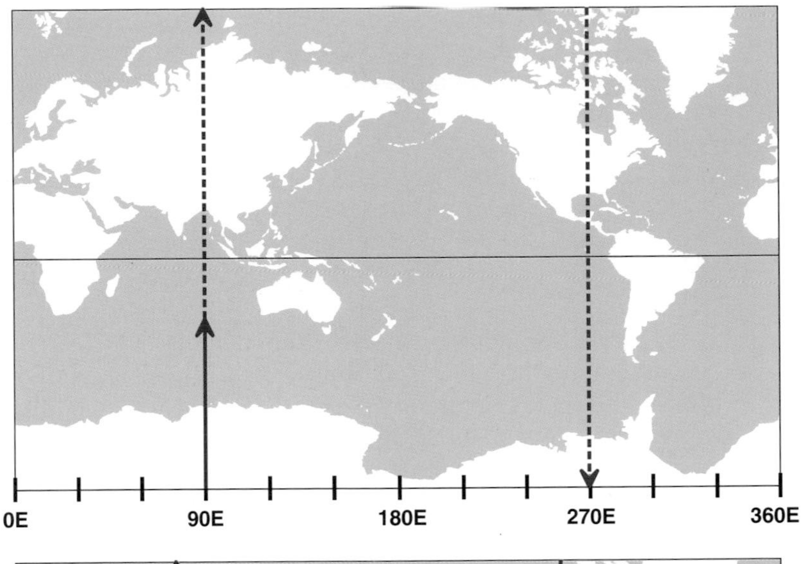

Row 1

A satellite on a polar orbit moves in the north-south direction and flies over both poles. This row shows the ground track of a satellite over Antarctica that is just beginning a two-hour orbit around the Earth. If the Earth does not spin, the satellite will move from south to north along the 90° East longitude line, fly over the North Pole, appear at the other side of the world, then move from north to south along the 270° East longitude line. The solid line shows the part of the orbit already completed, and the dotted line shows the parts that the satellite needs to fly on to complete the rest of the orbit.

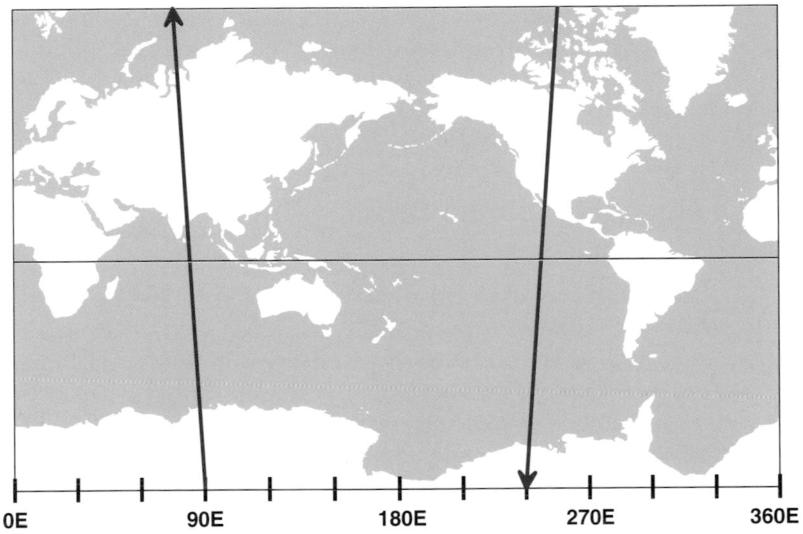

0E 90E 180E 270E 360E

Row 2

Now, add the Earth's spin into the picture. The Earth spins from west to east by 15° every hour. A satellite takes one hour to fly from the South to North Pole on a two-hour orbit, and another hour to fly back to the south from the north. Consequently, in this orbit, every time the satellite completes one half orbit and arrives at one of the poles, it will be over a location 15° in longitude to the west of its longitude while at the opposite pole. Thus, the first half orbit starts at the South Pole at 90° East, ends at the North Pole at 75° East, and the second half orbit starts at 255° East and ends at 240° East.

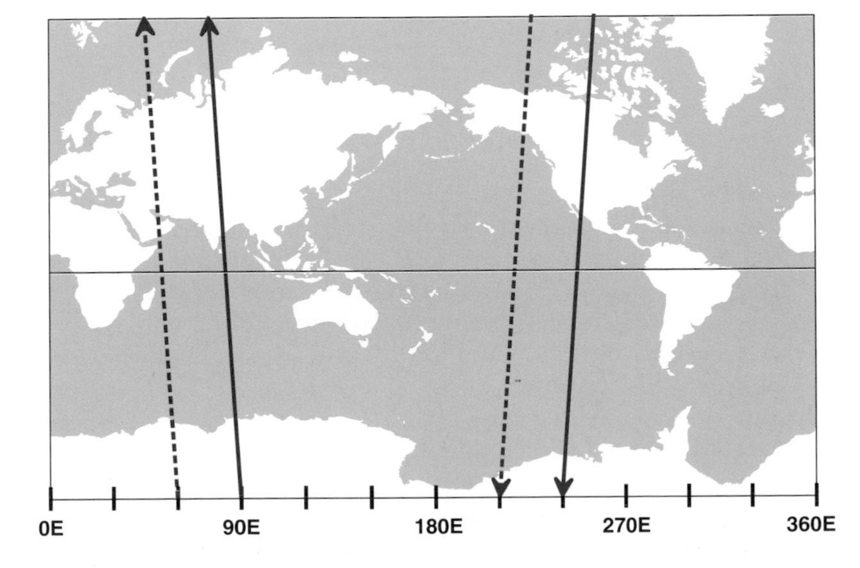

Row 3

Every new orbit in this example starts 30° to the west of the starting point of the previous orbit. The second orbit is shown by the dashed line on the left. It starts at 60° East longitude right over Antarctica and ends back at the South Pole, but over 210° degrees East. The ground tracks of several orbits will form a pattern of diagonal lines on the world map.

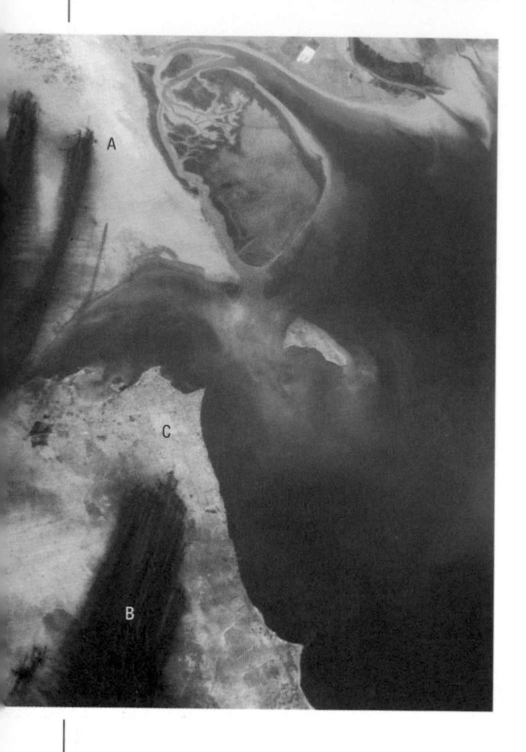

Above:

Near the end of the 1991 Gulf War with defeat imminent, the Iraqi Army set fire to more than 550 oil wells (A) in six major oil fields near Kuwait City, Kuwait (C). The smoke plumes from this human-caused environmental disaster (B) were visible to astronauts on Space Shuttle *Atlantis* more than 1,000 kilometers away. This photograph was taken in April 1991, several months after the American-led United Nations troops defeated Iraq.

The net effect of the Earth's rotation causes the south to north segments of the ground track to slant diagonally to the northwest, and the north to south segments to slant diagonally to the southwest. Ground tracks of circular polar orbits with longer periods experience a greater slant than circular polar orbits with shorter periods. In this example, each half orbit finishes at the opposite pole at a longitude 15° to the west of where it started, and each new orbit begins 30° to the west of where the previous orbit started. Eventually, the ground tracks will create a diagonal-mesh pattern across the entire world.

During the course of one orbit, a polar satellite will fly directly over both the North and South Poles, as well as all of the latitudes in between. Remember that because the maximum latitude that a satellite flies over equals the tilt angle of the orbit, only a polar orbit allows a satellite to eventually fly over every point on the Earth's surface. This unique characteristic allows a polar satellite to make extremely detailed observations of the entire Earth in a time span of between two to three weeks. Satellites that survey the land and seas to look for natural resources, some weather satellites, photographic satellites, and some military surveillance satellites use polar or near-polar orbits (inclinations between 80° and 100°).

2.4 Geosynchronous Orbits

Learning all the intricate details of satellite motion takes an extremely long time. Although the last three sections of this chapter provide a good introduction to the basics of orbital mechanics, the bad news is that the material barely scratches the surface of the subject. Fortunately, for those who do not wish to spend a lifetime studying orbits but are still interested in spaceflight, the good news is that the most useful type of Earth orbit is one of the easiest to understand and is also one of the most interesting to learn about. Spaceflight engineers call this type of orbit a *geosynchronous orbit*.

Before the Space Age

In order to appreciate the usefulness of geosynchronous orbits, consider the state of global communications before the advent of spaceflight. In those days, radio and television transmissions (collectively referred to as radio signals) were transmitted on a line-of-sight basis. In other words, they could not reach locations that did not lie on an unobstructed straight line connecting the transmitter to the receiver. This constraint restricted broadcasts from reaching beyond the horizon as seen from the transmission tower. Consequently, communications companies used tall transmission towers or placed their towers on top of tall mountains. Either method allowed the transmitter to "see" a larger area than possible from the ground. However, mountains do not exist everywhere, and increasing the tower height to increase the broadcast range becomes structurally infeasible after a certain point.

Historical Fact

Before the space age, communications with locations beyond the horizon required long stretches of cables or bouncing radio signals off of the Earth's upper atmosphere. This "bouncing" scheme was unreliable at best.

One solution to transmitting beyond the horizon involved using the upper atmosphere (ionosphere) to reflect radio signals to distant locations. Unfortunately, the clarity of this type of transmission depends on the conditions in the ionosphere, which can vary minute by minute. Today, some types of radio stations and amateur (HAM) radio operators still use ionospheric reflection to broadcast past the horizon. Another scheme involved placing transmission towers roughly 50 to 80 kilometers (31 to 50 miles) apart. In this "bucket brigade" method, one tower received the transmission from the previous tower, then rebroadcasted the signal to the next tower down the line. A big problem with this scheme was that it could not reach across the ocean or to ships at sea.

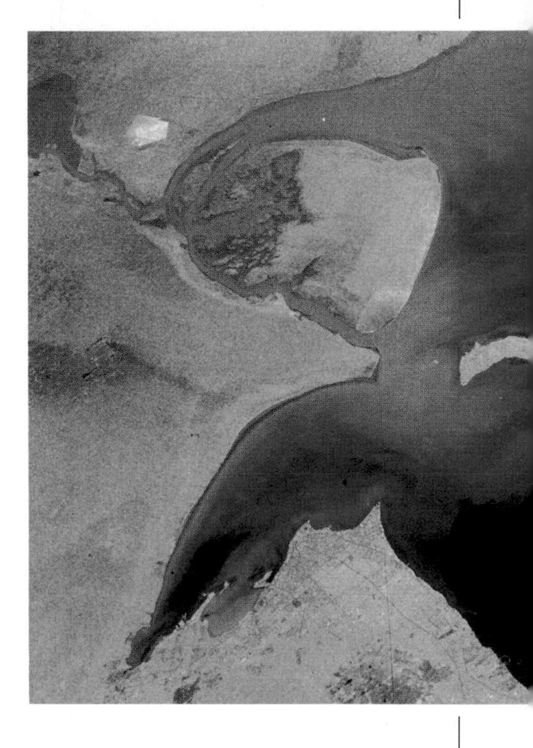

Above:
More than one year after the 1991 Gulf War, the oil fires in Kuwait have been extinguished, but the scars are still clearly visible. The oil-covered sands of the Kuwait oil fields show up in this August 1992 photograph as dark patches (south of Kuwait City, bottom middle of picture) surrounded by bright-colored, oil-free sands. Also visible within the oil fields are numerous oil lakes. They stand out as small, black circular areas. This photograph was taken by astronauts on Space Shuttle *Atlantis*.

Reflectors in Space

The advent of satellite technology in the late 1950s allowed communications engineers to move beyond the "ionospheric reflection" or "bucket brigade" methods of transmission. They initially thought, "Why not use a satellite as a reflector instead of the unreliable ionosphere?" U.S. Army engineers first tried this idea in the early 1950s by attempting to bounce radio signals off the Moon's surface. Although the experiment met with little success, the Moon inadvertently became the first communications satellite in history.

Historical Fact

Thanks to the U.S. Army, the Moon inadvertently became the first communications satellite in history.

Engineers from NASA tried a similar radio signal reflection experiment with a satellite called *Echo 1* in August of 1960. Essentially, *Echo* was a 30.5-meter- (100-foot-) wide aluminized Mylar balloon folded into a small package inflated after reaching orbit. Many people on Earth saw the giant balloon in space because it looked like a bright star moving across the night sky. NASA's reflection experiment met with a lot more success than the Army's Moon experiment. However, they found that the spherical balloon tended to scatter the radio signal in many different directions into space. Very little of the original signal was reflected to the intended destination on Earth.

The next logical step involved trying to use a satellite as an active relay instead of a passive reflector. This scheme works by sending the radio signal up to a receiver in the satellite, and then using the satellite to rebroadcast the signal to locations beyond the original ground transmitter's horizon. American Telephone and Telegraph (AT&T) funded the first active relay satellite experiment and NASA launched it in July of 1962. The satellite, called *Telstar 1*, made history 15 hours after launch by relaying the first transatlantic television broadcast. It transmitted a picture of the American flag proudly flying in the New England breeze to waiting viewers in France, England, and New Jersey. *Telstar* looked like a small metallic ball, measured about 87 centimeters in diameter (34 inches), and weighed only 77 kilograms (170 pounds).

Telstar 2, *Relay 1*, and *Relay 2* followed over the next eighteen months. The Relay satellites offered the additional ability to transmit either ten simultaneous phone calls or one television broadcast.

Historical Fact

America's first artificial communications satellite was a 30-meter-diameter balloon called *Echo 1*. It worked by reflecting radio signals beamed toward it back to the ground. Although the bouncing-signal scheme met with little practical success, *Echo* managed to reflect a pre-taped message from President Eisenhower across the nation.

The first American communications satellites that actively relayed television signals were placed in low Earth orbit. Engineers soon discovered that this idea was not very practical.

Scientific Fact!

A satellite in geosynchronous orbit over the equator appears to hover, motionless over a single location on the equator. However, a satellite in an inclined geosynchronous orbit traces out a figure eight-shaped ground track. The "eight" is centered on the equator. It is impossible to "hang" a satellite anywhere other than at the equator.

However, their relatively low orbital altitudes of about 6,400 kilometers (4,000 miles) allowed them to fly within sight of any one ground transmission or reception station for less than half an hour at a time. This constraint severely limited the practical usefulness of these early experimental communications satellites.

Hovering

NASA borrowed an old idea from science fiction writer Arthur C. Clarke, author of *2001: A Space Odyssey*, for its next improvement in global communications after *Telstar* and *Relay*. Clarke's novel idea first appeared in an article he wrote for a 1945 issue of the magazine *Wireless World*. In the article, Clarke wrote about placing communications satellites in orbits that allow them to hover stationary with respect to the ground. He theorized that a satellite in a motionless orbit would possess the ability to maintain a constant communications link between two or more ground stations within the satellite's field of view.

By now, you might be wondering how satellites can hover if they must remain in constant motion to avoid falling out of orbit. The key to understanding this paradox involves remembering that the orbital velocity of a satellite depends on the size of the orbit. Specifically, satellites at higher altitudes move slower and take longer to circle the world than satellites at lower altitudes. For example, at a typical low Earth orbit altitude of 300 kilometers (186 miles), a satellite takes about 90 minutes to complete an orbit. However, something special happens at an altitude of 35,786 kilometers (22,236 miles). At that height, a satellite moves slowly enough that it takes exactly one day to circle the Earth. Since a satellite in this type of orbit circles the Earth at the same rate that the Earth spins, the satellite will appear to hover motionless over the ground (see Figure 18 on page 65). Spaceflight engineers refer to these one-day orbits as geosynchronous.

Keep in mind that a geosynchronous satellite must also be circular (zero eccentricity) and equatorial (zero inclination) to appear motionless with respect to the ground. Remember that a satellite on an inclined orbit must spend part of its time north of the equator and part of its time south of the equator. In addition, because the center of the Earth must lie at the focus of an orbit, it is impossible to orbit a satellite exclusively in the Northern or Southern Hemispheres. Tilting a geosynchronous orbit with respect to the equator results in a figure-eight–shaped ground track with the center of the "eight" located at the equator. For these reasons, it is impossible to "hang" a satellite over places like San Francisco or Australia. The only valid hover points lie on the equator.

The ability to hover motionless with respect to the ground allows a communications satellite to provide an uninterrupted, twenty-four-hour communications link with any two ground stations within the satellite's field of view. Consequently, ground stations need not track low-flying, fast-moving satellites through space. All

Figure 18: *How a Geosynchronous Satellite Appears to Hover*

A geosynchronous satellite completes one orbit in exactly one day. Since its rate of motion through space matches the Earth's rotation rate exactly, a geosynchronous satellite in orbit above the Earth's equator will appear to hover motionless over a single point on the equator. This figure illustrates why this happens. Keep in mind that a satellite that is an inclined geosynchronous orbit does not appear to hover motionless. Instead, this satellite will trace out a figure-eight-shaped ground track with the center of the "eight" exactly on the Earth's equator.

Figure Not Drawn to Scale

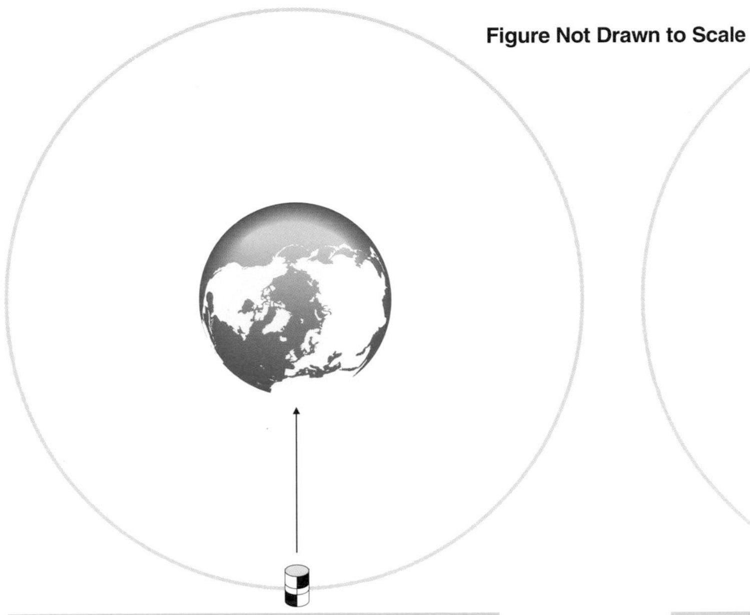

Frame A - Current Time Is 12:00 Noon

Communications satellite in geosynchronous orbit is currently over the west coast of Africa on the equator.

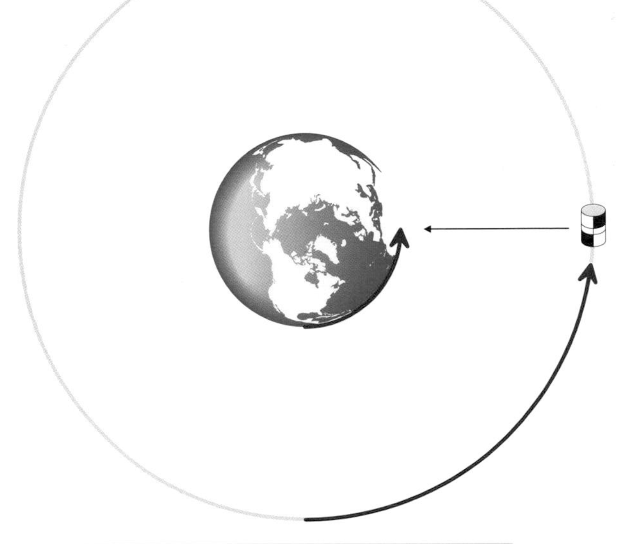

Frame B - Current Time Is 6:00 p.m.

Six hours later, the satellite has completed one quarter of an orbit, but the Earth has also turned one quarter of a revolution.

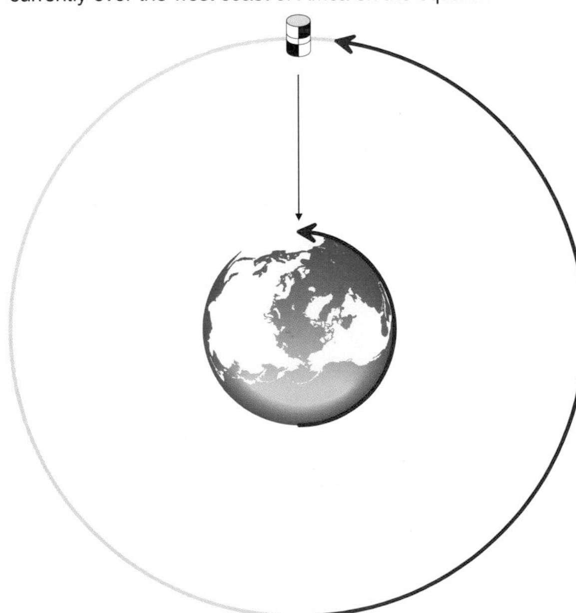

Frame C - Current Time Is 12:00 Midnight

Another six hours has gone by, and the satellite has now completed one half of an orbit. The Earth has also turned one half of a revolution.

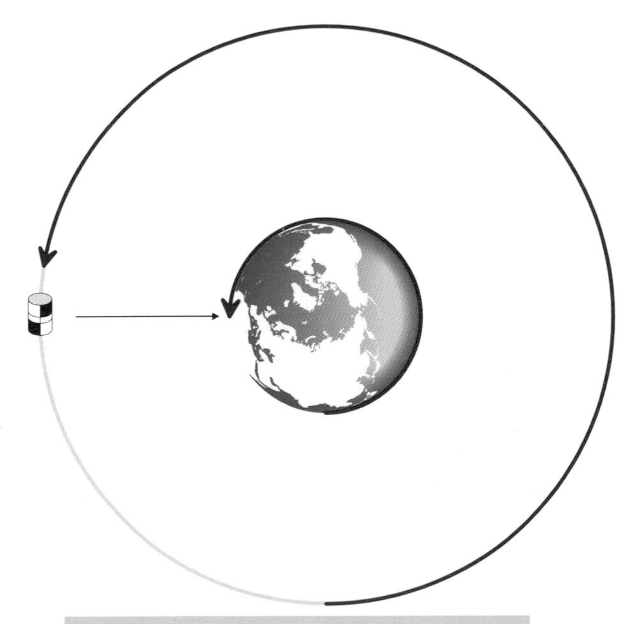

Frame D - Current Time Is 6:00 a.m.

Now, both Earth and satellite have completed three-fourths of a revolution in the 18 hours since Frame A. Notice that in all the frames, the satellite remained over West Africa.

they need to do is point their transmission or reception antenna toward the satellite and keep it fixed in that position. In addition, the 35,786-kilometer altitude provides a geosynchronous satellite with a clear line of sight to roughly one-third of the Earth's surface. This wide view allows a single geosynchronous satellite to provide a communications link between any two points within an 18,000-kilometer-(11,200-mile-) wide circle centered on the equator. Such a circle extends almost from the South Pole to the North Pole. Therefore, three geosynchronous satellites can provide a link between any two points on the world with the exception of high-latitude locations near the poles (see Figure 19 on page 67).

NASA launched the world's first geosynchronous communications satellite for television in August of 1964. This satellite, called *Syncom 3*, was placed at a hover point near the International Date Line in the Pacific Ocean. The launch occurred just in time for the satellite to relay the live television broadcast of the opening ceremony of the 1964 Summer Olympic Games in Tokyo, Japan. Viewers throughout the United States were introduced to the phrase, "Live via satellite."

An important concept requires clarification at this point. Satellites do not perform the actual television broadcast in most cases. Instead, they relay the signal from the camera at the remote location to the local television station. The local station then sends the signal to their transmission tower for broadcast to television sets in the local neighborhood. However, a growing minority of people around the world own their own satellite antenna dishes that allow them to receive television signals directly from satellites.

Communications Today

Today, more than 120 functional geosynchronous communications satellites orbit over many different locations on the equator. They belong to various governments, government agencies, and private communications companies. Some of these groups include NASA, the Department of Defense, the European Space Agency (ESA), Canada, Indonesia, Great Britain, AT&T, RCA, and Western Union. NASA and ESA put most of these satellites into orbit.

In addition to all the satellite operators listed previously, a unique satellite communications organization deserves special mention. The International Telecommunications Satellite Organization (INTELSAT) is a multinational organization that owns and operates the largest global satellite communications network. This consortium consists of more than 100 member nations. Each country pays a share of the operating costs based on that nation's percentage of the total communications traffic on the network. Traffic in the United States accounted for over half of INTELSAT's business in the 1960s. Now, with many more countries interested in satellite communications than before, the American share has dropped to about 25%, even

RULE OF THUMB

Next time you see a satellite antenna dish in somebody's backyard, notice which way it points. In the United States, they all point toward the south. The reason is that all television communications satellites "hover" in geosynchronous orbits over the equator.

Historical Fact

The "hotline" between the White House and the Kremlin in Moscow helped cool tensions during the cold war. Contrary to popular belief, the hotline is not a red telephone. Instead, the White House and Kremlin communicate through use of a teleprinter that prints out telegrams. Messages from the president's Oval Office travel north to Frederick, Maryland. There, at the Army's Fort Detrick station, the message is transmitted to a Russian Molniya communications satellite. Then, the message is beamed down to a Russian ground station south of Moscow before arriving at the Kremlin. In order to verify the operational status of this system, technicians checked the link once every hour during the cold war.

Figure 19: *How to Use Three Satellites to Like the Entire World* ◀

Geosynchronous satellites orbit at an altitude of 35,786 kilometers (22,236 miles) above the surface of the Earth. From that vantage point, a satellite has a clear line of sight to roughly one-third of the Earth's surface. A geosynchronous equatorial orbit (also called geostationary) is an ideal location for a communications satellite because it will appear to hover motionless over the equator. This allows one satellite to provide a constant radio link for communications between any two points in its field of view which extends almost from the North Pole to almost the South Pole. This figure shows how three geostationary satellites, spaced equally apart, can provide a link for communications between any two points on the world except for the polar regions.

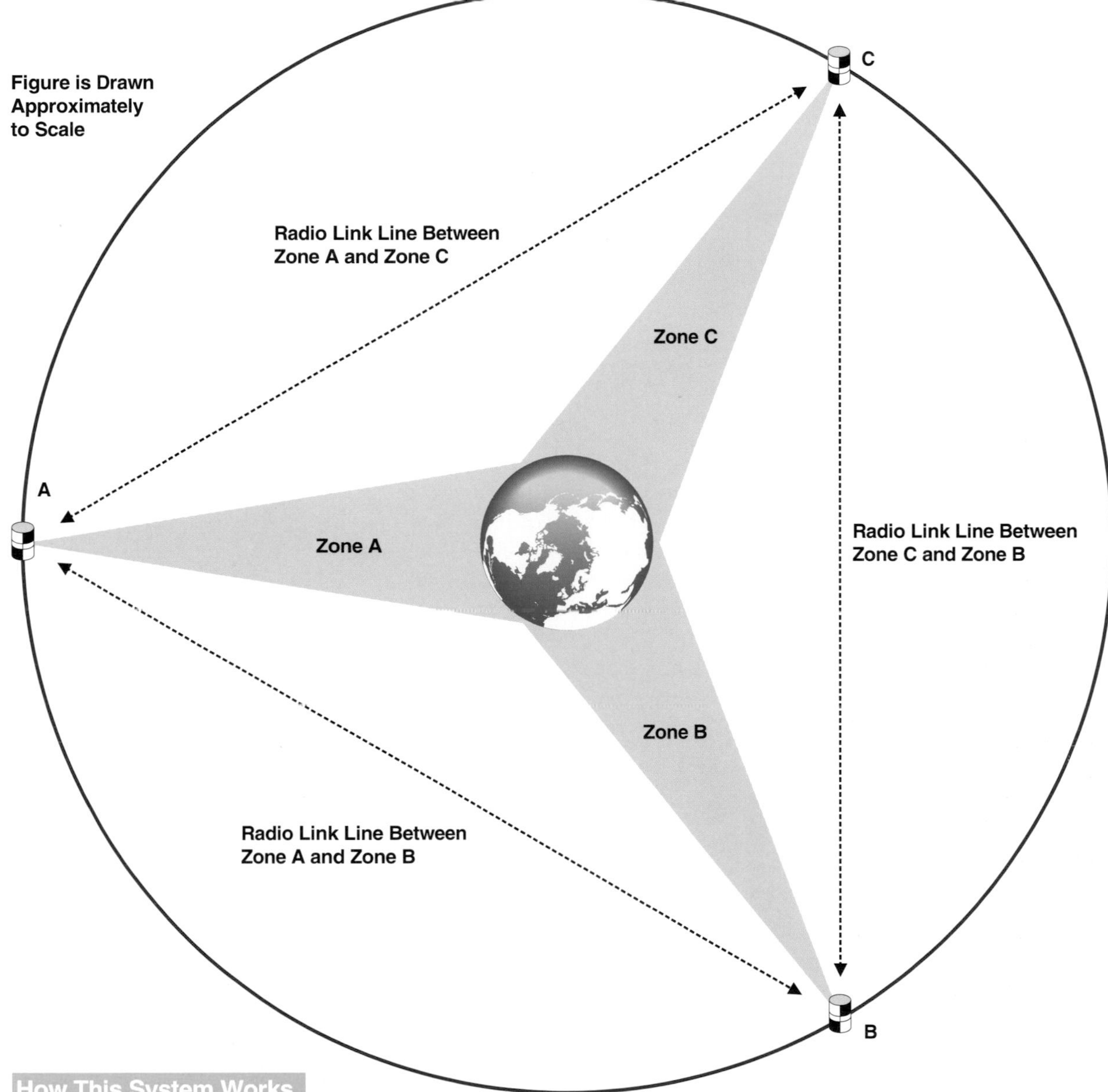

How This System Works

Satellite A provides communications links between any two points in Zone A (North, South, and Central America). Satellite B provides communications links between any two points in Zone B (Europe, Africa, Western Asia). Satellite C provides communications links between any two points in Zone C (Eastern Asia, Australia, Pacific Islands).

Communicating between two different zones requires the use of two different relay satellites. For example, talking from North America to Australia requires sending the signal to Satellite A. It then relays the signal to Satellite C. In turn, Satellite C relays the original signal back down to the Earth at Australia.

What's on Satellite TV?

ABC *(American Broadcasting Corporation)*
National network television
Telstar-4 at 89° W longitude
Satellite Channel 21

CBS *(Columbia Broadcasting System)*
National network television
Telstar-4 at 89° W longitude
Satellite Channel 10

CNN *(Cable News Network)*
Around the clock news reports
Galaxy-5 at 125° W longitude
Satellite Channel 5

Discovery Channel
Science and educational programs
Galaxy-5 at 125° W longitude
Satellite Channel 12

ESPN
Total sports network
Galaxy-5 at 125° W longitude
Satellite Channel 9

HBO *(Home Box Office)*
Premium movie channel
Galaxy-5 at 125° W longitude
Satellite Channel 18

MTV *(Music Television)*
Rock and roll music videos
Satcom-C4 at 135° W longitude
Satellite Channel 17

NASA Television
Live coverage of Space Shuttle missions
General Electric-2 at 85° W longitude
Satellite Channel 9

NBC *(National Broadcasting Corporation)*
National network television
Satcom-C1 at 137° W longitude
Satellite Channel 8

Showtime
Premium movie channel
Galaxy-9 at 123° W longitude
Satellite Channel 15

WGN
Superstation from Chicago, IL
Galaxy-5 at 125° W longitude
Satellite Channel 7

The third line of each row in this table lists the satellite that carries the television transmission for the listed station and the satellite's position over the equator in geosynchronous orbit. Each satellite carries 24 different channels. Keep in mind that stations may use multiple channels on the same or different satellites. Only one per station is listed.

though the total volume of the American communication traffic through INTELSAT has increased.

The first INTELSAT satellite, called *Early Bird* or *INTELSAT 1*, left Earth for geosynchronous orbit on a NASA Delta rocket in 1965. *Early Bird* was designed to last only for about one year, and carried 240 phone calls or one television broadcast at any one time. Although this capacity seems rather limited by today's standards, it increased the transatlantic phone capacity during the mid-1960s by more than 50%. Since then, NASA and ESA have placed over thirty satellites in geosynchronous orbit for INTELSAT. About seventeen still function today, including five new INTELSAT Type-6 satellites. Each one of these new satellites can relay 120,000 phone calls and three color television broadcasts simultaneously. Engineers expect them to function for over thirteen years, and carry over two-thirds of the world's annual international phone traffic. Another recent addition to the fleet, called *INTELSAT K*, can relay thirty-two simultaneous color television broadcasts between the two American continents and Europe.

Satellite communications is a booming but expensive business. For example, INTELSAT paid $750 million to build the five *INTELSAT 6* satellites, and $500 million to launch them. In addition, one of the five got stranded in low Earth orbit on the way to geosynchronous orbit when its upper-stage rocket engine failed to fire. INTELSAT subsequently paid about $93 million for NASA to send the Space Shuttle *Endeavour* and six astronauts into space to fix the rocket in May of 1992. Many people questioned INTELSAT's wisdom when they spent almost as much money on the rescue and repair mission as the price of a new satellite. However, consider that it would have taken over four years to build a replacement, and that each *INTELSAT 6* satellite is expected to generate close to $1 billion of revenue from usage fees over its thirteen-year operational lifetime.

How much does international phone conversation really cost? INTELSAT charges telephone companies about one dollar per day to rent one of their 120,000 phone circuits on an *INTELSAT 6* satellite. The phone company then charges their customers roughly one dollar a minute to call a country across the ocean. Although this price markup may seem unfair, keep in mind that the phone company provides the infrastructure (electronic network) to link phones together on the ground. Without this network, simple local calls would not be possible. Scientists and spaceflight engineers are currently developing experimental systems that may eventually allow two phones to communicate directly through a satellite without the help of the local phone company.

Broadcast Frequencies

Experts expect the worldwide demand for satellite communications to dramatically increase during the next decade. More than 120 functional satellites currently occupy positions in geosynchronous orbit. A good number of dead satellites also litter the skies

above the equator. Fortunately, the control technology exists to allow geosynchronous satellites to orbit half a degree in true anomaly apart from each other without fear of collision. This separation distance physically amounts to 364 kilometers (226 miles) at an altitude of 35,786 kilometers in space, and 55 kilometers (34 miles) in terms of the hover positions on the ground. Such a scheme theoretically provides space for 720 satellites around the world. About 150 of these slots are in positions that allow satellites to provide links between ground stations in the United States.

In reality, the number of total slots available in geosynchronous orbit falls well below 720 because of the satellites' transmission requirements. Current international regulations dictate that communications satellites utilize C-band frequencies that extend from 3.7 gigahertz to 6.4 gigahertz. Radio frequencies represent the different "channels" or "stations" that satellites broadcast on. Higher frequencies correspond to higher channels. For example, FM radio station 99.5 (0.0995 gigahertz) broadcasts on a lower frequency than station 105.7 (0.1057 gigahertz). Frequencies corresponding to satellite broadcasts would correspond to stations 3,700 to 6,400 on an FM radio if the dial went that high.

Most satellites receive transmissions from the ground at 6 gigahertz and transmit to the ground at 4 gigahertz. They use lower frequencies to transmit to the ground because lower frequency signals transmit better through clouds and rain. A single C-band signal can carry twelve television broadcasts simultaneously. On the other hand, phone conversations contain significantly less data to electronically encode than a television transmission. The latest techniques in data compression allow satellites to transmit up to 10,000 phone calls in the same amount of "signal space" required to send a single television broadcast. Another electronic data compression trick, called signal polarization, allows C-band satellites to double the number of simultaneous television transmissions from twelve to twenty-four.

The problem with all satellites using the same frequency is that signals from two adjacent satellites tend to interfere with each other. In order to minimize interference, satellites in geostationary orbit usually stay about 1° to 2° in true anomaly apart from each other instead of half a degree. This separation requirement limits the total number of available orbital "slots" to between 180 and 360. Of these 360, no more than 60 of them lie over locations that allow a satellite to have a clear line of sight to the United States. A United Nations committee oversees the distribution of the rights to use the slots. Some of the newer satellites use the Ku instead of the C-band. These newly United Nations–approved frequencies extend from 12 to 18 gigahertz and should help alleviate signal interference and overcrowding in the C-band.

RULE OF THUMB

Overcrowding in geosynchronous orbit is becoming a major problem that must be solved soon. If communications satellites are placed too close together in geosynchronous orbit, their transmissions will start to interfere with each other. In general, they must be spaced hundreds of kilometers apart in orbit even though the control technology exists to place them closer together.

Scientific Fact!

Another reason that satellites broadcast at a frequency of 4 gigahertz and receive at 6 gigahertz is that broadcasting at the lower frequency requires less power. It is much easier for ground stations to broadcast at a higher power setting than satellites.

Satellite Lifetime and Overcrowding

The Sun and Moon constantly play a game of gravitational "tug-of-war" with satellites in geosynchronous orbit. These external forces tend to cause satellites to drift away from their designated hover points and tumble out of control at the same time. Ground controllers must periodically fire a satellite's maneuvering thrusters (small rocket engines) to place it back into position and to de-tumble it. Without these maintenance maneuvers, a satellite may eventually collide with another, or lose the ability to maintain a communications link with the ground. As a result, a communications satellite's useful lifetime lasts only as long as its maneuvering propellant supply. Modern satellites usually carry enough propellant to last for between five and thirteen years. After that, spaceflight engineers consider the satellite to be dead and useless.

Current "spaceflight etiquette" calls for the owners to remove a dead communications satellite from its geosynchronous slot so that another may take its place. Otherwise, all the slots will eventually become littered with dead satellites. In low Earth orbit, a satellite normally re-enters the atmosphere and burns up after a few months to a few years because atmospheric drag slowly lowers the orbital altitude. Unfortunately, drag does not affect satellites orbiting at extremely high altitudes. Consequently, a dead geosynchronous satellite will continue to circle the Earth forever.

The accepted method to dispose of a geosynchronous satellite involves using its last drops of propellant to boost it above geosynchronous altitude. Dead satellites in these so-called graveyard orbits circle the Earth every two to five days and do not interfere with the launch and operation of new geosynchronous satellites. However, somebody will eventually need to figure out how to dispose of the satellites flying in the "graveyard" orbits. ❑

Opposite Page:
Space Shuttle astronauts Joe Allen (left) and Dale Gardner (on the right, and standing on the end of the Shuttle's robot arm) advertise the successful salvage of a Westar communications satellite in November 1984. On this mission, the Shuttle *Discovery* carried two satellites into space, and returned two broken ones (*Westar* and *Palapa*) back to Earth. Both the *Westar* and *Palapa* were left stranded in low Earth orbit when their rocket engines, used for transfer to geosynchronous orbit, misfired. Unfortunately, there is no current way to recover malfunctioning or dead satellites in geosynchronous orbit because the Shuttle cannot fly that high.

Dancing in the Dark

Chapter 3: How to Perform Space Maneuvers

Imagine sitting at the controls of a rescue spacecraft. Your mission objectives involve saving astronauts currently onboard a broken spacecraft. A quick glance out the window shows that the broken spacecraft circles the Earth in the same orbit as yours, but several kilometers ahead of your current position. How do you maneuver your spacecraft to catch up? A commonsense solution involves speeding up to catch up. As you will see later on in this chapter, the real answer is surprising.

This chapter explains the basic orbital maneuvers that satellites and astronauts use to change orbits. The chapter also demonstrates that planning orbital maneuvers requires careful thinking. However, do not be intimidated. Space maneuvers, like the basic orbital mechanics concepts presented in the previous chapter, can be understood with a little geometry, and without resorting to a single mathematical equation.

Opposite Page:
This photograph shot by NASA astronauts in low Earth orbit shows the Florida peninsula and the Gulf of Mexico (right side of picture) illuminated by the mid-morning sun.

3.1 How to Change an Orbit

Satellites use small rocket engines called thrusters to change their orbit. These engines work like and look like smaller versions of the large engines that booster rockets use to send satellites into orbit. Thrusters create a new orbit out of the old by changing the satellite's flight direction, velocity, or both. Spaceflight engineers typically refer to this velocity or direction change as a ΔV (pronounced delta vee). The capital Greek letter delta means "a change" in mathematical jargon, and the "V" stands for velocity.

Impulsive Burn Assumption

A satellite in the process of changing its orbit keeps moving during the course of a several-second to several-minute rocket burn maneuver. The constant movement of the satellite complicates orbital change calculations because the effect of a burn critically depends on a satellite's exact position on the orbit. Because a satellite always remains in a state of constant motion, its position will change slightly in between the time of ignition (burn start) and cutoff (burn end). Consequently, most spaceflight engineers simplify burn calculations with only minimal sacrifice in numerical accuracy by assuming that all ΔVs occur instantaneously. This assumption, called the impulsive burn assumption, works because the burn time represents only a tiny fraction of the orbital period. In the interest of maximum accuracy for navigation, computers that control rocket burns in space do not use this assumption.

Above:

Astronauts on Space Shuttle *Endeavour* took this photograph of the eastern end of the ancient world in June 1993. The lush Nile River delta appears as a dark area near the bottom of the picture. To the right, the Suez Canal shows up as a thin, near-vertical line connecting the Mediterranean Sea (middle of the photo) with the Red Sea (not shown). On the eastern (right) edge of the photograph, the border between Egypt and Israel is visible as a diagonal line separating a lighter-colored area (sand in the Gaza Strip) from a darker region (farmland in Israel). The island of Cyprus is also visible near the north shore of the Mediterranean.

RULE OF THUMB

The new orbit that results from an orbital-change maneuver *must* intersect or touch the old orbit at the location that the maneuver occurred. This rule is the most fundamental rule governing the design of space maneuvers.

The impulsive burn assumption also gives rise to one of the most important rules governing spaceflight maneuvers. This rule states that a satellite on a new orbit created by a ΔV passes through the burn point on every successive orbit if no subsequent rocket burns occur. In other words, the original and new orbital paths must intersect or touch at the exact altitude and J2000 coordinate location of the rocket burn.

Components of Motion

Components of motion provide powerful tools that help in understanding the theory of orbital maneuvers and how ΔVs affect orbits. A *component breakdown* allows the description of any arbitrary direction of motion as a combination of motions along convenient directions perpendicular to each other. For example, an object moving exactly northeast has equal components of motion in both the north and east directions. This is another way of saying that the object simultaneously moves due north and due east at the same velocity. However, the sizes of components often differ. Consider the fact that an object moving slightly south of west has both south and west components of motion. In this case, the size of the west component greatly exceeds the size of the south component because an object moving slightly south of west travels mainly in the westward direction.

On an orbit, the horizontal-radial component system replaces north, south, east, and west (see Figure 20 on page 75). The horizontal component measures the velocity of a satellite in a direction parallel to the surface of the Earth, while the radial component measures velocity in a direction perpendicular to the surface. In other words, horizontal velocity describes how fast a satellite moves forward along the ground track, and radial velocity characterizes how fast a satellite gains or loses altitude. The topographical map direction of forward depends on a satellite's current location on the ground track. This map direction constantly changes as a satellite snakes back and forth between its maximum northern and southern latitudes.

The relative size of the horizontal and radial components of velocity depends on the eccentricity of the orbit. A satellite on a circular orbit (zero eccentricity) has no radial component of velocity because the altitude of the orbit never changes. Furthermore, the size of the horizontal component remains constant and depends on the orbital altitude. Satellites in higher circular orbits move slower. Consequently, they have smaller horizontal components than satellites in lower circular orbits.

On an elliptical orbit, the size of the radial velocity component constantly changes as a satellite moves along the orbital path (see Figure 20 on page 75). The amount of radial velocity varies from none at perigee, increases to a maximum outward (upward) at 90° true anomaly, decreases back to none at apogee, increases to a maximum inward (downward) at 270° true anomaly, and then decreases

Figure 20: *Using Orbital Components to Describe Motion*

Posigrade Horizontal (PH):	Direction that points forward along the orbital path, and parallel to the surface of the Earth	Outward Radial (OR):	Direction that points directly away from the center of the Earth, and perpendicular to the surface
Retrograde Horizontal (RH):	Direction that points backward along the orbital path, and parallel to the surface of the Earth	Inward Radial (IR):	Direction that points directly at the center of the Earth, perpendicular to the surface

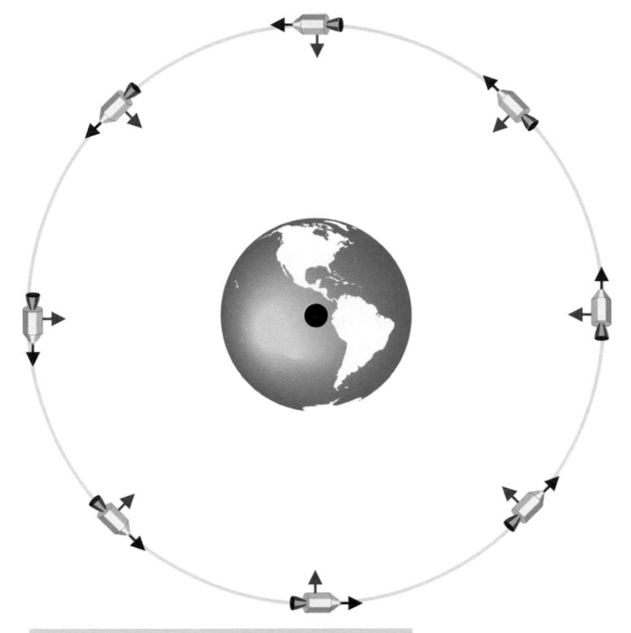

Directions on a Circular Orbit

This drawing above shows the direction components on a posigrade circular orbit. The arrows indicate direction only and are not proportional in size to the velocity. Notice that horizontal is always along (tangent to) the orbital path.

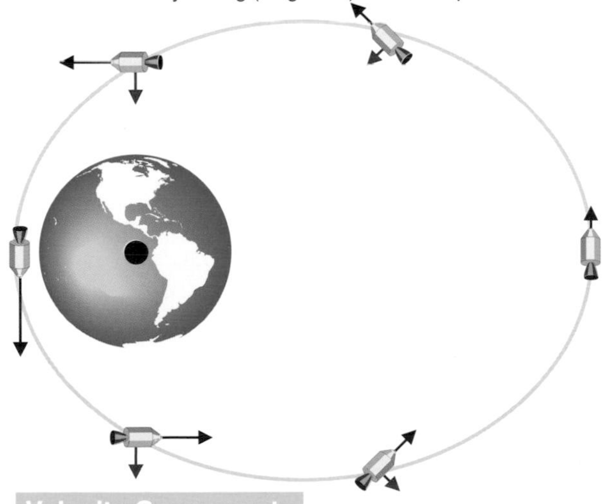

Velocity Components

The drawing above illustrates how fast a spacecraft moves in the horizontal (black) direction as compared to radial (red) direction at various points on an elliptical orbit. The size of the arrows shows the relative speed in each direction. Notice that there is no radial velocity at perigee and apogee.

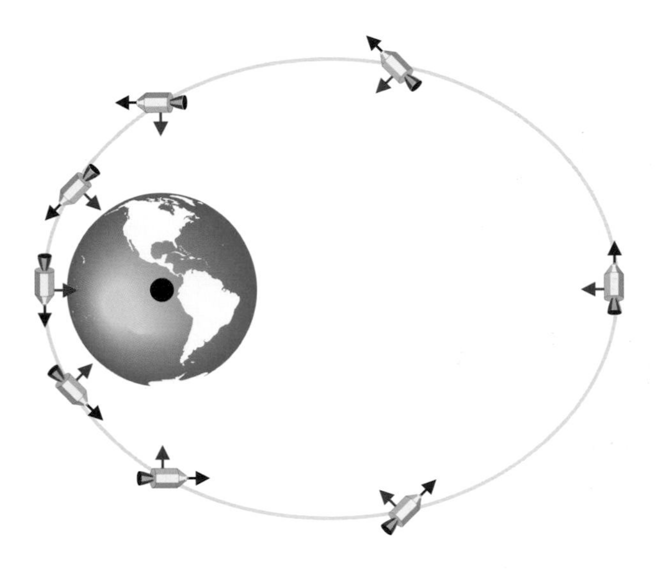

Directions on an Elliptical Orbit

This drawing above shows the direction components on a posigrade elliptical orbit. The arrows indicate direction only and are not proportional in size to the velocity. Notice that horizontal is not necessarily along the orbital path. That only happens at perigee and apogee.

Legend to the Two Drawings Above

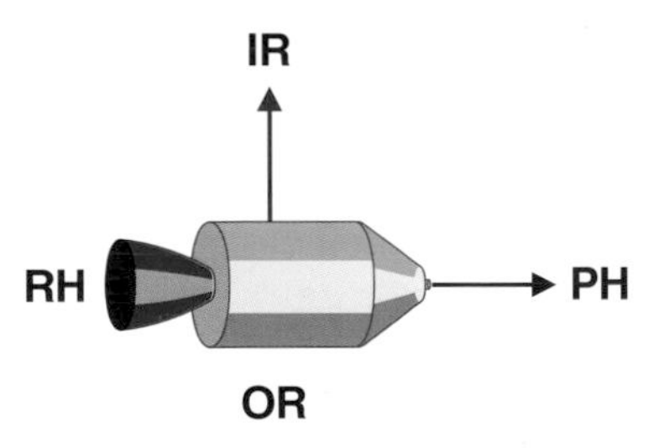

In the two "Directions" drawings above, the black arrow indicates the posigrade horizontal direction, and the red arrow indicates the inward radial direction. Notice that the horizontal directions are always perpendicular to a line connecting the center of the spacecraft to the center of the Earth, and are always parallel to the surface of the Earth directly underneath the spacecraft. In the "Velocity" drawing to the left, the red arrow indicates either outward or inward radial depending on the spacecraft's position.

How Posigrade Burns Affect Circular Orbits

Orbit Parameter	Result
ΔV = 5% Increase	
Orbit Period	18% increase
Apogee Velocity	15% decrease
Apogee Radius	23% increase
Perigee Velocity	5% increase
Perigee Radius	remains the same
ΔV = 10% Increase	
Orbit Period	42% increase
Apogee Velocity	28% decrease
Apogee Radius	53% increase
Perigee Velocity	10% increase
Perigee Radius	remains the same
ΔV = 20% Increase	
Orbit Period	139% increase
Apogee Velocity	53% decrease
Apogee Radius	157% increase
Perigee Velocity	20% increase
Perigee Radius	remains the same
ΔV = 30% Increase	
Orbit Period	479% increase
Apogee Velocity	76% decrease
Apogee Radius	445% increase
Perigee Velocity	30% increase
Perigee Radius	remains the same
ΔV = 40% Increase	
Orbit Period	12400% increase
Apogee Velocity	97% decrease
Apogee Radius	4800% increase
Perigee Velocity	40% increase
Perigee Radius	remains the same

Although the terms perigee and apogee apply specifically to Earth orbits, the data in this table is valid for satellites in orbit around any body (Sun, Earth, Moon, Mars, etc.) at any altitude. This table assumes that the satellite starts in a circular orbit and performs a posigrade horizontal burn to increase its velocity by the percentage shown. The percentage increase or decrease in velocity, altitude, and period are relative to the values on the original circular orbit. Remember, radius is the distance from the center of the body being orbited, not the surface.

back to none at perigee again. This pattern coincides with the fact that a satellite gains altitude on the trip to apogee from perigee, and then loses altitude during the return to perigee from apogee.

A satellite on an elliptical orbit also experiences a changing horizontal velocity component. The rate of forward motion (size of the component) starts at a maximum at perigee, decreases to a minimum at apogee, then increases again as a satellite heads back to perigee from apogee. This pattern coincides with the fact that satellites move fastest at perigee and slowest at apogee. An important fact to realize is that at perigee and apogee, satellites only move in the horizontal direction, since the radial component vanishes at these two points.

The new orbit that results from a ΔV depends on how much a rocket burn changes the horizontal and radial components of velocity, and the specific location on the orbit that the burn occurs. In general, the calculations to determine the new orbit are quite involved, even with the use of the impulsive burn assumption. However, many useful orbital change maneuvers that employ a velocity change in only the horizontal or radial direction can be understood without the use of mathematical equations.

Horizontal Burns

Horizontal burns increase or decrease a satellite's forward velocity. The exact effect depends on which way the rocket engine fires. Specifically, a rocket engine fires in the direction opposite that of the desired velocity change or direction of motion. As a result, an increase in forward velocity occurs when thrusters fire in the reverse direction to "push the satellite forward from behind it." Conversely, firing the thrusters in the forward direction slows a satellite by "pushing the satellite backward from ahead of it." On a posigrade orbit, these types of horizontal burns are respectively called posigrade and retrograde burns.

A posigrade burn that increases a satellite's forward velocity affects an orbit in a similar fashion to what happens when a ball rolls off a tabletop. An increase in the speed with which a ball rolls across a table increases how far the ball lands from the table after falling off of the edge. Similarly, increasing a satellite's forward velocity increases how far the satellite falls away from the curved surface of the Earth on the orbit. This increase allows a satellite to gain altitude.

The top of Figure 21 on page 77 shows that a posigrade burn on a circular orbit raises the altitude of every point on the orbital path except for the burn point. This altitude gain increases the semi-major axis length and eccentricity of the original orbit. In other words, a posigrade burn on a circular orbit increases the orbit's size and stretches the circle into an ellipse at the same time. The perigee of the new elliptical orbit always occurs at the burn point because a posi-

Figure 21: *How Posigrade Horizontal Burns Affect Orbits* ◄

What Happens: The burn raises the altitude of every point on the orbit except for the burn point. Therefore, the orbit becomes larger. The shape change depends on where the burn occurs. Also, the burn point must lie on both the pre-burn and post-burn orbits.

► Solid arrow indicates position of the burn and points in the direction of the velocity change

▷ Hollow arrow indicates position of the burn and points in the direction that the rocket engines fire to perform the burn

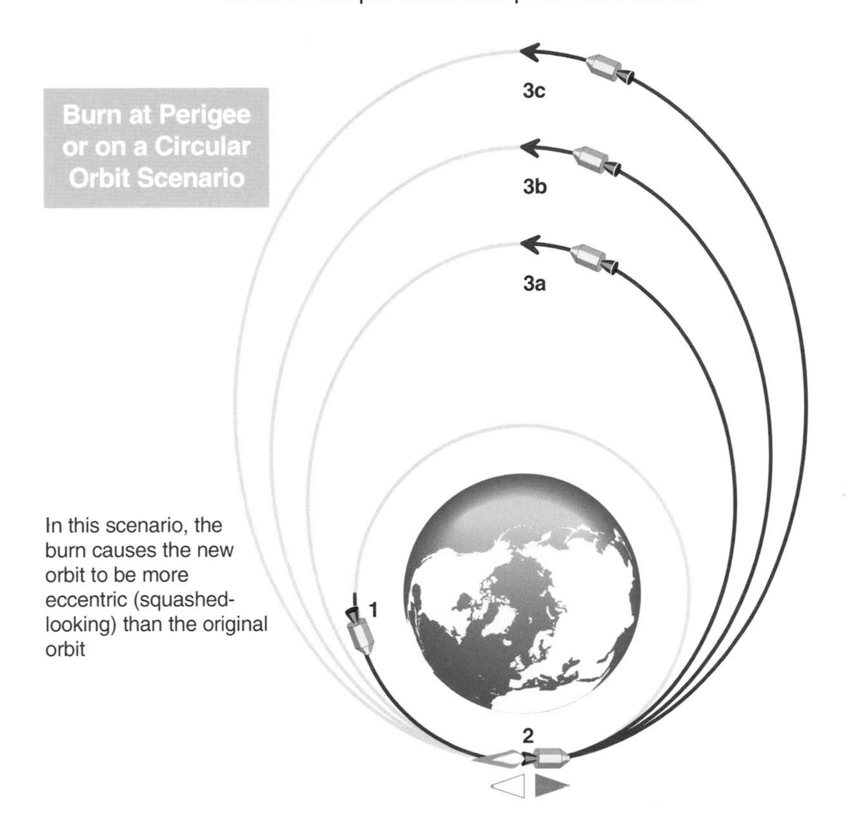

In this scenario, the burn causes the new orbit to be more eccentric (squashed-looking) than the original orbit

1 Satellite starts on a circular orbit.

2 Posigrade burn (velocity increase) occurs here. Burn point becomes perigee of new orbit because the burn raises every point on the old orbit except for the burn point.

3 Which orbit the satellite winds up in after the burn depends on the amount of propellant used in the burn. Subsequently, it also depends on the amount of velocity increase supplied by the burn. It takes more propellant to reach orbit 3c than 3b, and more to reach 3b than 3a.

If the satellite starts on orbit 3a and performs a velocity increase at perigee, it will wind up in either 3b or 3c depending on the amount of propellant used in the burn.

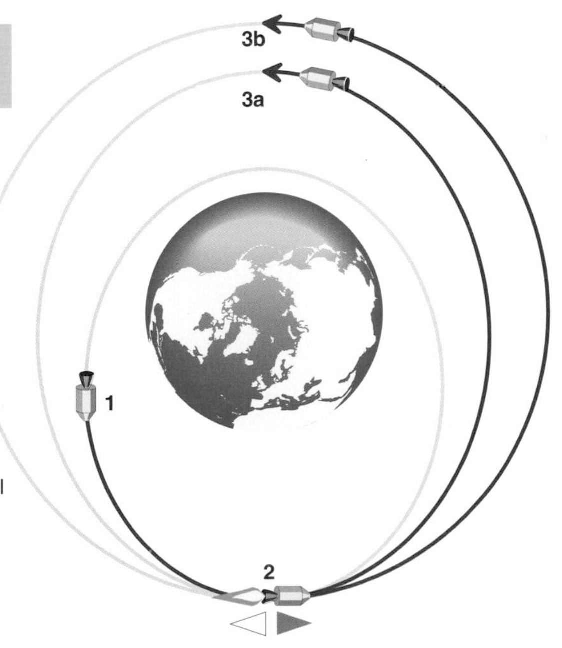

In this scenario, the burn causes the new orbit to be less eccentric and more circular than the original orbit

1 Satellite starts on the lower elliptical orbit.

2 Posigrade burn (velocity increase) occurs here at apogee. This burn increases the altitude of every point on the orbit except for the burn point. As a result, the new perigee is higher than the perigee of the old orbit.

3 Which orbit the satellite winds up in after the burn depends on the amount of velocity increase at the time of burn. It takes more propellant to reach 3b than 3a. Notice that in the case of 3b, the burn supplies enough velocity increase to raise the perigee altitude to match the apogee altitude. This type of maneuver is called circularization.

▶ Figure 22: *How Retrograde Horizontal Burns Affect Orbits*

What Happens: The burn lowers the altitude of every point on the orbit except for the burn point. Therefore, the orbit becomes smaller. The shape change depends on where the burn occurs. Also, the burn point must lie on both the pre-burn and post-burn orbits.

Solid arrow indicates position of the burn and points in the direction of the velocity change

Hollow arrow indicates position of the burn and points in the direction that the rocket engines fire to perform the burn

Burn at Apogee or on a Circular Orbit Scenario

3b

3a

1

2

In this scenario, the burn causes the new orbit to be more eccentric than the original orbit

1 Satellite starts on upper circular orbit.

2 Retrograde burn (velocity slow down) occurs here. This burn decreases the altitude of every point on the orbit except for the burn point. As a result, the burn point turns into the apogee of the new orbit.

3 Which orbit the satellite winds up in after the burn depends on the amount of velocity decrease supplied by the burn. It takes more propellant to reach 3a than to reach 3b.

If the satellite starts on orbit 3b and performs a retrograde burn at apogee, it will wind up on orbit 3a.

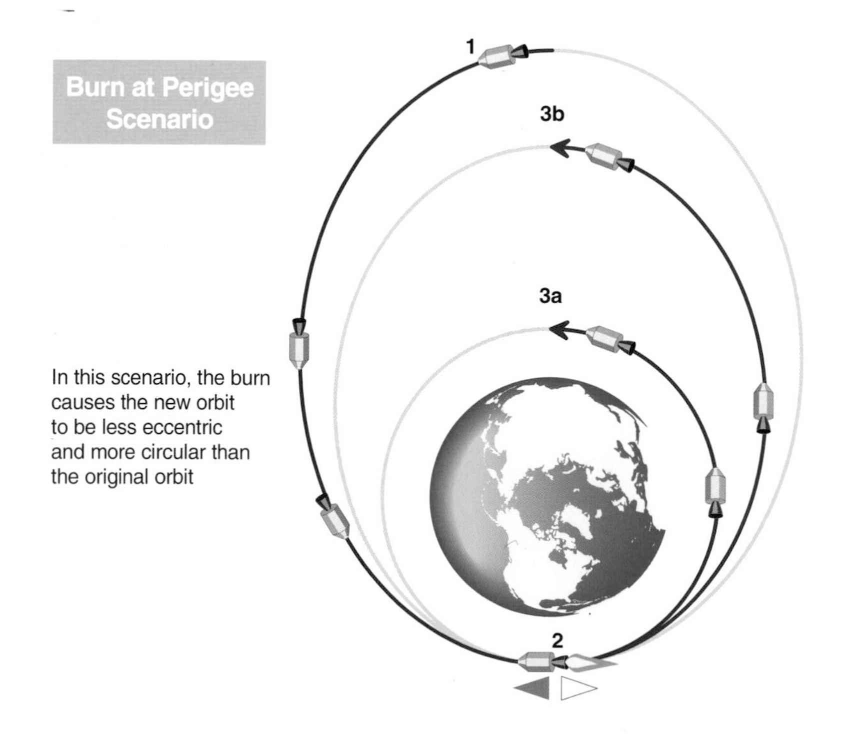

1 Satellite starts on outer (upper) elliptical orbit.

2 Retrograde burn (velocity slow down) occurs here at the perigee of the orbit. The burn lowers the altitude of every point on the orbit except for the burn point. As a result, the new apogee will lie at a lower altitude.

3 Which orbit the satellite winds up on depends on the amount of velocity decrease supplied by the burn. It takes more propellant to reach orbit 3a than to reach 3b.

Notice that in both the retrograde burn at perigee and apogee scenarios, the satellite must turn around backward and point in the direction opposite of its motion in order to fire the rocket.

grade ΔV raises the altitude of all the other points on the original orbital path.

A posigrade burn that occurs at the perigee of an elliptical orbit works in the same fashion as a posigrade burn on a circular orbit. The ΔV increases the altitude of every point on the orbital path except for the burn point. This change increases both the semi-major axis length and eccentricity of the orbit by raising the apogee to a higher altitude and keeping the perigee at the same location (at the burn point). The burn works analogously to riding a bike up a hill in that pedaling faster at the bottom of a hill allows the rider to coast higher up the hill.

In this case and the circular orbit case, the amount of size and squash increase depends on the strength of the burn. A stronger burn produces more ΔV than a weaker burn. Consequently, a stronger burn raises the apogee altitude more. However, keep in mind that increasing the ΔV also increases the amount of propellant that the thrusters require to perform the burn.

A satellite that performs a posigrade burn at the apogee point of its orbit also experiences an orbit change that involves every point on the orbit shifting to a higher altitude, except for the burn point. This shift keeps the apogee location constant (at the burn point) and elevates the perigee to a higher altitude. As a result, the size of the orbit increases and the shape takes on a more circular appearance. Notice that at the bottom of Figure 21 on page 77, the amount of increase in perigee altitude depends on the strength of the burn. A strong enough burn can provide enough posigrade ΔV to circularize the ellipse by raising the perigee altitude to the apogee altitude.

Retrograde burns work in exactly the opposite manner than their posigrade counterparts. A satellite that decreases its forward velocity on a circular orbit lowers the altitude of every point along the orbital path except for the burn point. The top of Figure 22 on page 78 shows that a retrograde burn on a circular orbit shrinks the semi-major axis length and increases the eccentricity. In other words, the burn simultaneously decreases the size of the orbit and squashes the circle into an ellipse. The apogee of the new elliptical orbit always occurs at the burn point because a retrograde ΔV lowers the altitude of all the other points on the original orbit.

A retrograde burn that occurs at the apogee of an elliptical orbit also decreases an orbit's semi-major axis length and increases its eccentricity. This change occurs because the apogee location stays constant (at the burn point) while the ΔV lowers the altitude of all the other points on the orbit. The net result is that the apogee altitude remains constant and the perigee altitude decreases. In both this and the retrograde burn on a circular orbit case, the amount of decrease in perigee altitude depends on the strength of the burn. Larger ΔVs alter the orbit to a greater degree than smaller ΔVs, but require more propellant.

How Retrograde Burns Affect Circular Orbits

Orbit Parameter	Result
ΔV = 5% Decrease	
Orbit Period	13% decrease
Perigee Velocity	22% increase
Perigee Radius	18% decrease
Apogee Velocity	5% decrease
Apogee Radius	remains the same
ΔV = 10% Decrease	
Orbit Period	23% decrease
Perigee Velocity	47% increase
Perigee Radius	32% decrease
Apogee Velocity	10% decrease
Apogee Radius	remains the same
ΔV = 20% Decrease	
Orbit Period	37% decrease
Perigee Velocity	113% increase
Perigee Radius	53% decrease
Apogee Velocity	20% decrease
Apogee Radius	remains the same
ΔV = 30% Decrease	
Orbit Period	46% decrease
Perigee Velocity	208% increase
Perigee Radius	68% decrease
Apogee Velocity	30% decrease
Apogee Radius	remains the same
ΔV = 40% Decrease	
Orbit Period	52% decrease
Perigee Velocity	356% increase
Perigee Radius	78% decrease
Apogee Velocity	40% decrease
Apogee Radius	remains the same

Although the terms *perigee* and *apogee* apply specifically to Earth orbits, the data in this table is valid for satellites in orbit around any body (Sun, Earth, Moon, Mars, etc.) at any altitude. This table assumes that the satellite starts in a circular orbit and performs a retrograde horizontal burn to decrease its velocity by the percentage shown. The percentage increase or decrease in velocity, altitude, and period are relative to the values on the original circular orbit. Remember, radius is the distance from the center of the body being orbited, not the surface.

Above:

This picture, taken by astronauts on Shuttle *Discovery* in April 1990, shows the northwestern part of Mauritania, Africa. The view points northwest over the Sahara to the Atlantic.

Where to Circularize?

Eccentricity Value	Relative ΔV Needed to Circularize at Apogee
0.1	95% of the ΔV needed to circularize at perigee
0.3	86% of the ΔV needed to circularize at perigee
0.5	75% of the ΔV needed to circularize at perigee
0.7	63% of the ΔV needed to circularize at perigee
0.9	41% of the ΔV needed to circularize at perigee

Although the terms *perigee* and *apogee* apply specifically to Earth orbits, the comparisons in this table apply for satellites in orbit around any body (Sun, Earth, Moon, Mars, etc.) at any altitude.This table shows that using horizontal burns, it is always takes less ΔV to circularize an orbit at apogee than at perigee.

A retrograde burn that occurs at the perigee of an elliptical orbit works slightly differently than the two other retrograde burn cases. In this case, the ΔV shifts the apogee to a lower altitude and keeps the perigee at the same location (at the burn point). Consequently, the size of the orbit decreases and the shape takes on a more circular appearance. A strong enough burn that lowers the apogee altitude to match the perigee altitude circularizes the elliptical orbit.

The choice of where to perform a circularization maneuver using a horizontal burn depends on the desired altitude of the new orbit. Circularization of an elliptical orbit at perigee results in a circular orbit with an altitude equal to the perigee altitude, while the same maneuver at apogee creates an orbit with an altitude equal to the apogee altitude. In other words, circularizing at apogee creates a higher circular orbit than circularizing at perigee. Sometimes, energy considerations also play a role in the choice of burn locations. The reason is that for any given orbit, circularization at apogee always requires less energy than at perigee.

One of the most important concepts to remember from this section deals with the differences between velocities on circular and elliptical orbits. Since a velocity increase on a circular orbit creates an elliptical orbit, a satellite at the perigee of an elliptical orbit moves faster than a satellite on a circular orbit with an altitude equal to the elliptical orbit's perigee altitude. Conversely, a satellite at the apogee of an elliptical orbit always travels slower than a satellite on a circular orbit with an altitude equal to the elliptical orbit's apogee altitude. These two facts provide the basis to understand other space maneuvers that this book presents, and also to understand some of the choices that spaceflight engineers face when planning a space mission.

Satellites can perform horizontal burns at locations on the orbit other than perigee or apogee. In general, these types of maneuvers create new orbits out of the old by altering the semi-major axis length, eccentricity, and by changing the position of perigee and apogee. These changes are difficult to predict without the use of mathematical equations. However, remember that both the original and new orbit must touch or intersect at the burn point.

Radial Burns

Radial burns alter a satellite's velocity in the upward or downward direction. The exact effect of the burn depends on the direction that the thrusters fire. An outward radial burn increases a satellite's upward velocity. This type of burn works by firing the rockets downward toward the Earth's surface to "push the satellite up from under it." Conversely, an inward radial burn works by firing the rockets in the upward direction to "push the satellite down from above it." Such a burn increases a satellite's downward velocity.

Figure 23: *How Radial Burns Affect Orbits* ◄

What Happens: A radial burn occurs when a satellite fires its thrusters directly toward or away from the center of the Earth. The burn turns a circular orbit into an ellipse and increases its size. The results in this figure have been exaggerated for clarity.

Solid arrow indicates position of the burn and points in the direction of the velocity change

Hollow arrow indicates position of the burn and points in the direction that the rocket engines fire to perform the burn

Outward Burn Scenario

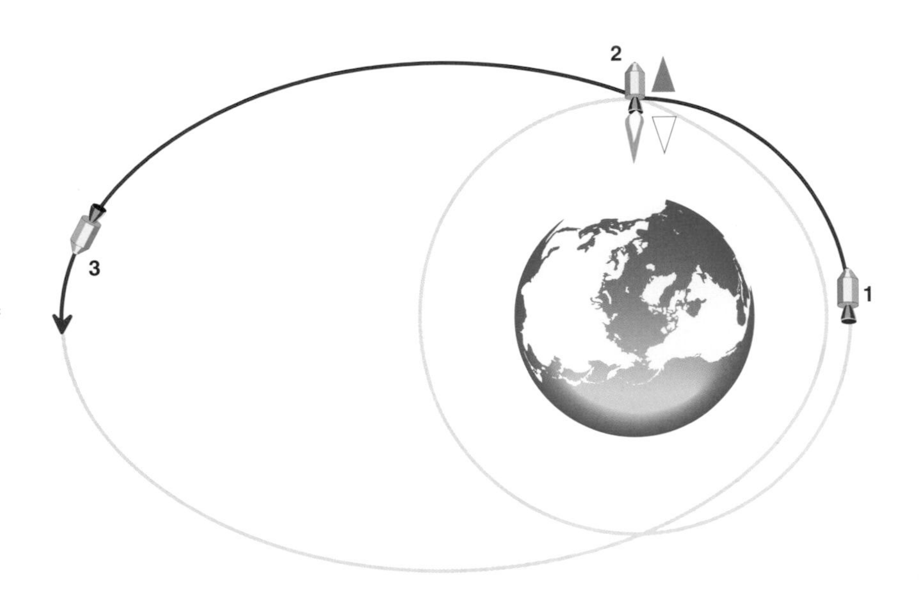

1 Satellite starts on the circular orbit.

2 Here, the satellite fires its thrusters directly toward the center of the Earth to increase its velocity in the direction directly away from the center of the Earth. The burn point becomes the 90° true anomaly point on the resulting elliptical orbit.

3 The satellite coasts up toward apogee of the new orbit after the burn. Notice that the perigee of the new orbit is lower than the original circular orbit altitude.

Inward Burn Scenario

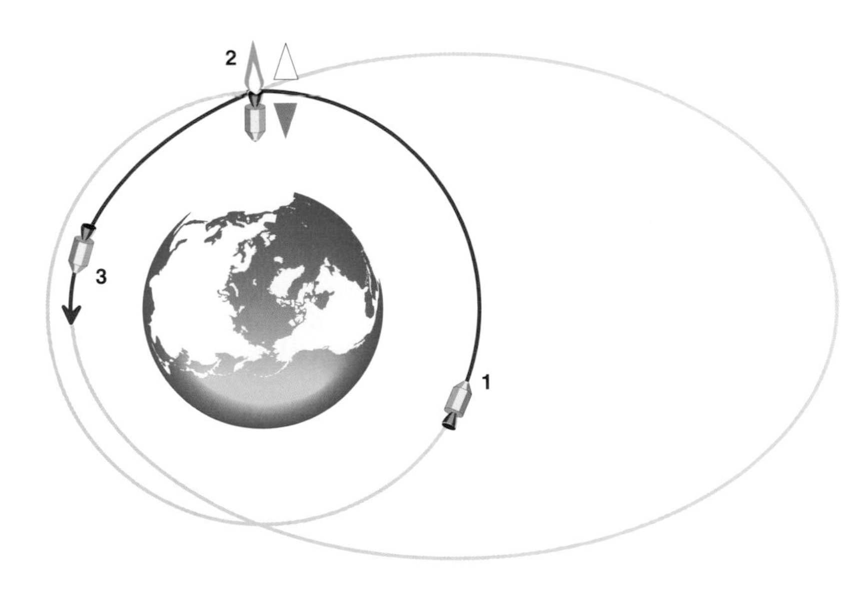

1 Satellite starts on the circular orbit.

2 Here, the satellite fires its thrusters directly away from the center of the Earth to increase its velocity in the direction directly toward the center of the Earth. The burn point turns into the 270° true anomaly point on the resulting elliptical orbit.

3 The satellite coasts downward to the perigee of the new orbit after the burn. Notice that the perigee of the resulting orbit is lower than the original circular orbit altitude.

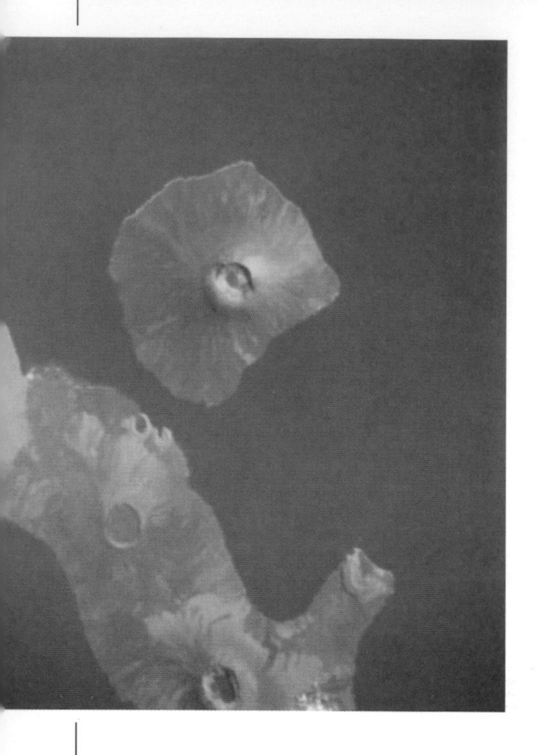

Above:
The tops of the volcanos of the Galápagos Islands poke out from the surface of the Pacific Ocean. Lava flows that occurred most recently show up as dark areas crossing the older, lightly vegetated flows (lighter areas). An abrupt steepening of the slope, which characterizes the Galápagos volcanos, is especially noticeable at Wolf Volcano (bottom center). Volcano Fernandina, one of the more active ones in this set, sits alone as an island in the middle of the photo. This picture was taken by astronauts on Space Shuttle *Atlantis* in October 1985.

RULE OF THUMB

An inclination (plane) change is one of the most expensive space maneuvers in terms of propellant needed. As such, the best place to perform a plane change is at the launchpad. In other words, it is best to launch the satellite into an orbit as close to the desired inclination as possible. Most satellites only carry enough propellant to change their inclination by a few degrees, at most.

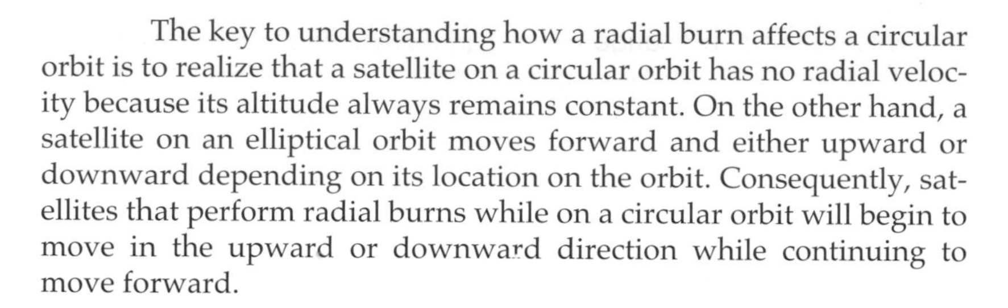

The key to understanding how a radial burn affects a circular orbit is to realize that a satellite on a circular orbit has no radial velocity because its altitude always remains constant. On the other hand, a satellite on an elliptical orbit moves forward and either upward or downward depending on its location on the orbit. Consequently, satellites that perform radial burns while on a circular orbit will begin to move in the upward or downward direction while continuing to move forward.

Remember that a satellite on its way to apogee from perigee moves both forward and upward. Because a satellite that performs an outward radial burn on a circular orbit will begin to move upward in addition to forward, the satellite will find itself on an elliptical orbit somewhere after perigee of the new orbit, but before apogee. Conversely, a satellite on its way to perigee from apogee moves both forward and downward at the same time. Therefore, an inward radial burn on a circular orbit will move a satellite onto an elliptical orbit somewhere after apogee of the new orbit, but before perigee.

Other types of radial burn maneuvers exist. A satellite on an elliptical orbit can use a radial burn to circularize its orbit. This maneuver requires "zeroing out" the radial velocity by firing the rocket engines to provide a ΔV of equal magnitude to the satellite's radial velocity, but in the opposite direction. For example, in order to cancel an outward (upward) radial velocity component, the thrusters must fire upward to provide a downward ΔV. The cancellation of the radial velocity leaves the satelite traveling only in the horizontal direction. However, forward motion without upward or downward motion occurs at the perigee and apogee of elliptical orbits as well as on circular orbits. The only two places on an elliptical orbit where canceling the radial velocity results in circularization are the 90° and 270° true anomaly points. A radial ΔV that cancels a satellite's radial velocity anywhere else turns the burn point into the perigee or apogee of a new elliptical orbit.

Plane Change Maneuver

A plane change maneuver changes the inclination of an orbit without affecting the orbit's orientation in three-dimensional space. Figure 24 on page 83 illustrates the theory behind this type of maneuver. The figure shows three circular orbits with inclinations of 0°, 45°, and 90°. Notice that orbits that differ only in their inclination values only intersect at the equator. Therefore, a simple plane change maneuver must occur as a satellite crosses the equator (at the ascending or descending node) because the burn point must lie on both the original and new orbits. For a plane change from an equatorial to an inclined orbit, the maneuver can occur anywhere on the original orbit because the entire orbit lies over the equator. In this case, the burn point turns into the ascending or descending node of the new orbit.

The burn direction and strength for a plane change depends on the desired magnitude of the inclination change. For example,

A simple plane change alters the inclination of an orbit without affecting its size and shape. The amount of propellant required depends on the amount of inclination change, and the altitude where the change maneuver occurs. In general, a satellite only has enough propellant to perform a change of a few degrees at a low-Earth orbit altitude. But, the same amount of propellant used near geosynchronous altitude can change the inclination by 30° or more. To use the same amount to go from equatorial to polar, you would have to be farther out than the Moon.

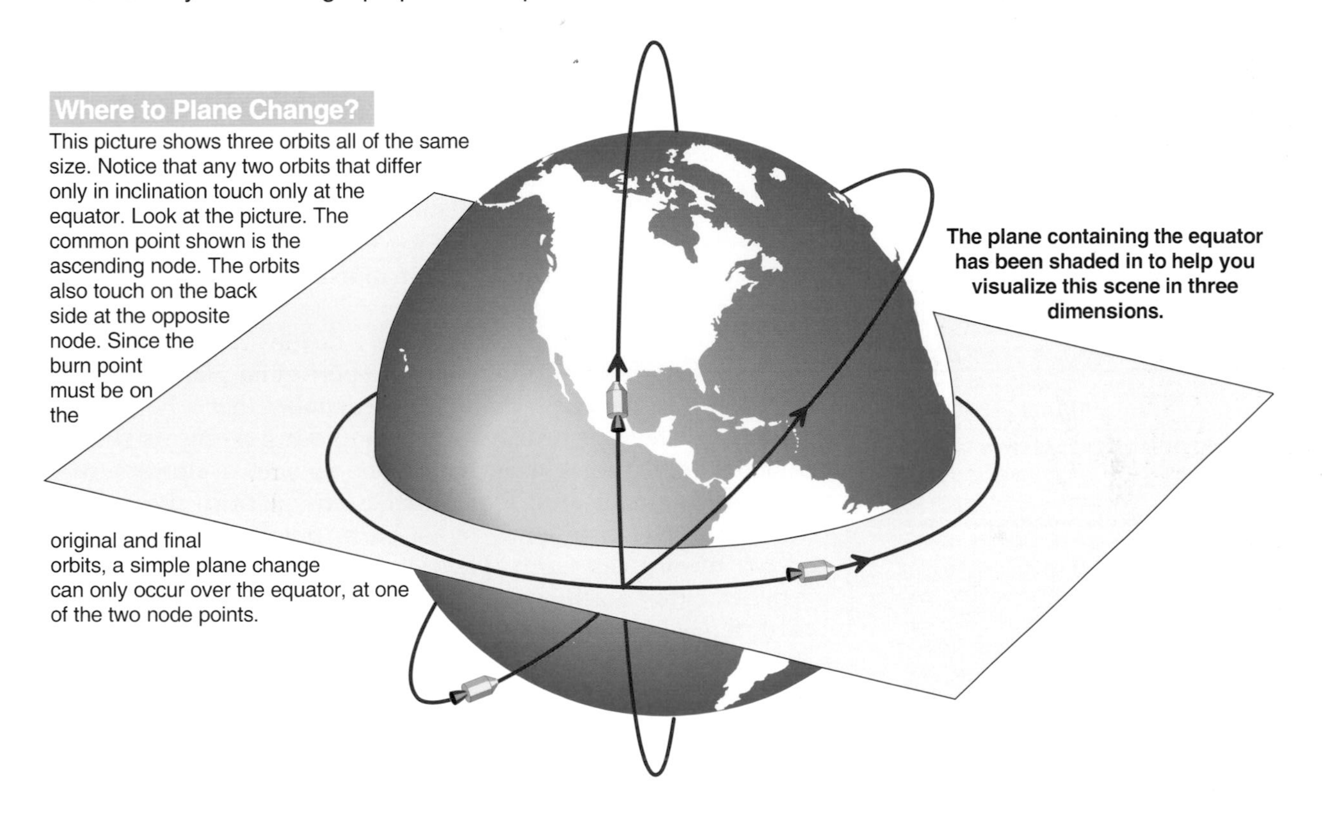

Which Way Should the Engines Fire?

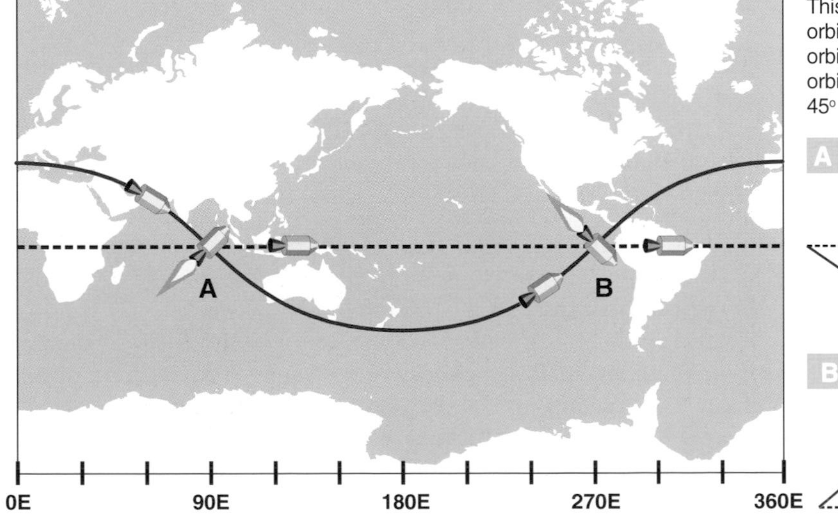

This drawing shows the ground tracks of two orbits. The solid line shows a 45° inclination orbit, and the dashed line is an equatorial orbit. How does a spacecraft change from the 45° orbit to the equatorial orbit?

A - at the Descending Node

Initial direction of flight is to the southeast. Final desired motion is eastward. Engines must fire in the southwest direction in order to provide a northeast velocity change since east is northeast of southwest.

B - at the Ascending Node

Initial direction of flight is to the northeast. Final desired motion is eastward. Engines must fire in the northwest direction in order to provide a southeast velocity change since east is southeast of northeast.

Cost of an Inclination Change

Change in Inclination	Amount of ΔV Required
5°	9% of current velocity 674 m/s at 300 km 289 m/s at 30,000 km
10°	17% of current velocity 1,347 m/s at 300 km 577 m/s at 30,000 km
30°	52% of current velocity 3,999 m/s at 300 km 1,714 m/s at 30,000 km
45°	77% of current velocity 5,913 m/s at 300 km 2,534 m/s at 30,000 km
60°	100% of current velocity 7,726 m/s at 300 km 3,310 m/s at 30,000 km
90°	141% of current velocity 10,926 m/s at 300 km 4,681 m/s at 30,000 km

The data in the first line of each row is general data valid for satellites in a circular orbit around any body (Sun, Earth, Moon, Mars, etc.) at any altitude. The bottom two lines of each rows give specific ΔV values for two altitudes in circular Earth orbit. Notice that for inclination changes of 60° or greater, a satellite must impart a velocity change greater than or equal to its current velocity. Note: "m/s" stands for meters per second.

RULE OF THUMB

Remember, a satellite that performs a plane change neither speeds up nor slows down. The velocity change simply alters the satellite's direction of motion.

consider a plane change between a posigrade equatorial orbit and another posigrade orbit inclined to the equator. A satellite on the original orbit travels due east along the equator at all times. If the satellite fires its thrusters southward, it will receive a northward push and will begin to move northward as well as eastward. The new orbit will be inclined to the equator with the ascending node (south to north equatorial crossing) at the burn point.

Conversely, if the satellite fires its thrusters northward, it will receive a southward push and will begin to move southward as well as eastward. In this case, the new orbit will be inclined to the equator with the descending node (north to south equatorial crossing) at the burn point. The amount of the inclination change in both cases depends on the ΔV provided by the thrusters. A stronger burn changes a satellite's orbital inclination by a larger amount than a weaker burn.

Keep in mind that a burn to perform a plane change is neither a horizontal nor radial burn. Remember that a horizontal burn affects a satellite's forward speed without altering its direction of motion, and a radial burn changes the upward or downward speed. On an equatorial orbit, horizontal burns point in the eastward or westward direction, while radial burns point up toward the sky or down toward the Earth. The burns in the two previous examples pointed either north or south. Think of the differences as comparable to different ways to drive a car. A horizontal burn works analogously to driving a car and stepping on the gas or stepping on the brakes without turning the steering wheel. On the other hand, a plane change works analogously to turning the steering wheel.

Plane changes require a tremendous amount of propellant as compared to other types of space maneuvers. How much is a lot? Consider the fact that most satellites only carry enough propellant to change their inclination by less than two degrees. However, the amount of propellant required for a plane change decreases with increasing altitude. A good analogy to explain this concept is that a person running at full speed requires more effort to "cut to the side" and suddenly change his or her direction of motion than a person walking slowly. Similarly, a satellite in a higher orbit moves much slower than one in a lower orbit. Consequently, the satellite in the higher orbit also requires less effort and propellant to change its direction of motion.

Perturbation Sources

All of the space maneuvers discussed up to this point require the use of rocket engines. However, some orbit changes do not. The reason is that many different forces in space other than the Earth's gravity tend to push and pull on satellites. Although a satellite's orbit tends to stay fairly constant in the short run, these forces gradually perturb the orbit to create small, unplanned orbit changes. Spaceflight engineers must understand these changes so they can either

predict a satellite's future location accurately, or plan *orbital maintenance* maneuvers to counteract the perturbing forces.

What forces tend to perturb satellite motion? Atmospheric drag is a big culprit. Gas molecules from the atmosphere, although extremely few in number, still exist at altitudes of greater than 300 kilometers (186 miles). Every collision with a gas molecule acts as a tiny retrograde ΔV. These tiny but constant collisions slow a satellite's horizontal velocity and lower the orbital altitude over a period of time. Eventually, this lowering effect causes satellites to fall into the upper atmosphere. There, they burn up due to frictional heating from colliding with air molecules at near orbital velocities.

The science of predicting the exact effects of atmospheric drag at very high altitudes is inexact at best. However, spaceflight engineers do know that satellites with large solar panels, or large cross-sectional areas tend to suffer much more from drag effects than those with smaller cross-sectional areas. In general, drag effects will cause a satellite with an orbital altitude of less than 300 kilometers to fall back into the atmosphere in a few weeks to a few months. Satellites can keep from falling out of orbit by periodically performing orbital maintenance maneuvers that raise their orbit.

In addition to atmospheric drag, solar radiation pressure also perturbs orbits. This pressure results from the constant stream of photons (sunlight) that strike a satellite's body. Satellites with large solar panels and/or large cross-sectional areas are extremely susceptible to this type of perturbation. Another major source of orbital perturbation comes from the gravitational pull of celestial bodies such as the Moon or the Sun. Gravitational perturbations tend to affect satellites in high altitude orbits more than those that orbit at lower altitudes. Both solar and gravitational perturbations act slowly over a long period of time. Their exact effects are extremely difficult to model.

What is a "Delta V" Budget?

When spaceflight engineers design a satellite, they must carefully plan all of the orbit maneuvers ahead of time, and then decide how much propellant to place on the satellite. The reason is that rockets cannot lift an unlimited amount of mass into space. Every kilogram of propellant that a satellite stores in its fuel tanks results in one less kilogram of equipment (antennas, computers, scientific instruments) that the satellite can carry into orbit. Typically, a satellite carries only the minimum amount of propellant needed to complete its maneuvers successfully, plus a little extra for orbit maintenance maneuvers (see previous section on perturbations) and emergency situations.

The ΔV budget of a satellite provides an easy way for spaceflight engineers to monitor the total amount of propellant a satellite carries. For example, suppose a satellite must perform three orbit

Above:
Snow covers the Kaibab Plateau, located to the north (lighter-gray area to the right) of the Colorado River in the Grand Canyon. In the photo, the river appears as a thick, dark groove that appears to cut the canyon in half. The Grand Canyon measures close to 450 kilometers (280 miles) in length, over 1.6 kilometers (1 mile) wide, and about 1.6 kilometers deep. Most visitors visit the Coconino Plateau on the southern rim (gray area to the left) of the canyon. This picture was taken by astronauts on shuttle *Discovery* in February 1994.

Historical Fact

The most dramatic example of atmospheric drag claiming the life of a satellite was when *Skylab* fell out of orbit on 11 July 1979. It was a large, abandoned space station that NASA astronauts had used to conduct science in the early 1970s. Most of the space station burned up in the upper atmosphere. However, a few large chunks fell into the Indian Ocean and on Australia in the "outback." Fortunately, there was no loss of life or property. NASA's new space station, scheduled for completion early in the twenty-first century, will use small thrusters to compensate for atmospheric drag.

Designing a Spacecraft

Spacecraft Parameters	Total ΔV Budget
Propellant 10% of total	
ISP = 275 seconds	284 m/s
ISP = 300 seconds	310 m/s
ISP = 325 seconds	336 m/s
Propellant 25% of total	
ISP = 275 seconds	776m/s
ISP = 300 seconds	846 m/s
ISP = 325 seconds	917 m/s
Propellant 50% of total	
ISP = 275 seconds	1,869 m/s
ISP = 300 seconds	2,039 m/s
ISP = 325 seconds	2,209 m/s
Propellant 75% of total	
ISP = 275 seconds	3,379 m/s
ISP = 300 seconds	4,078 m/s
ISP = 325 seconds	4,418 m/s

The first line in each row lists the amount of propellant carried by the spacecraft as a percentage of the spacecraft's total mass. The next three lines in each row list the spacecraft's total ΔV budget for three different values of propellant ISP. Remember, ISP measures the efficiency of a propellant (refer back to Chapter 1). Note: "m/s" stands for meters per second.

RULE OF THUMB

It takes at least two burns to transfer a satellite in between two orbits that do not intersect. A Hohmann transfer provides the most energy-efficient method to accomplish this task. Amazingly enough, the idea was proposed by Dr. Walter Hohmann thirty-two years before the space age began.

change maneuvers during the course of its mission that include a 10-meter-per-second speed up, a 30-meter-per-second slowdown, and a 15-meter-per-second inclination change. In this case, the satellite must carry enough propellant to change its velocity by at least a total of 55 meters per second. The physical amount of propellant a satellite carries depends on this total ΔV. In general, a higher ΔV budget requires more propellant. In addition, for a given amount of total ΔV, a heavier satellite, or one that uses less efficient propellant must carry more propellant than a lighter satellite, or one that uses more efficient propellant.

In order to understand why spaceflight engineers use this system to keep track of propellant, consider planning a day's worth of errands while driving a car. If an errand list includes driving 10 kilometers to the store, 20 kilometers to school, and then 15 kilometers back home, most people would think in terms of having enough gas in the tank to drive at least a total of 45 kilometers. By comparison, thinking in terms of a specific number of liters in the tank is much more difficult. On the ground, the distance a car can travel on one tank of gas measures the total performance potential of the car. In space, engineers measure the total performance potential of a satellite by how many orbit changes it can perform on one tank of propellant.

3.2 How to Transfer Between Orbits

Use of a single rocket burn in space restricts orbit changes to a transfer between two orbits that touch or intersect at the burn point. However, spaceflight engineers frequently transfer satellites between two orbits that do not intersect. These types of orbit changes require a maneuver sequence that utilizes at least two rocket burns and an intermediate transfer orbit to move the satellite between the original and destination orbits. The first burn transfers the satellite from the original orbit to the intermediate orbit. Later on, the second burn takes the satellite off the intermediate orbit and onto the final destination orbit.

Hohmann Transfer

In 1925, a German engineer named Dr. Walter Hohmann derived the most energy-efficient method to transfer a satellite between two circular orbits with the same inclination but different altitudes. The amazing aspect about his discovery is that it occurred thirty-two years before spaceflight was possible. He proved that the optimal transfer trajectory from a ΔV perspective is an elliptical orbit with a perigee at the lower orbit and apogee at the higher orbit. A satellite can use this type of transfer ellipse to move from the lower altitude orbit to the higher altitude orbit or vice versa.

Essentially, a Hohmann transfer involves a combination of two horizontal burn maneuvers (see Figure 25 on page 87). A transfer

Figure 25: *Moving Between Orbits Using a Hohmann Transfer* ◄

Characteristics: Provides the most energy-efficient method to transfer a satellite between two circular orbits with the same inclination. Requires two burns to complete. Satellite moves through 180° of true anomaly while on the transfer orbit.

► Solid arrow indicates position of the burn and points in the direction of the velocity change

▷ Hollow arrow indicates position of the burn and points in the direction that the rocket engines fire to perform the burn

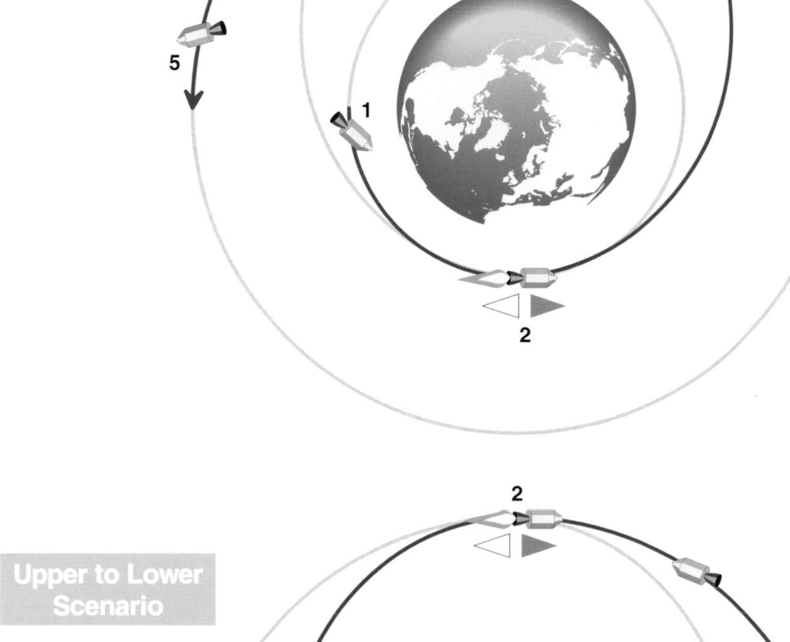

1 Satellite begins on lower orbit.

2 Posigrade burn (velocity increase) occurs here. Burn point becomes perigee of the transfer orbit.

3 Satellite coasts upward to the apogee of the transfer orbit, slowing down in the process.

4 Another posigrade burn occurs here at the apogee of the transfer orbit. This velocity increase circularizes the transfer orbit at the altitude of the upper orbit.

5 Satellite now orbits on the upper circular orbit.

6 Without the second burn, the satellite will remain on the transfer orbit and coast back down to the perigee.

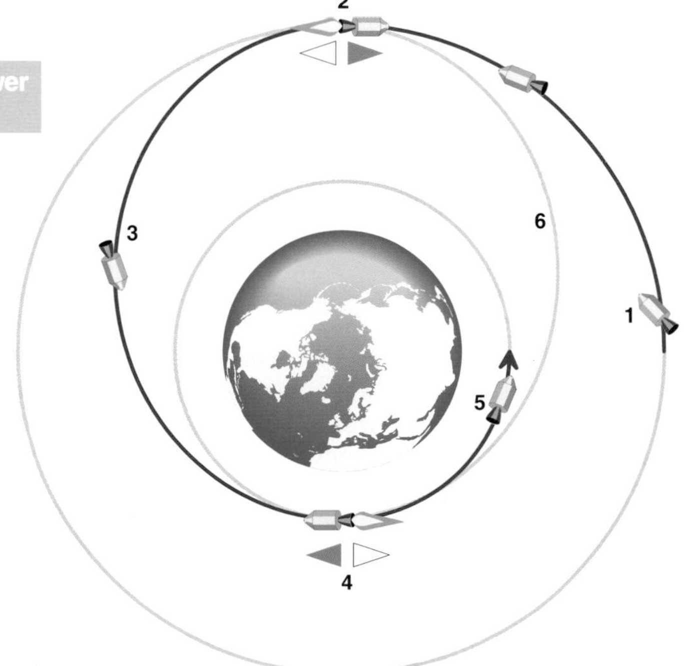

1 Satellite begins on upper orbit

2 Retrograde burn (velocity decrease) occurs here. Burn point becomes apogee of the transfer orbit.

3 Satellite coasts downward to the perigee of the transfer orbit, gaining velocity in the process.

4 Another retrograde burn occurs here at the perigee of the transfer orbit. This velocity decrease circularizes the transfer orbit at the altitude of the lower orbit

5 Satellite now orbits on the lower circular orbit.

6 Without the second burn, the satellite will remain on the transfer orbit and coast back up to apogee.

Planning a Hohmann Transfer

Transfer Statistic	Value
New Orbit is 2x Radius of Old	
ΔV to enter Hohmann	15.5%
ΔV to exit Hohmann	13.0%
ΔV Total	28.5%
Transfer Time	91.9%
New Orbit is 3x Radius of Old	
ΔV to enter Hohmann	22.5%
ΔV to exit Hohmann	16.9%
ΔV Total	39.4%
Transfer Time	141.0%
New Orbit is 4x Radius of Old	
ΔV to enter Hohmann	26.5%
ΔV to exit Hohmann	18.4%
ΔV Total	44.9%
Transfer Time	198.0%
New Orbit is 5x Radius of Old	
ΔV to enter Hohmann	29.1%
ΔV to exit Hohmann	18.9%
ΔV Total	48.0%
Transfer Time	259.0%
New Orbit is 10x Radius of Old	
ΔV to enter Hohmann	34.8%
ΔV to exit Hohmann	18.1%
ΔV Total	52.9%
Transfer Time	645.0%

Values for ΔVs and Hohmann transfer times are given as a percentage of the value on the original circular orbit. For example, if the transfer time is listed as 400%, then the time it takes to complete the Hohmann transfer is four times longer than the satellite's period on the original circular orbit. This table applies for satellites in orbit around any body (Sun, Earth, Moon, Mars, etc.). Remember, radius measures the distance from the satellite to the center of the body being orbited, not the surface of the body.

from the lower to the upper orbit begins when a satellite fires its thrusters to increase its velocity (posigrade burn). This burn takes the initial lower circular orbit and turns it into an elliptical orbit with a perigee at the burn point and apogee at the altitude of the upper orbit. The satellite then coasts upward from the perigee at the burn point toward the apogee of the elliptical transfer orbit. At apogee, the thrusters perform a second velocity increase (posigrade burn) to circularize the transfer orbit by raising the perigee of the transfer orbit to match the apogee altitude.

The theory behind the maneuver sequence described in the previous paragraph appears to result in a paradox. Consider the fact that it takes velocity increase burns on two separate occasions to effect a Hohmann transfer from a lower altitude orbit to a higher altitude orbit. However, the satellite's final velocity is slower than the initial velocity because it winds up in a higher orbit than the original. How is this apparent contradiction possible? In reality, no paradox exists. Remember that the first velocity increase allows the satellite to climb to the altitude of the higher orbit. In the process, the satellite loses a lot of velocity. By the time the satellite reaches the top of the transfer ellipse, its velocity is slower than that required to maintain a circular orbit at that altitude. Consequently, the satellite must perform another velocity increase burn.

The transfer from the upper orbit to the lower works in a similar fashion. A satellite on the upper orbit starts the transfer by firing its engines to slow down (retrograde burn). This burn takes the initial upper circular orbit and turns it into an ellipse with an apogee at the burn point and perigee at the altitude of the lower orbit. The satellite then coasts downward from apogee at the burn point to the perigee of the transfer orbit. At perigee, the thrusters perform a second velocity decrease (retrograde burn) to circularize the transfer orbit by lowering the apogee of the transfer orbit to match the perigee altitude. The transfer time equals half the period of the elliptical transfer orbit and is the same for both the lower to upper and upper to lower case.

On an upper to lower transfer, a satellite performs burns to slow down on two separate occasions, but winds up traveling faster than when it started because the final orbit is lower than the original orbit. In this case, the first slowdown allows the satellite to descend to the lower orbit on the transfer ellipse. The satellite gains a lot of velocity descending to the lower orbit similar to a ball gaining velocity by rolling down a hill. By the time the satellite reaches the bottom of the ellipse, it is moving too fast to maintain a circular orbit at that altitude. As a result, the satellite must perform a second slowdown burn.

A Hohmann transfer requires less propellant than any other type of transfer between two circular orbits that do not intersect. However, the energy minimization comes at the expense of the transfer time. This maneuver takes more time to complete than most other types of multiple burn transfers. The slow time can be partially attrib-

uted to the fact that a satellite on a Hohmann ellipse must travel through 180° of true anomaly during the course of the transfer.

Faster Transfer Techniques

Hohmann transfers minimize propellant expenditures at the expense of a long transfer time. Another alternative exists that requires more propellant but greatly reduces the transfer time. Consider the transfer of a satellite from a lower-altitude circular orbit to a higher-altitude circular orbit. A satellite begins the faster transfer by burning more propellant than that required to turn the lower orbit into a Hohmann transfer ellipse. As in the Hohmann case, the burn converts the lower circular orbit into an elliptical-transfer orbit with a perigee at the burn point. However, the stronger burn elevates the apogee of the transfer orbit above the altitude of the upper circular orbit.

After the burn, the satellite coasts upward from the perigee at the burn point toward the apogee of the newly created transfer orbit. Figure 26 on page 90 shows that the satellite reaches the altitude of the upper orbit before reaching apogee of the transfer orbit. At that point, the satellite circularizes the transfer orbit by performing a combination of a radial and a horizontal burn. The decrease in the transfer time over the Hohmann scenario occurs because the geometry of this technique allows a satellite to travel a shorter distance but still reach the upper orbit's altitude.

The increase in the total propellant expenditure as compared to the Hohmann transfer occurs in two places. One increase occurs at the initial burn to create the transfer ellipse. This ellipse is larger than a Hohmann transfer ellipse. Therefore, it takes more ΔV to create. The other increase occurs at the final burn to circularize the transfer orbit. In general, a circularization maneuver that requires a combination horizontal-radial burn takes more propellant to accomplish than a circularization maneuver that requires a single horizontal burn.

General Transfers

Orbital transfers in real life almost always involve more complicated scenarios than the circular to circular with the same inclination case. In general, the sizes, shapes, inclinations, and orientations of the original and destination orbits may differ. The bottom line is that a transfer ellipse must have a focus at the center of the Earth, and must also intersect both the original and destination orbits. Typically, a satellite may need to perform a combination of a horizontal burn, a radial burn, and a plane change to get on and off the transfer orbit.

LEO to GEO Transfers

One important physical limitation of launching rockets is that a rocket cannot launch a satellite directly into an orbit with a tilt

Above:
On the morning of 19 September 1994, two volcanic cones of the Rabaul volcano in Papua, New Guinea began to erupt with little warning. The eruption sent a plume over 18.3 kilometers (60,000 feet) into the sky and forced the evacuation of over 50,000 people from the Rabaul harbor area. Astronauts on Space Shuttle *Discovery* took this picture of the plume 24 hours after the initial eruption. The plume flattens out at the top and is carried in a westerly direction by prevailing winds. A layer of ash (gray area under the plume) is being distributed by lower level winds.

RULE OF THUMB

Using a Hohmann transfer saves propellant at the expense of time. Using more propellant than required for a Hohmann transfer can reduce the transfer time.

► Figure 26: *A Look at Other Types of Orbital Transfers*

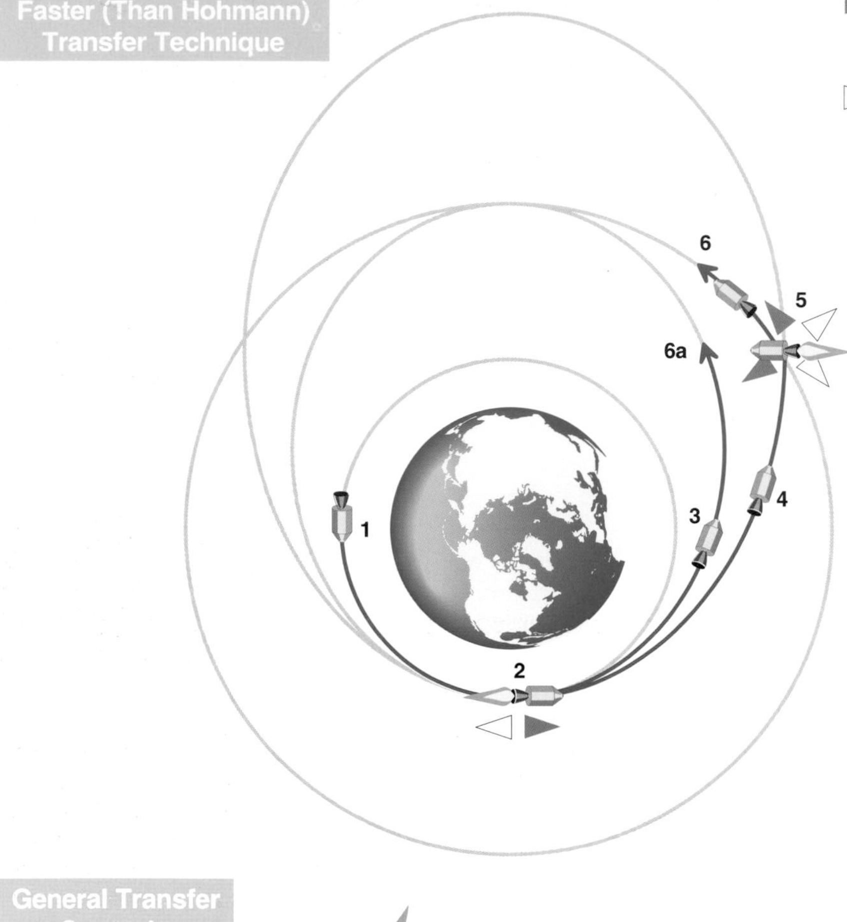

Solid arrow indicates position of the burn and points in the direction of the velocity change

Hollow arrow indicates position of the burn and points in the direction that the rocket engines fire to perform the burn

1 Spacecraft begins on the lower circular orbit with the goal of reaching the upper circular orbit.

2 Posigrade burn (velocity increase) occurs here.

3 If the burn barely raises the apogee to the altitude of the upper orbit, then the spacecraft enters a Hohmann transfer.

4 However, using more propellant will put the spacecraft on a larger (than Hohmann) transfer orbit with an apogee at a higher altitude than the upper orbit.

5 A combination of a posigrade and an inward radial burn allows the spacecraft to turn the corner and enter the upper circular orbit.

6 The fast transfer is complete. If a Hohmann transfer had been used, the spacecraft would still be on the transfer orbit at position marked 6a.

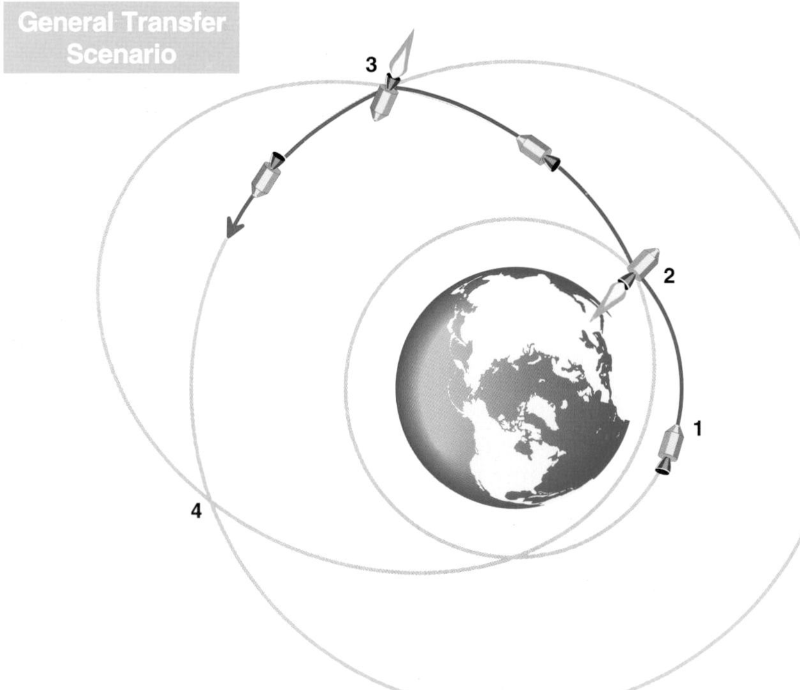

1 Spacecraft begins on the lower circular orbit with the goal of reaching the upper circular orbit.

2 A rocket burn here moves the spacecraft onto the transfer orbit. Notice that the transfer ellipse can be of any size, eccentricity, and orientation as long as it intersects both the original and destination orbits.

3 Rocket burn here takes the spacecraft off the transfer orbit and onto the destination (upper circular) orbit. A general transfer uses a lot of propellant because it may require the spacecraft to use two horizontal-radial combination burns.

4 The spacecraft could have also waited until here to exit the transfer orbit.

angle less than the latitude of the launch site. One important consequence of this limitation is that American rockets cannot send satellites directly into equatorial orbits because the nation's launch site at the Kennedy Space Center in Florida lies at a latitude 28.5° north of the equator. This restriction is unfortunate because a geosynchronous orbit high above the equator provides an ideal location for communications and weather satellites. In order to overcome the launch site latitude constraint, NASA uses a technique called the low Earth orbit (LEO) to geosynchronous orbit (GEO) transfer. This type of transfer is interesting from a mission design perspective because it combines a Hohmann transfer with a plane change.

An expendable rocket or the Space Shuttle must initially place the GEO-bound satellite in a posigrade low Earth orbit as close to the equatorial plane as physically possible. Consequently, all LEO to GEO missions starting from the Kennedy Space Center begin with a launch into a low orbit with an inclination of 28.5°. GEO-bound satellites always arrive in LEO attached to an auxiliary rocket called an upper stage. This rocket provides the propulsion to reach GEO and functions as the satellite's "booster" stage for the Hohmann transfer from LEO to GEO. The majority of upper stage designs in use today utilize solid propellants.

At the appointed time, the upper stage's engine performs a posigrade burn to move the satellite onto a transfer ellipse with a perigee at the burn point and apogee at geosynchronous altitude. The satellite then coasts upward toward apogee, jettisons the upper stage somewhere along the way, and arrives at geosynchronous altitude 5 hours and 16 minutes after the initial burn. At apogee, a small solid rocket engine attached to the satellite, called an apogee kick motor, fires to create a posigrade ΔV that circularizes the transfer orbit at geosynchronous altitude. The apogee kick motor typically stays attached to the satellite after the burn, but serves no further purpose.

The problem with this mission plan is that both the transfer orbit and the geosynchronous orbit have the same inclination as the initial low Earth orbit. A 28.5° inclination change must occur sometime during the transfer in order to allow the satellite to reach an equatorial geosynchronous orbit. This change typically occurs simultaneously with the circularization maneuver at the apogee of the transfer orbit because plane changes at higher altitudes require less propellant. In reality, the optimal mathematical solution that minimizes propellant consumption calls for the upper stage to perform a 2.2° inclination change to insert the satellite into the perigee of a 26.3° inclination transfer orbit. The apogee kick motor then performs the rest of the 26.3° inclination change at the time of circularization at apogee.

The apogee of the Hohmann ellipse must occur at an equatorial crossing (either the ascending or descending node) because it is physically impossible for a satellite to transfer from an inclined orbit to an equatorial orbit at any place other than the equatorial plane.

Above:

In September 1991, astronauts on *Discovery* took this photograph centered on Kashmir, India. The view from the Space Shuttle is looking north into the Takla Makan Desert of eastern China. Notice that a layer of haze covers almost all of western India. At the time this picture was taken, the haze extended all the way past China and into the Siberian plains of Russia (not shown in this picture).

Scientific Fact!

A Hohmann transfer from low Earth to geosynchronous orbit takes 5 hours and 16 minutes to complete.

▶ Figure 27: *How to Reach Geosynchronous Orbit from Florida*

Characteristics: Provides a way for NASA to send satellites to an equatorial geosynchronous orbit (0° inclination) even though the latitude of the Kennedy Space Center limits achievable orbits to greater than or equal to 28.5° in inclination.

Solid arrow indicates position of the burn and points in the direction of the velocity change

Hollow arrow indicates position of the burn and points in the direction that the rocket engines fire to perform the burn

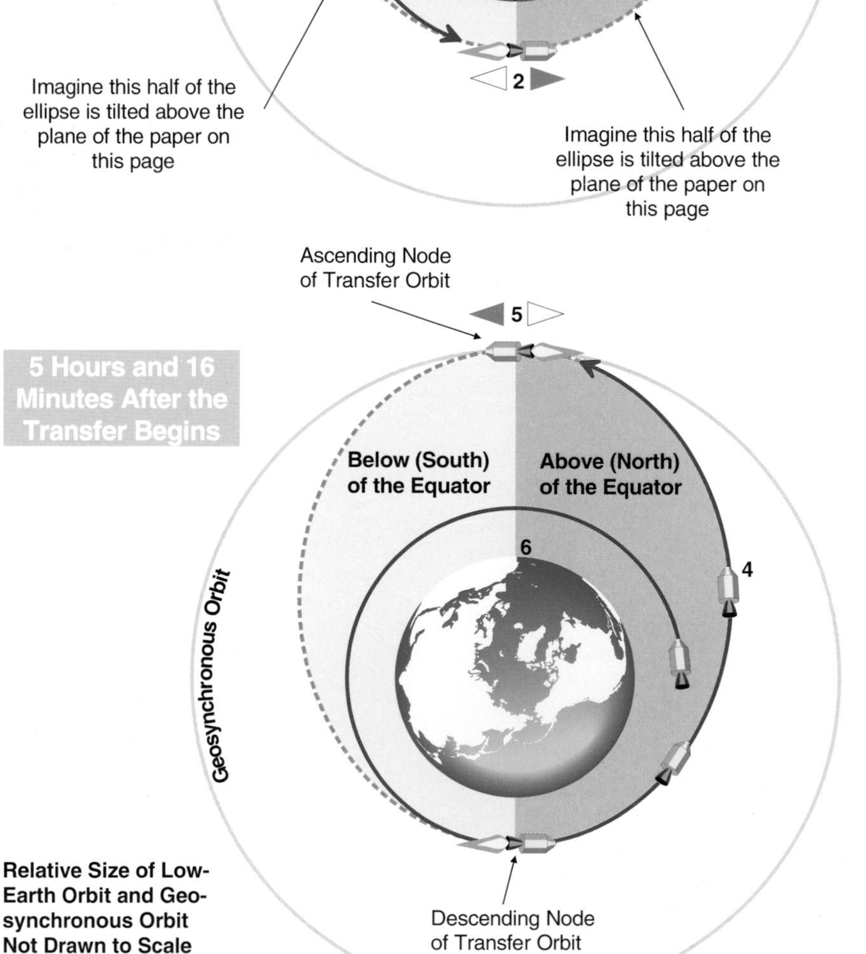

1 The satellite begins on a 28.5° inclination low-Earth orbit. The goal is to place the satellite into an equatorial geosynchronous orbit over the west coast of Africa.

2 Posigrade burn (velocity increase) at the ascending node of the low-Earth orbit, combined with a slight inclination change, places the satellite onto a 26.3° inclination Hohmann transfer orbit. Insertion onto the transfer orbit occurs at the perigee and the ascending node of the new orbit.

3 Notice that the insertion onto the transfer orbit happens when the satellite is 100° behind the west coast of Africa. The two thin-black lines on the Earth show this angle. The satellite must begin 100° behind because the Earth rotates through only 80° in the time the satellite takes to move through 180° on the Hohmann transfer.

4 After the initial burn to create the Hohmann transfer orbit, the satellite coasts upward to the apogee of the transfer orbit which occurs high above the Earth at geosynchronous altitude.

5 At the apogee (which is also the descending node), the satellite performs another posigrade burn to circularize the transfer ellipse at geosynchronous altitude. Also, the satellite simultaneously performs a 26.3° inclination change maneuver to switch to an equatorial orbit (0° inclination). This burn happens 5 hours 16 minutes after the first burn (event 2).

6 Notice that the satellite arrives at geosynchronous orbit over West Africa as desired.

Therefore, the perigee must occur at the other node point because the nodes lie 180° apart and a satellite moves through 180° of true anomaly on a Hohmann transfer. These two constraints lead to two possible LEO to GEO transfer scenarios. The first involves a burn at the ascending node in LEO to create the transfer orbit, a transfer that occurs over the Northern Hemisphere, and a circularization burn at the descending node of the transfer orbit at GEO altitude. Scenario number two involves a burn at the descending node in LEO to create the transfer orbit, a transfer that occurs over the Southern Hemisphere, and a circularization burn at the ascending node of the transfer orbit at GEO altitude. Figure 27 on page 92 illustrates the first case.

One interesting question remains. When should the upper stage engine fire to insert the satellite into the Hohmann transfer orbit? Remember that a satellite in a geosynchronous orbit appears to hover over a single location on the equator. This location, also called the *final station*, determines what parts of the Earth's surface the satellite can "see." In order to place the satellite over the proper hover point, the Hohmann transfer must begin at a time that allows the satellite to fly over the hover point at the end of the transfer. The fact that the Earth constantly rotates as the satellite moves on the transfer orbit complicates this problem.

The key to understanding how to place the satellite over the proper hover point (final station) at the end of the transfer involves thinking about angles in the J2000 coordinate system. All locations on the surface of the Earth rotate around the center of the J2000 coordinate system (located at the center of the Earth) at a rate of 360° every 24 hours, or 15° every hour. This rotation rate causes the hover point to rotate counterclockwise (eastward) through an angle of about 80° in the 5 hours and 16 minutes that the satellite spends on the transfer orbit. In the same amount of time, the satellite moves counterclockwise through 180° as it travels from the perigee to the apogee of the Hohmann ellipse. The fact that the satellite moves from west to east through 100° more than the hover point during the transfer means that the transfer must begin 100° to the west of the hover point. Figure 27 on page 92 shows how this works.

All of these constraints dictate that the upper stage and satellite must wait in low Earth orbit until they cross the equator over a location 100° in longitude to the west of the hover point. On every orbit, a satellite in low Earth orbit flies over a point on the equator between 20° to 30° to the west of the location on the previous orbit (see the section in the last chapter on ground tracks). Unfortunately, as many as forty to seventy orbits can pass in between opportunities where one of the equatorial crossings falls at the proper position to allow an insertion into the Hohmann transfer orbit that reaches geosynchronous orbit. However, spaceflight engineers have some flexibility in solving this problem because satellites typically carry enough propellant to correct their final hover point by up to 5° in longitude once they reach geosynchronous orbit.

Above:

Astronauts on Space Shuttle *Endeavour* took this photograph as they flew over the southern tip of Italy in October 1994. In the picture, Italy appears as the long peninsula (from the top to the bottom) that looks like a boot. The island of Sicily lies at the left bottom side of the image.

Scientific Fact!

Due in part to the 28.5° inclination change that must be performed, it takes more propellant to launch a satellite into geosynchronous orbit from Florida than to send the same satellite to the Moon.

Above:

In May 1991, astronauts on Space Shuttle *Discovery* shot this picture of the author's home town of San Diego, California. Those familiar with Southern California should be able to recognize Point Loma (cloud-covered peninsula), Coronado Island and the Naval Air Station (in the middle of the harbor), Mission Bay (square water area to the north of Coronado), and the Miramar Naval Air Station (diagonally up and to the left of Mission Bay). Those with a good eye might be able to spot the San Diego border with Tijuana, Mexico. The border shows up as an extremely thin diagonal line near the bottom center of the photograph, about half an inch from the edge. Hint: at the ocean, the border separates a darker area (United States) from a lighter area (Mexico).

Historical Fact

The first successful orbital rendezvous mission occurred on 15 December 1965. On that day, NASA astronauts Frank Borman and Jim Lovell in a spacecraft called *Gemini 7* rendezvoused with Wally Schirra and Thomas Stafford in *Gemini 6*.

3.3 How to Rendezvous in Orbit

A rendezvous in orbit involves two or more satellites matching their orbits, velocities, and positions. This type of maneuver primarily occurs in situations where two spacecraft need to dock with each other to transfer astronauts or supplies, or where one spacecraft needs to reach the other in order to perform a repair. A successful rendezvous requires a lot of planning and a good intuition into how orbital mechanics works. Many different solutions exist for any given rendezvous scenario. The problem involves finding the most feasible solution given the amount of time and propellant available for the maneuver.

Single Catch-up Rendezvous

A single catch-up scenario occurs when both spacecraft start on the same orbit but at different positions. One of the satellites, called the chaser, maneuvers to perform the rendezvous. The other, called the target, maintains a constant orbit. The catch-up scenario can involve the chaser starting either ahead of or behind the target. The following example shows how to perform a rendezvous when the chaser starts behind the target. This scenario represents the ideal situation in most cases.

Suppose that the Space Shuttle flies in the same circular orbit but lags behind a satellite that the astronauts want to catch. A commonsense solution that works on the highway involves "speeding up by stepping on the gas." This solution does not work in space because a speed-up involves a posigrade burn that increases the size of the orbit and lengthens the period. The new elliptical orbit forces the orbiter to take a longer time to circle the Earth and come back to the burn point than the satellite takes to complete one transit around the original circular orbit. Therefore, the astronauts fall further behind the satellite on every successive pass through the burn point.

Instead, the astronauts must "hit the brakes" and slow down. A retrograde burn slows the orbiter at the burn point, decreases the orbit size, and shortens the period. This smaller elliptical orbit allows the shuttle to circle the Earth and come back to the burn point in a shorter time than the satellite takes to complete one transit around the original circular orbit. As a result, the shuttle moves closer to the satellite on every successive pass through the burn point (see Figure 28 on page 96).

Another aspect of this rendezvous example concerns how long the chaser takes to catch the target. Consider the following scenario. Both the Space Shuttle and a satellite initially circle the Earth in an orbit that takes 92 minutes to complete. Suppose that the satellite initially leads the shuttle by four minutes. In other words, the satellite reaches any given point on the orbit four minutes ahead of the shuttle. What options allow the astronauts to catch up to the satellite?

Figure 28 on page 96 shows one option. A retrograde burn allows the shuttle to drop from the original 92-minute circular orbit into a smaller elliptical orbit with a 91-minute period. During the first 91 minutes after the burn, the shuttle makes one trip around the world on the new orbit and returns back to the burn point. In this time, the satellite also travels around the world, but winds up one minute short of reaching its original position. As a result, the astronauts come one minute closer to the satellite every 91 minutes. This rate of catch-up allows the shuttle to make up the initial four-minute difference after four orbits on the new elliptical orbit. The chase lasts for four 91-minute orbits which amounts to slightly more than six hours.

The right half of Figure 29 on page 97 shows a faster option. A stronger retrograde burn allows the shuttle to drop into a smaller, elliptical orbit with a shorter period than in the first case. This time, the burn drops the shuttle from the 92-minute orbit into a 90-minute orbit. Every time the astronauts make one trip around the world on the new orbit and arrive back at the burn point, they close the gap by two minutes. Therefore, the orbiter makes up the initial four-minute difference after two orbits on the new elliptical orbit. The chase lasts for two 90-minute orbits (three hours). In both cases, the shuttle must recircularize its orbit at the end of the chase to perform the rendezvous.

Both the slow and fast chase options represent valid solutions to the single catch-up rendezvous problem. The choice of options depends on the mission situation. For example, a slow option that saves propellant makes sense if the chaser carries non-perishable supplies to a space station. However, saving propellant at the expense of time makes no sense if the chaser is a rescue party on the way to rendezvous with and save astronauts aboard a damaged spacecraft.

Rendezvous Using a Hohmann Transfer

A different rendezvous scenario exists if both the chaser and target start in different orbits. Consider the following scenario. The Space Shuttle currently circles the Earth at a different altitude than a broken satellite that the astronauts need to repair. Mission control instructs the astronauts that they need to pay for the propellant for the rendezvous out of their salaries because of cuts in NASA's budget. Therefore, the astronauts choose to use a Hohmann transfer in the rendezvous process.

One possible solution to this problem involves a Hohmann transfer to reach the broken satellite's orbit, then a single catch-up rendezvous to meet the satellite. However, a better solution uses careful timing to allow the astronauts to meet the broken satellite at the exact instant the shuttle finishes the Hohmann transfer. The way to understand this timing involves thinking about how fast the

Scientific Fact!

Understanding how to rendezvous two spacecraft in orbit requires thinking opposite to what common sense would seem to dictate. For example, in order to catch up with a satellite ahead of you in the same orbit, you must slow down. In order to let a satellite behind you in the same orbit catch up, you must speed up.

RULE OF THUMB

The amount of time required to complete a rendezvous depends on the amount of propellant used. More propellant usually results in a shorter rendezvous time. The average NASA Space Shuttle rendezvous mission takes about two to three days to complete.

Historical Fact

A NASA astronaut, Dr. Edwin "Buzz" Aldrin Jr., pioneered the research work in the 1960s that allowed NASA to solve the problem of how to rendezvous spacecraft. The information was crucial in allowing America to win the "race to the Moon." Insiders nicknamed him "Dr. Rendezvous." Eventually, Aldrin became the second person to walk on the Moon.

► Figure 28: *Catching Another Spacecraft in Orbit, Slower Option*

Scenario: The Space Shuttle and its astronauts are in the same orbit (circular, with a 92-minute period) as a satellite that they want to catch. However, they find themselves behind the satellite by four minutes. In other words, the satellite at any given point on the orbit is four minutes ahead of the space shuttle. How can the astronauts catch the satellite? This figure looks at two possible options.

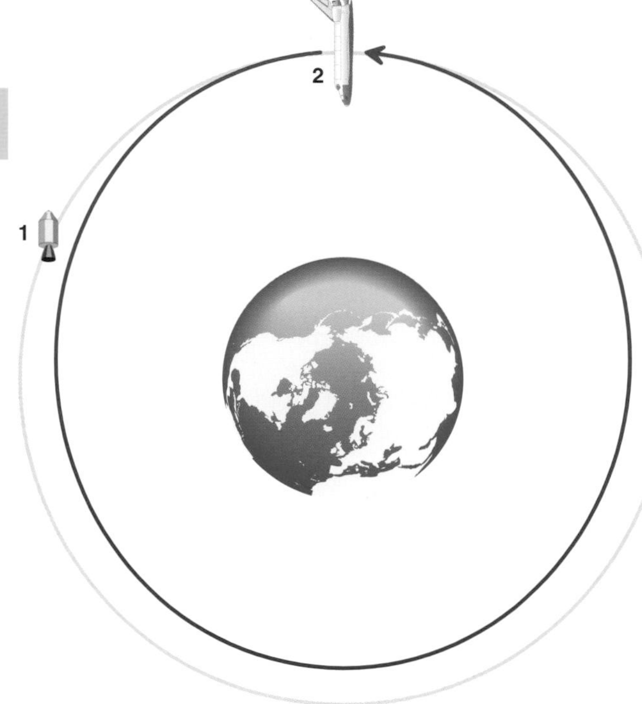

1 Both the satellite and the shuttle begin on the outer circular orbit (gray line) with a 92-minute period. The shuttle starts four minutes behind the satellite.

2 To catch up, the shuttle fires its thrusters to slow down. This action drops it into an elliptical orbit with apogee at the burn point, and perigee slightly lower than the original circular orbit. This new orbit allows the shuttle to go around the world in 91 minutes (one minute quicker than the satellite).

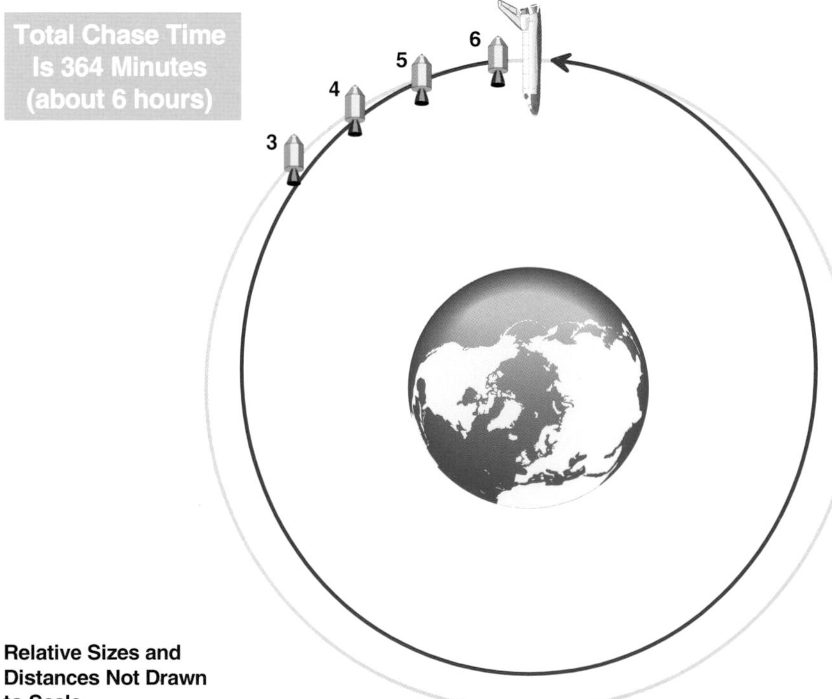

3 91 minutes after the burn, the shuttle comes back to the apogee of its new orbit at the circular orbit altitude. However, the satellite stayed on the original orbit which has a 92-minute period. As a result, the satellite is now only three minutes ahead of the shuttle instead of four.

4 Another 91 minutes elapses, and the satellite is now only two minutes ahead of the shuttle.

5 Yet another 91 minutes elapses, and the satellite is now only one minute ahead of the shuttle.

6 A final 91-minute interval passes and the shuttle is now caught up with the satellite. The total chase took the shuttle four orbits that lasted 91 minutes each. To complete the rendezvous, the shuttle must return to the original circular orbit.

Relative Sizes and Distances Not Drawn to Scale

Figure 29: *Catching Another Spacecraft in Orbit, Faster Option*

Which option the astronauts decide to use to catch up with the satellite (either the slower or the faster method) depends on how much propellant they have (or are willing to use), and the time criticality of the rendezvous. It takes more propellant to drop to the 90-minute orbit than the 91-minute orbit. However, the lower orbit (faster option) allows the shuttle to catch up with the satellite in half the time of the slower option.

Faster Catch Up Option

1 Both the satellite and the shuttle begin on the outer circular orbit (gray line) with a 92-minute period. The shuttle starts four minutes behind the satellite.

2 To catch up, the shuttle fires its thrusters to slow down. This action drops it into an elliptical orbit with apogee at the burn point, and perigee slightly lower than the original circular orbit. This new orbit allows the shuttle to go around the world in 90 minutes (two minutes faster than the satellite).

Notice that this catch-up orbit is slightly lower than the one on the facing page.

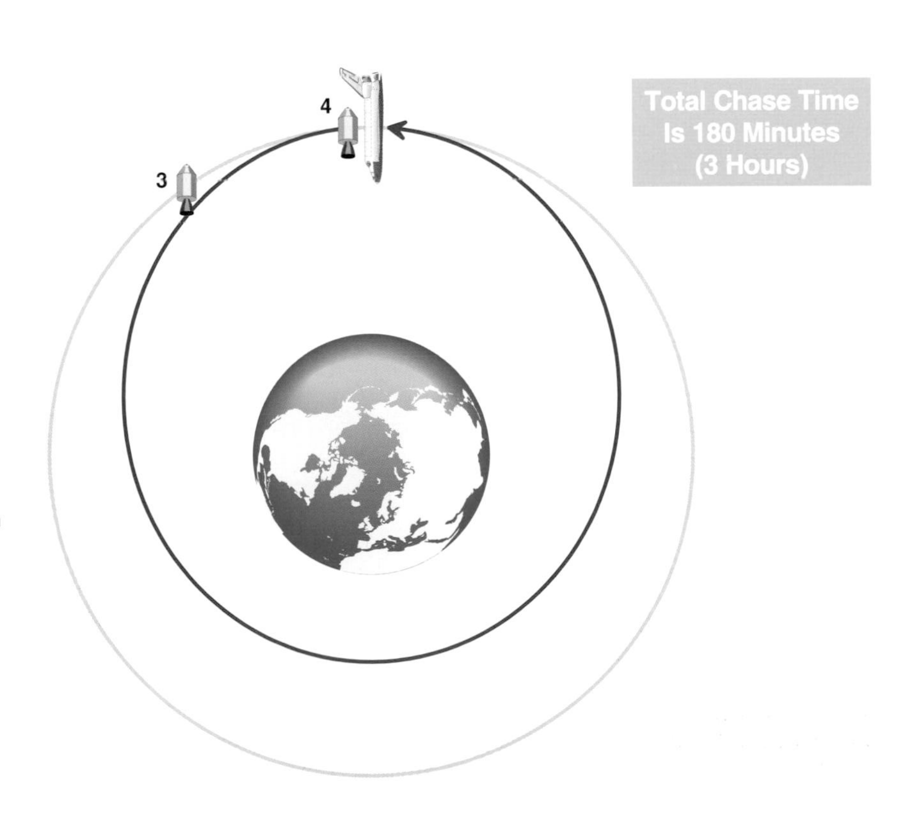

3 90 minutes after the burn, the shuttle comes back to the apogee of its new orbit at the circular orbit altitude. However, the satellite stayed on the original orbit which has a 92-minute period. As a result, the satellite is now only two minutes ahead of the shuttle instead of four.

4 Another 90 minutes elapses and the shuttle has gone around its new orbit one more time. It has now caught up with the satellite. The total chase only took the shuttle two orbits that lasted for 90 minutes each. To complete the rendezvous, the shuttle must return to the original circular orbit.

IMPORTANT
The size differences between the 90-minute catch-up orbit on this page and the 91-minute catch-up orbit on the facing page are exaggerated for clarity

Above:

A great number of circular, center-pivot irrigation plots appear in the west-looking view of northern Saudi Arabia. These small plots of irrigated farmland show up as dots near the bottom of the photo. The darker area near the middle, left edge of the photograph is the Nejd Plateau area. Water from this high country flows east toward the agricultural region where it is pumped up from underground aquifers. The north end of the Red Sea and Sinai Peninsula is visible near the top of the photograph. This picture was taken on Space Shuttle *Columbia* in March 1994.

RULE OF THUMB

When astronauts in the Space Shuttle rendezvous with a satellite to repair it, they sometimes perform 10 or more orbital change maneuvers during the course of the rendezvous.

shuttle moves on the Hohmann transfer relative to how fast the broken satellite moves on its orbit.

The first possibility is that the broken satellite lies at a higher altitude than the shuttle's initial orbit (see upper half of Figure 30 on page 99). Notice that in this case, the entire Hohmann ellipse, except for the apogee, lies below the broken satellite's altitude. As a result, the shuttle moves faster along the transfer orbit than the broken satellite moves on its circular orbit. The reason is that higher-altitude satellites have longer periods and move slower. This velocity differential means that the shuttle must start the Hohmann transfer behind the satellite in order for both to meet at the same place at the same time.

The other possibility occurs when the broken satellite orbits at a lower altitude than the shuttle's initial orbit (see lower half of Figure 30 on page 99). Here, the entire transfer orbit, except for the perigee, lies above the satellite's altitude. In this case, the shuttle travels slower along the Hohmann transfer ellipse than the broken satellite moves along its circular orbit. As a result, the shuttle must start the Hohmann transfer ahead of the satellite in order for both to reach the same place at the same time. The exact starting differences in both this and the first possibility depend on the initial altitude of both the orbiter and the broken satellite.

General Rendezvous

Real rendezvous missions typically utilize many different rocket burns and many different maneuvers that may include, but are not limited to, single catch-ups, orbital transfers, inclination changes, and node position-change maneuvers. Some scenarios employ a double rendezvous where both spacecraft are active. In this situation, instead of designating one satellite as the chaser and the other as the target, both spacecraft chase each other to simultaneously arrive at a predesignated point in space. Planning the sequence and the strength of the burns for a successful rendezvous in real life is a complicated task.

3.4 Satellite Navigation

Successfully executing a space maneuver in orbit requires fast reflexes and knowing the exact position and velocity of the satellite to be maneuvered. Consider the effect of a rocket burn timed one tenth of a second late. A satellite in low Earth orbit moves at 7,700 meters (25,262 feet) per second. In that one-tenth of a second, a satellite will move 770 meters, almost an entire kilometer. A position error of this magnitude can ruin an entire rendezvous mission.

Recently, the Department of Defense has developed a novel system that will allow satellites to navigate in Earth orbit with an

Figure 30: *Catching a Spacecraft in a Different Orbit* ◄

Scenario: The Space Shuttle and its astronauts are in a different orbit (lower or higher) than a satellite that they want to catch.

Solid arrow indicates position of the burn and points in the direction of the velocity change

Hollow arrow indicates position of the burn and points in the direction that the rocket engines fire to perform the burn

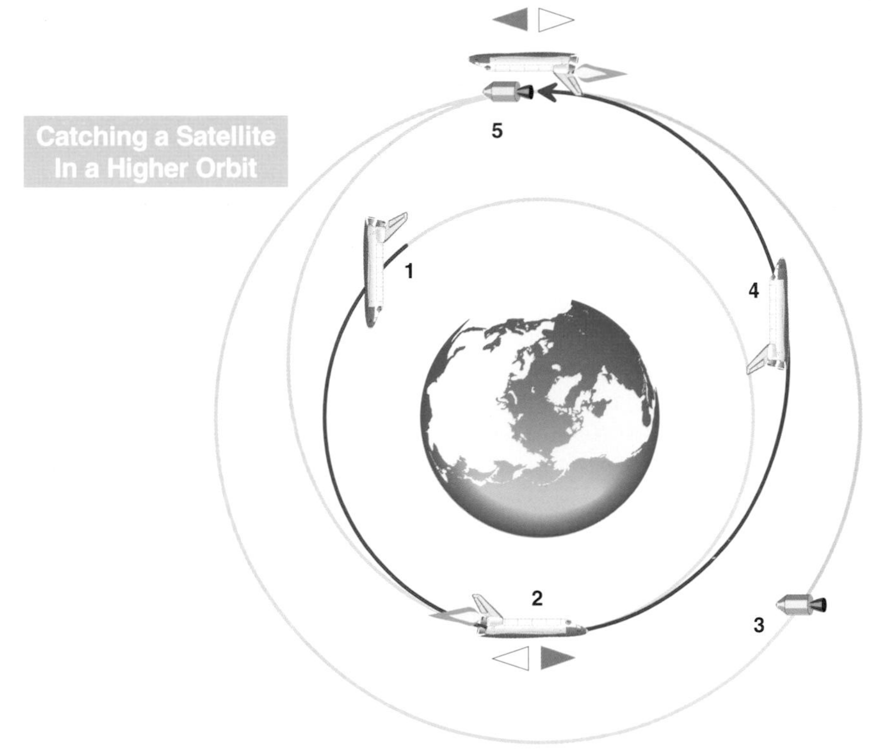

1 Space Shuttle begins on the lower circular orbit.

2 A posigrade burn (velocity speed-up) occurs here. This burn puts the shuttle onto a Hohmann transfer to the upper circular orbit where the satellite is orbiting.

3 This is the satellite's position at the time the shuttle performs its initial burn to enter the Hohmann transfer. Since the shuttle moves faster on the transfer orbit than the satellite on the upper orbit, the shuttle must enter the transfer orbit behind the satellite to avoid arriving at the rendezvous point before the satellite.

4 Shuttle coasts up to the apogee of the transfer orbit.

5 Shuttle and satellite arrive here at the same time. Another posigrade burn by the shuttle circularizes its orbit at the satellite's altitude.

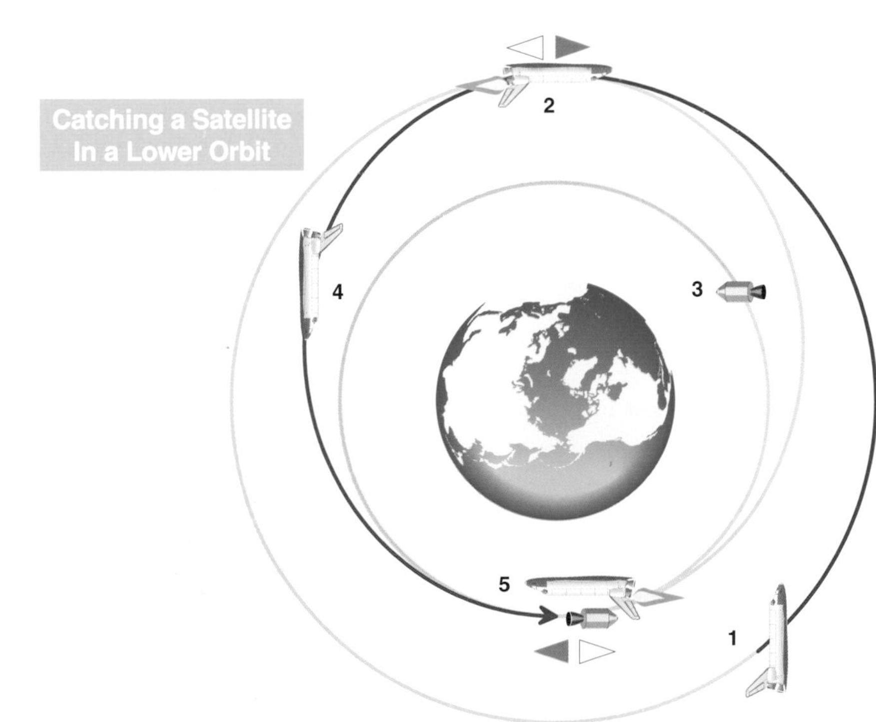

1 Shuttle begins on the upper circular orbit.

2 Retrograde burn (velocity slow-down) occurs here. The burn puts the shuttle onto a Hohmann transfer to the lower circular orbit where the satellite is orbiting.

3 This is the satellite's position at the time the shuttle performs its initial burn to start the Hohmann transfer. Since the shuttle moves slower on the transfer than the satellite on the lower orbit, the shuttle must enter the transfer ahead of the satellite to avoid arriving at the rendezvous point after the satellite.

4 Shuttle coasts down to the perigee of the transfer orbit.

5 Shuttle and satellite arrive here at the same time. Another retrograde burn by the shuttle circularizes its orbit at the satellite's altitude.

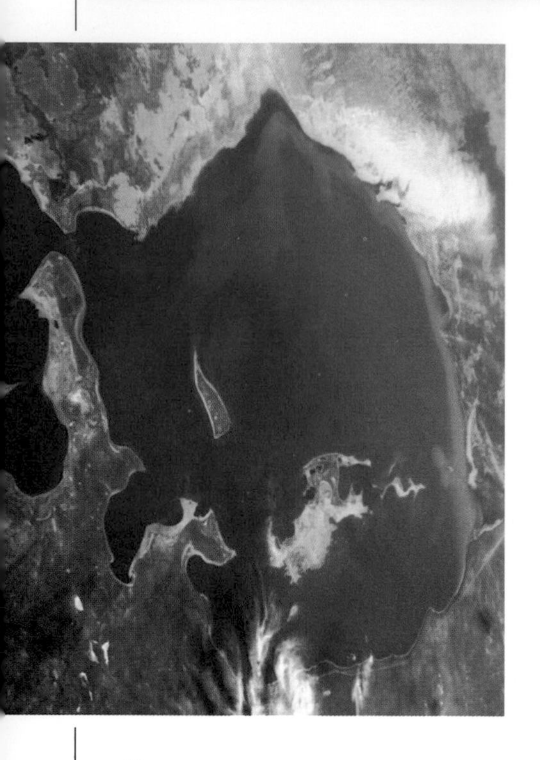

Above:

The Aral Sea, located to the east of the Caspian Sea in what used to be the former Soviet Union, is fed almost entirely by the inflow of two rivers. Since the 1960s, the lake's water level has dropped 14 meters because the water from the rivers has been diverted to irrigate fields in central Asia. The declining lake level has exposed the lake bottom (light-colored area near the top), caused 20 species of fish to become extinct, and bankrupted a once thriving fishing industry. This photograph was taken by astronauts on Space Shuttle *Challenger* in August 1985.

Historical Fact

Global Positioning System (GPS) uses a fleet of 21 satellites to allow an unlimited number of users to locate their position on the ground, in the air, or in orbit. This program began in 1973 with the "Block-1" first-generation GPS satellites. The U.S. Air Force launched 10 of them, and some still function today. In 1989, the Air Force launched the first of 28 new "Block-2" GPS satellites purchased from Rockwell International for $1.2 billion.

unprecedented level of accuracy. However, the most amazing aspect of this new navigation system is that it works anywhere and that anybody can use it. This system, called the Navstar Global Positioning System (GPS), uses a fleet of twenty-one satellites to help people navigate on the ground, at sea, in the air, and in orbit. GPS allows an unlimited number of users, both military and civilian, to instantly determine their exact latitude, longitude, and altitude with an accuracy of 30.5 meters (100 feet) or better. Essentially, GPS represents an advance in navigation technology as significant as the invention of the map and globe centuries ago.

In the near future, GPS will find its way into many aspects of everyday life. This system will be used in the nation's air traffic control network to guide passenger jets near crowded airports, to help lost campers find their way home, to aid motorists in guiding their cars through unfamiliar sections of town, to allow ships to safely navigate near coastal waters in the fog, and much more. The possibilities are endless.

How Satellite Ranging Works

People and satellites that want to take advantage of this unique navigation system must use a special device called a GPS receiver. This gadget monitors radio signals broadcast from GPS satellites. A computer inside the device relies on these signals to locate the exact position of the receiver. The first step in the locating process involves using the broadcast signals to figure out how much distance separates the receiver from a GPS satellite.

Remember the classic high school algebra problem that states, "If a car moves at 100 kilometers an hour, how far does it travel in four hours?" The answer to this question involves using the equation that states, "Distance traveled equals velocity multiplied by the travel time." Essentially, the computer inside the GPS receiver performs the same type of calculation. Radio signals travel at the speed of light which amounts to 300,000 kilometers (186,282 miles) per second. Therefore, the computer figures out the number of kilometers that separates the satellite from the receiver by first measuring the number of seconds the signal takes to reach the receiver, then it multiplies this time by 300,000.

Figuring out the time it takes a signal to propagate from the satellite requires a little more effort than simple multiplication. Consider the following analogy. Imagine that you and a friend each stand at opposite rims of a wide canyon. At precisely 12 noon, your friend starts to sequentially shout the letters, "a...b...c...d...e...f...," across the canyon. You also begin to recite the alphabet mentally to yourself at the same rate as your friend. Sound takes some time to travel across the canyon. Consequently, you might be up to the letter "d" by the time you hear the letter "a" from your friend's voice. Therefore, since both of you started to recite the alphabet at the same instant, the time that your friend's voice takes to reach you equals the time differential

between the moment you say the letter "a" and the moment you hear an "a" arrive from across the canyon. You can now figure out the distance across the canyon by multiplying this time differential by the speed of sound.

GPS timing works in a similar fashion. Both the satellite and the receiver contain circuits that generate a stream of energy pulses that form a continuous sequence of electronic code. Furthermore, the code sequences generated by the satellite and the receiver are identical and synchronized. The satellite uses a radio transmitter to broadcast this code on a continuous basis. In the voice across the canyon analogy, you were the GPS receiver, the letters of the alphabet represented the electronic code, and your friend was the satellite.

A GPS receiver listens to the code broadcast from the satellite. Since the signal typically takes some time to reach the receiver, the computer inside the receiver must look back into its memory and examine its clock to see how long ago it generated the same sequence of code. This differential equals the amount of time that the signal took to propagate from the satellite to the receiver. The computer multiplies this time differential by the speed of light in order to figure out the distance to the satellite.

Each one of the 21 satellites in the GPS fleet broadcasts a different code sequence to allow receivers to distinguish between the various satellites. This differentiation is necessary because a receiver must determine the distance between it and several different satellites to locate an unknown position. The next section explains why.

Triangulation

How does knowing the distance to a satellite help to reveal a GPS receiver's unknown location? The answer lies in the use of an old geometrical technique called triangulation. Consider the following example in order to understand how this technique works. For the time being, assume that the universe is two-dimensional (flat).

Imagine that you get lost on vacation, but just happen to know the exact locations of three different GPS satellites in space. Furthermore, you observe that 17,700 kilometers separates your position from one of the satellites. This bit of seemingly trivial information narrows your possible locations from the entire universe to the edge of a circle that surrounds the satellite. In fact, you can plot all of your possible locations by drawing a circle of radius 17,700 kilometers around the satellite's position on a map (see Figure 31 on page 102).

Now, what happens if a further observation reveals that 19,300 kilometers separates you from a second satellite? This new data puts you simultaneously on a circle of radius 19,300 kilometers surrounding the second satellite, and on the original circle of radius 17,700 kilometers around the first satellite. The figure shows that this

Above:
The declining lake level in this photograph of the Aral Sea is dramatic when compared to the view of the same body of water taken seven years earlier (see photograph on opposite page). The north end of the lake (upper left) has been totally cut off from the southern half. Also, the islands in the Aral Sea have become substantially larger due to the declining water level. Because the lake's surface area has decreased by over 40% since 1960, this body of water no longer exerts a moderating influence on air temperature in the region. Consequently, the length of the growing season has been substantially reduced. This photograph was taken by astronauts on Space Shuttle *Endeavour* in September 1992.

RULE OF THUMB

Taking advantage of GPS for extremely accurate navigation requires using a device called a GPS receiver. These devices vary in size and shape. Thanks to the miracle of miniaturization, some receivers can fit in the palm of your hand.

► Figure 31: *How GPS Triangulation Works*

GPS stands for Global Positioning System. This system uses a fleet of 21 satellites that allow anybody with a device called a GPS receiver to determine their exact location on the Earth, or in space around the Earth. Basically, the receiver is pre-programmed with the knowledge to locate all of the GPS satellites. The satellites broadcast signals that allow a GPS receiver to determine the distance between the satellite and the receiver. By knowing the distance between it and several GPS satellites, a receiver can triangulate its position. This figure shows how triangulation works.

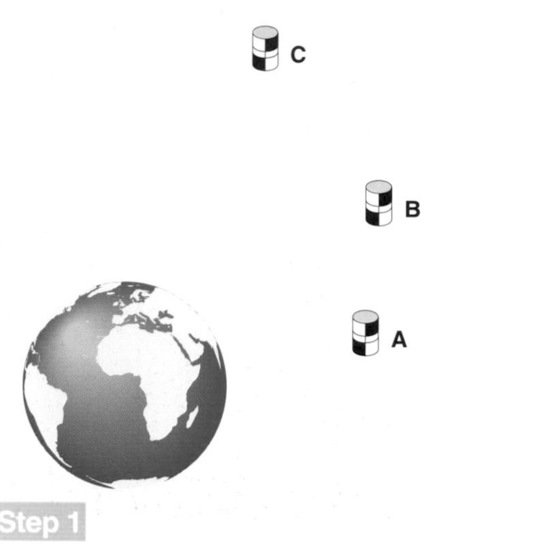

Step 1

You are somewhere on the Earth, or in Earth orbit, but do not know where. However, suppose that you know that you are 17,700 km from GPS satellite A, 19,300 km from GPS satellite B, and 29,000 km from GPS satellite C. Furthermore, you know the exact position of all three satellites. How do you use this information to determine your location? (note, the distances are not drawn to scale in this figure).

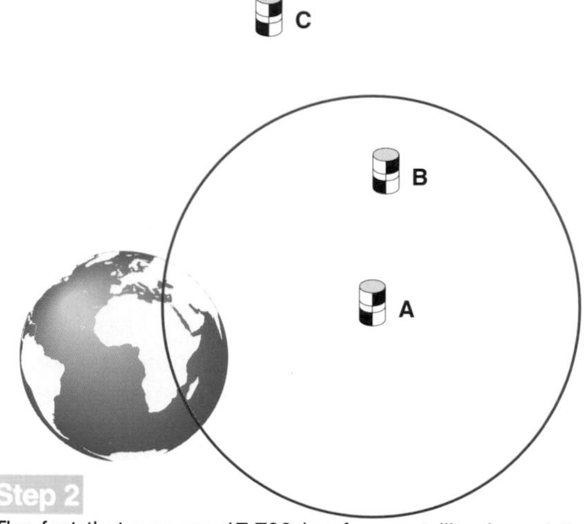

Step 2

The fact that you are 17,700 km from satellite A, and that you know A's position means that you could be anywhere on a circle of radius 17,700 km surrounding satellite A. This circle is drawn above using a red solid line.

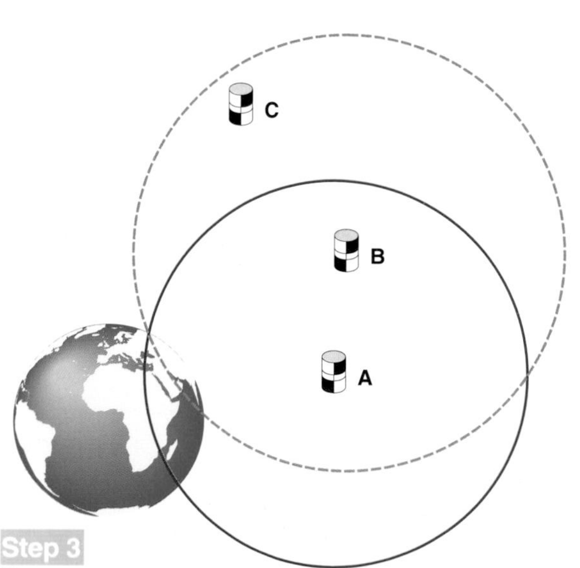

Step 3

Since you also know that 19,300 km separates you from satellite B, that means that you are also on a circle of radius 19,300 km surrounding B as well as the first circle. Therefore, you must be at one of the two points where the two circles intersect. However, which one of the two? The circle around B is drawn with a green dashed line.

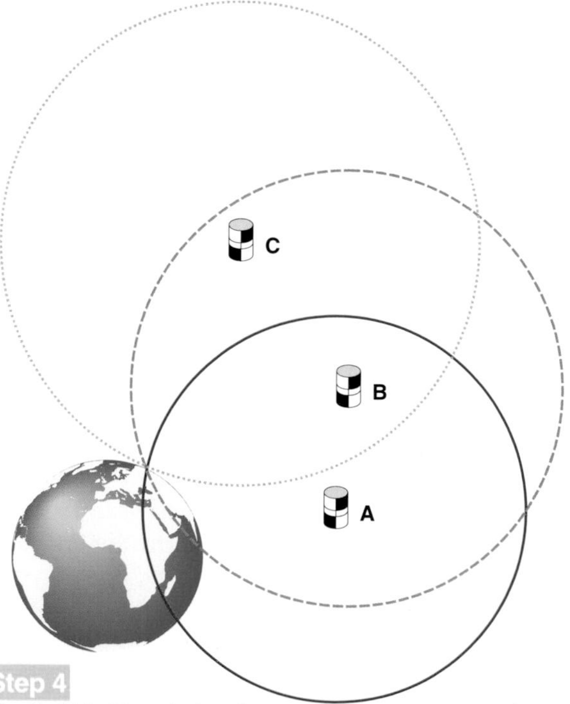

Step 4

The final bit of knowledge, that we are 20,900 km from satellite C, places us on a circle of radius 20,900 km surrounding C in addition to the other two circles. This new circle is shown by a brown dotted line. Notice that the 3 circles intersect in only one place. That place is our location - in this case, northern Russia. A GPS receiver does this type of analysis to find its location.

second bit of information limits your possible location to the two points on the map where the two circles intersect.

Which one of these two points represents the correct answer? Suppose that one more observation determines a 29,000-kilometer separation distance between you and the third satellite. Therefore, your position must exist on a circle of radius 29,000 kilometers surrounding the third satellite, and on the other two circles all at the same time. The third observation geometrically eliminates one of the two previous possibilities because three overlapping circles on a map intersect at only one point. Your location lies at this point.

In reality, the receiver's computer faces a slightly different problem because the universe is not flat. This extra dimension means that if 17,700 kilometers separates an unknown position from a satellite, then that position lies anywhere on a sphere of radius 17,700 kilometers surrounding the satellite instead of a circle. As a result, the existence of three satellite distance measurements creates a situation where the computer must mathematically solve for the intersection point of three spheres instead of three circles.

Above:
The crew of Space Shuttle *Endeavour* photographed the plume of a volcanic eruption on the Kamchatka Peninsula in October 1994. This peninsula is located on the eastern Pacific coastline of Russia, slightly north of Japan. The plume stretches for several hundred kilometers.

Clock Errors

The ability of a GPS receiver to accurately locate its position on the Earth critically depends on precise synchronization between its clock and the clocks on the satellites. Remember that the clock on the receiver allows it to measure the signal propagation time from the satellite by seeing how long ago it generated the same sequence of code as the satellite. As a result, the clock provides the means to calculate the distance from the satellite. Any small error in the synchronization will result in an inaccurate measurement. Consider the fact that radio signals travel at 300,000 kilometers a second. At this speed, a 0.0000003-second (3 parts in 10 million) difference in clock synchronization between the receiver and satellite will result a 100-meter (328-foot) error in the distance measurement.

RULE OF THUMB

GPS requires super-precise timing in order to be accurate. As such, all GPS satellites carry extremely accurate atomic clocks. However, the clock inside a GPS receiver is not as accurate as the atomic clock on the satellite. That is why the receiver needs to measure signals from four GPS satellites. Three of the measurements allow the receiver to triangulate its unknown position. The fourth measurement allows the receiver to mathematically determine the difference between its clock and the clocks on the GPS satellites.

All GPS satellites employ a super-accurate atomic clock that loses less than one second every 100,000 years. Despite the name, these types of clocks do not run on nuclear power. The "atomic" comes from the fact that the clock bases its timing on the stable and predictable vibrational characteristics of the cesium atom. However, these clocks weigh a lot and cost a fortune. Most receivers use much smaller and less accurate clocks than their atomic counterparts. In fact, some of the clocks in the cheaper receivers function not much better than an expensive quartz wristwatch.

How do GPS receivers compensate for cheap clocks? One of the most important laws of mathematics states, "Solving for a group of unknown quantities requires a number of observations equal to or greater than the number of unknown quantities." In the theoretical GPS case, the receiver's latitude, longitude, and altitude coordinates represent three unknowns, and the three satellite distance

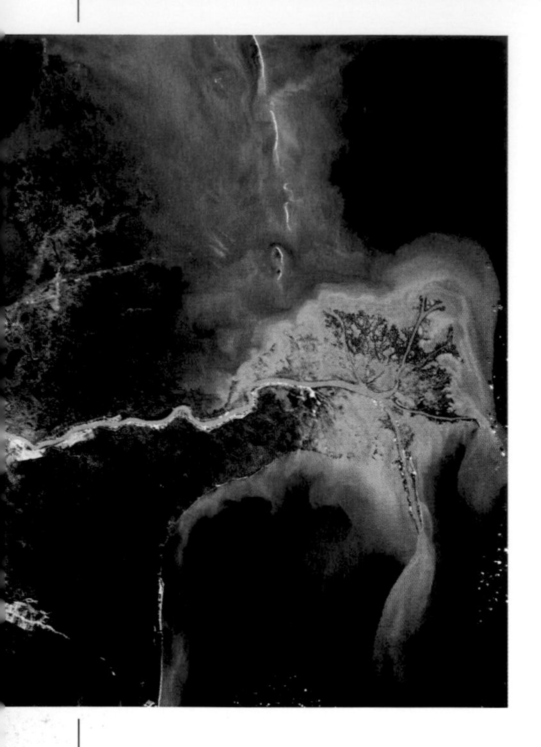

Above:

After a journey from the northern reaches of Minnesota, the muddy, freshwater from the Mississippi River empties into the Gulf of Mexico. The river delta and all of the mud and silt dumped into the Gulf are visible as different shades of gray at the bottom of the picture. The city of New Orleans, Louisiana, is visible as a bright patch at the left edge of the photograph, near the center. This picture was taken by astronauts on Space Shuttle *Discovery* in January 1985.

measurements represent three independent observations. In practice, the receiver's clock always lags or leads the satellites' clocks by a small amount because the receiver's clock is not as accurate. This time differential, called *clock skew*, represents a fourth unknown. The skew time varies from receiver to receiver just as different types of wristwatches have different accuracies.

The mathematical trick to figure out the skew involves taking four satellite distance measurements instead of three. This way, the computer can solve for all four unknowns. Finding the intersection point of four spheres allows the computer to figure out the receiver's unknown latitude, longitude, and altitude coordinates, as well as to determine the clock skew at the same time. The computer then multiplies the skew time by the speed of light to find out how much distance error the mis-synchronization of the clocks caused.

This entire scheme requires perfect synchronization between the clocks on all 21 satellites in the GPS fleet. As a result, the Department of Defense updates the time on all the satellites twice a day. The update information is based on the time kept by an atomic clock in Colorado Springs, Colorado. This atomic clock, the most accurate ever designed, keeps the time for the entire world, or at least for those who are not enemies of the United States. In short, remember that a GPS receiver must take four satellite distance measurements in order to locate the user's unknown location.

Locating a GPS Satellite

Successful triangulation requires that a GPS receiver know the exact location of the satellites used in the distance measurements. The reason is that the computer cannot begin to solve for the receiver's unknown position without knowing where to place the four spheres on its electronic map. Consequently, Department of Defense engineers constantly track the satellites to determine their precise locations. Then, they transmit the satellites' locations to the satellites. In turn, the satellites broadcast this information constantly. A GPS receiver takes this information and then solves orbital equations based on Kepler's laws to locate the satellites.

Historical Fact

Thanks to GPS, major American landmarks have moved. In reality, the landmarks are in the same place physically, but have moved on current maps after GPS determined their "true" position. For example, the Empire State Building in New York moved by about 40 meters, and the dome on top of the U.S. Capitol shifted location by roughly 30 meters. In addition, some state and national boundaries moved by several meters.

Navigational Accuracy

All of the GPS theory presented up until this point assumed that radio signals travel at a constant 300,000 kilometers per second. In reality, this speed varies depending on what type of medium the radio signal travels through. The 300,000 kilometers per second figure assumes that the signal propagates in a perfect vacuum. Other mediums, like the Earth's atmosphere, contain particles such as electrons, air molecules, and water vapor that tend to slow the propagation speed of the radio signal by a tiny amount. This slowdown increases the time the signal code takes to reach the receiver. Consequently, the computer will overestimate the distance to the satellite.

Ionospheric delays represent the single largest source of naturally occurring errors in a GPS distance calculation. The ionosphere is an extremely thin region of the atmosphere that exists above the altitude of 80 kilometers (50 miles). Here, ultraviolet radiation from the Sun begins to run into atmospheric molecules. The energy from the radiation knocks electrons away from the atoms of gas. In turn, these free electrons tend to slow the propagation speed of radio signals that pass through the ionosphere.

GPS satellites provide a way for the receiver to correct for ionospheric delays by broadcasting the electronic code simultaneously on two different frequencies (radio channels). Such a scheme works because the amount of radio signal slowdown in the ionosphere also depends on the frequency of the signal. As a result, the same code transmitted on two different frequencies arrives at the receiver at two different times. A GPS receiver's computer contains a program that can eliminate the ionospheric timing error by examining this time differential. Unfortunately, a receiver that simultaneously monitors broadcast codes on two different frequencies to correct for ionospheric errors costs much more than a receiver that scans only one frequency at a time.

The accuracy of GPS also depends on what kind of broadcast code the receiver listens to. Every GPS satellite broadcasts two types of codes. Non-military users almost always use the C/A or "clear acquisition" code. This version, sometimes called the civilian code, allows receivers to determine their location with an accuracy of between 31 and 70 meters (100 to 200 feet). On the other hand, military receivers listen to the P or "protected" code. One of the differences between the two is that GPS satellites broadcast the P code on two different frequencies, which is not the case with the C/A. As a result, receivers that use the P code can correct for ionospheric delay errors when determining the distance to the satellite. Corrections of this kind allow P code receivers to determine their position with an accuracy of about 9 meters (30 feet).

The Defense Department may choose to activate the GPS *anti-spoofing* mode in a time of war. This action encrypts the P into a Y code. Only authorized receivers who posses a circuit key to translate the Y code back into the P code will be able to use the P code. Furthermore, they may also choose to activate *selected availability*. This mode degrades the C/A accuracy to 107 meters (350 feet) by intentionally broadcasting inaccurate data regarding the satellite's location, and by intentionally mis-synchronizing the clocks on the satellites. Anti-spoofing and selected availability may inconvenience American civilian users, but it prevents the enemies of the United States from using GPS to their advantage. ❑

Above:

Trimble Navigation's GeoExplorer II GPS receiver is one of the many models of receivers available for purchase by the general public. It runs on 4 AA batteries, weighs 14 ounces, determines position within 2 to 5 meters, updates once every 0.7 seconds, and can store up to 9,000 three-dimensional GPS positions.

(Photograph provided by Trimble Navigation in Sunnyvale, California)

Moonstruck

▶ Chapter 4: The Story of the Race to the Moon

Between 1968 and 1972, twenty-four different Americans on nine separate occasions embarked on the greatest and most hazardous adventures ever envisioned. All of them set sail from the costal swamps of Florida on top of the most powerful machine ever built by humans. Those responsible for its design called it Saturn. Fully assembled, Saturn stood taller than a football field is long, could withstand heat and stress greater than ever imagined possible, and its pieces fit together with a precision better than the finest watch. Its power more than dwarfed that of 1,000 locomotives, and it made the proverbial speeding bullet look slow in comparison. Saturn, a gargantuan monument to human ingenuity, served no purpose other than to fulfill the centuries-old, seemingly unattainable dream of landing people on the Moon. This chapter tells the story of how the fantasy of walking on the Moon, thought of as impossible from the dawn of civilization, became the single greatest human accomplishment in history.

Opposite Page:

20 July 1969 – On the first Moon landing mission, astronaut Edwin "Buzz" Aldrin walks on the surface of the Moon at the Sea of Tranquility.

4.1 The Cold War Connection

Ironically, America's journey to the Moon began with the roar of a rocket launch deep in the heart of the former Soviet Union at a place called Kazakhstan. From there, America's cold war archenemy stunned the world by launching the first artificial satellite into space on 4 October 1957. *Sputnik,* as the Soviets called it, looked like a small aluminum ball less than 56 centimeters (22 inches) in diameter. The entire payload consisted of a detector to measure electron concentration in outer space, and a primitive radio transmitter that broadcasted meaningless electronic "chirps" to anybody listening with a radio receiver.

Nothing aboard *Sputnik* physically threatened the United States. However, the height of the cold war between America and the Soviet Union occurred during the 1950s and 1960s. Many Americans in the 1950s considered the Soviet Union as a technologically primitive society of farmers dominated by a totalitarian government eager to conquer the world. As a result, *Sputnik* essentially represented a political slap in the face for America in terms of international prestige. The popular consensus was that if the Soviets could orbit a spacecraft, then they could easily use the same rockets to drop their nuclear weapons on the United States without warning. Some even went so far as to describe the *Sputnik* launch as "the Pearl Harbor of the Cold War."

Fortunately, President Eisenhower had authorized the Department of the Navy to proceed with its Vanguard scientific

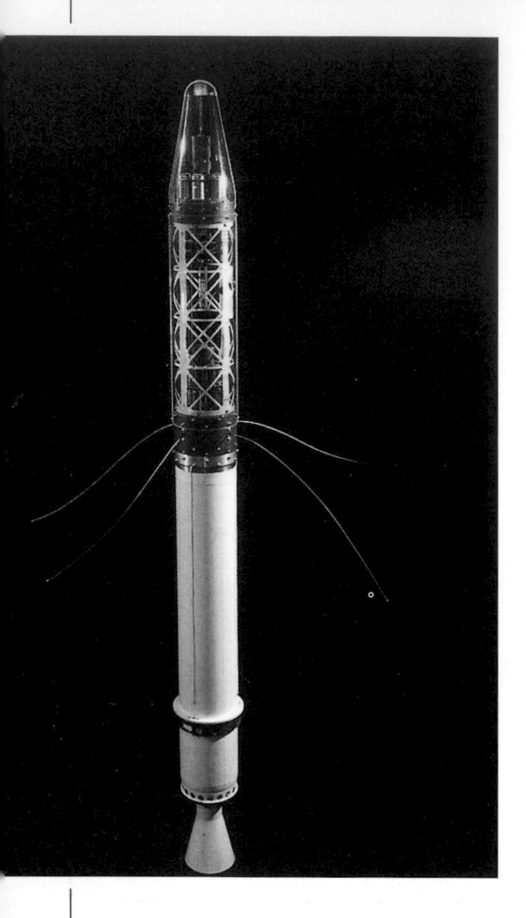

Above:
America's first satellite, called *Explorer 1*, was built by NASA's Jet Propulsion Laboratory in only 90 days back in 1957. The little spacecraft measured only 2 meters long and weighed less than 14 kilograms.

Historical Fact

The first living creature to travel in space was not a human. It was a Russian dog named Laika.

satellite project in 1955. He approved the Vanguard project not to beat the Soviets into space, but as part of the International Geophysical Year of international cooperation to study the Earth. Little did Eisenhower know that the Vanguard project would ultimately cause the United States much embarrassment in the years to come. However, progress went well with the project after its inception, and everybody naturally assumed that the Navy would launch the world's first artificial satellite into space. Despite the emotional setback associated with the *Sputnik* launch, the Navy pushed on in its effort to launch the *Vanguard* satellite.

The United States, still recovering from the shock of the *Sputnik* launch, suffered another blow one month later when the Soviets lifted *Sputnik 2* into orbit. This latest addition to the Soviet space fleet weighed 508 kilograms (1,120 pounds, six times heavier than *Sputnik*) and carried a dog named Laika. She eventually died after a few days in orbit when her oxygen supply ran out, but *Sputnik 2* continued to circle the Earth for five more months. American intelligence analysts speculated that the "dog in space" stunt represented a logical precursor to sending a human into orbit. Of course, this analysis only served to make Americans even more nervous.

Vanguard finally stood ready for launch about one month after Laika reached orbit. America placed all of its hopes for redemption on the tiny *Vanguard* satellite, which weighed only 1.5 kilograms (3.25 pounds). The White House invited news reporters from all over the world to the Cape Canaveral launch site in Florida to witness this historic event on 6 December 1957. American hopes soared high. Unfortunately, *Vanguard* did not. The rocket lost power about one second after launch, fell back onto the launchpad, and exploded into a spectacular ball of fire. Searchers looking through the wreckage found the tiny satellite lying several hundred feet away from the pad. The electronic transmitter continued to chirp as if nothing out of the ordinary had occurred.

Headlines such as "Washington Humiliated" in newspapers the next day expressed the general sentiment of embarrassment all throughout the country. Several days later, a Soviet delegate to the United Nations gloated to the American UN ambassador by asking if the United States would like to accept any UN aid destined for underdeveloped countries. America's doubt of itself also showed in the public criticism of everything from the education system to the military. Some major newspapers even printed editorials about the advantages of tight totalitarian control by the government.

A German rocket scientist named Wernher von Braun saved America's reputation. He and his team of scientists from the Army's rocket laboratory in Huntsville, Alabama, designed a rocket called Jupiter-C that could have easily launched a satellite into orbit more than a year before the first *Sputnik*. However, for various political reasons, they did not receive permission from Eisenhower to proceed. In fact, von Braun's team did not receive permission to proceed until

after the launch of *Sputnik 2*. Eisenhower lived to regret that decision, as history now records that the first two satellites were orbited by the Soviet Union. Nevertheless, the Army and von Braun rapidly readied their Jupiter rocket after the presidential go-ahead, and a team at the Jet Propulsion Laboratory in Pasadena, California, worked non-stop at a feverish pace for two months to build the satellite. On 31 January 1958, they restored America's hope by boosting the tiny 14-kilogram (31-pound) *Explorer 1* satellite into orbit from Cape Canaveral.

Vanguard suffered yet another extremely embarrassing launchpad explosion in February 1958 before finally reaching orbit in March. Eisenhower then announced to the world that the United States had launched as many satellites into orbit as their cold war rivals. He also claimed that the American satellites provided scientific data substantially more valuable than their *Sputnik* counterparts. Indeed, *Explorer* contained instruments to detect cosmic rays. In addition, its Geiger counter discovered dangerous zones of radiation surrounding the Earth called the Van Allen Radiation Belts.

Soviet premier Nikita Khrushchev remained unimpressed. He publicly ridiculed *Vanguard's* and *Explorer's* tiny size by comparing the two satellites to grapefruits and oranges. The Soviets also responded by launching the 1,327-kilogram (2,926-pound) *Sputnik 3* satellite during May of 1958. This new and impressively heavy spacecraft carried many scientific instruments. The launch allowed Khrushchev to dismiss claims of American scientific superiority in space. President Eisenhower, forced into a space race that he did not want, could not begin to understand why the Soviets wanted to spend an exorbitant amount of money on what he considered a glorified international propaganda campaign.

Khrushchev also enjoyed boasting to the world about the power of his rockets and the size of his large satellites. This "size gap" initially benefited the Soviets, but ultimately doomed them because they were never forced to learn the art of miniaturizing satellite components like their American counterparts. Miniaturization of electronics eventually played a key role in aiding space exploration. *Vanguard* and *Explorer* also had the "last laugh" on Khrushchev. All three *Sputniks* fell victim to atmospheric drag and dropped out of orbit several months after launch. *Explorer* circled the world 58,000 times before falling out of orbit in 1970. *Vanguard* still continues to orbit the Earth today, and will continue to do so for at least the next two hundred years.

4.2 First Steps Into Space

John F. Kennedy succeeded Eisenhower as president of the United States in 1961. Like his predecessor, Kennedy doubted the value of massive spending for space ventures. A single event changed his mind. On Wednesday, 12 April 1961, the Soviet Union stunned the world again, and created a political crisis in America by

Key Players During the Early Years of Space Exploration

von Braun, Wernher
German-born rocket scientist who designed the booster that launched America's first satellite into orbit in 1958

Gagarin, Yuri
Soviet Air Force major credited as the first human to fly in space

Glenn, John
Marine Lt. Colonel, NASA astronaut, and the first American to orbit the Earth

Kennedy, John F.
American president who committed the United States to landing astronauts on the Moon

Korolev, Sergei
Chief rocket designer for the Soviet Union responsible for *Sputnik* and the USSR's brilliant success early in the "space age"

Khrushchev, Nikita
Soviet leader whose foreign policy set the stage for the "space race" between the USSR and USA

Pickering, William
Lead the Jet Propulsion Laboratory team that designed America's first satellite

Shepard, Alan
Navy commander, NASA astronaut, and the first American to fly in space

Van Allen, James
American scientist who designed the instrument on America's first satellite that made the first scientific discovery from space

First 10 Humans in Space

#1 - Gagarin, Yuri
Launched on 12 April 1961 by USSR
1 orbit

#2 - Shepard, Alan
Launched on 5 May 1961 by USA
Sub-orbital, Mercury 3 Mission *(Freedom 7)*

#3 - Grissom, Gus
Launched on 21 July 1961 by USA
Sub-orbital, Mercury 4 Mission *(Liberty Bell 7)*

#4 - Titov, Gherman
Launched on 6 August 1961 by USSR
17 orbits

#5 - Glenn, John
Launched on 20 February 1962 by USA
3 orbits, Mercury 6 Mission *(Friendship 7)*

#6 - Carpenter, Scott
Launched on 24 May 1962 by USA
3 orbits, Mercury 7 Mission *(Aurora 7)*

#7 - Nikolayev, Andrian
Launched on 11 August 1962 by USSR
64 orbits, Vostok 3 Mission *(Falcon)*

#8 - Popovich, Pavel
Launched on 12 August 1962 by USSR
48 orbits, Vostok 4 Mission *(Golden Eagle)*

#9 - Schirra, Walter
Launched on 3 October 1962 by USA
6 orbits, Mercury 8 Mission *(Sigma 7)*

#10 - Cooper, Gordon
Launched on 15 May 1963 by USA
22 orbits, Mercury 9 Mission *(Faith 7)*

Mercury astronaut Deke Slayton was scheduled to fly in space in 1962, but NASA doctors grounded him because of a perceived heart problem.

launching the first human into orbit. Major Yuri Gagarin of the Soviet Air Force left the Earth from Baikonur Spaceport in Kazakhstan. He circled the Earth one time in a spacecraft shaped like a stubby ice cream cone with a scoop of ice cream on top. Gagarin sat in the "cream" part and the equipment rode in the "cone" section. He said upon returning, "I looked and looked and looked, but I didn't see God."

The Soviet's latest accomplishment turned into as much of an embarrassment for the Kennedy administration as Eisenhower's problems with *Sputnik* and *Vanguard*. Most Americans were tired of seeing the United States "second," and the media began to ask Kennedy what he was going to do about the situation. He replied, "However tired anybody may be, and no one is more tired than I am, it is a fact that it's going to take some time and I think we have to recognize it."

The new president also tried to avoid the space race issue by attempting to find other areas in science where the United States could be first. However, Kennedy learned from the world's reaction to Gagarin's orbital trip that it did not matter if there existed other areas in which America could be first. As Vice President Lyndon B. Johnson summarized the situation, "To be first in space is to be first, period. Second in space is second in everything." He recommended to Kennedy that America pursue a program to land astronauts on the Moon.

Gagarin's flight did not catch America completely off guard. Fortunately for Kennedy, America had already been working on sending humans into space since 1958. In July of that year, Eisenhower signed a bill into law that created the National Aeronautics and Space Administration (NASA) despite his continuing doubts of the value of space exploration. Eisenhower did not fully endorse participating in a "space race" with the Russians. But on the other hand, he felt that he could not ignore it from a political standpoint.

Project Mercury

One of NASA's charter goals involved promoting civilian interests and American leadership in space. As a result, NASA immediately started Project Mercury with the goal of launching American astronauts into space. Engineers working on this project designed a small cone-shaped spacecraft that looked like a bell 3 meters (10 feet) high and 2 meters (6.6 feet) in diameter at the base (see Figure 32 on page 111). They called it a *capsule* because the interior of the spacecraft barely contained enough room for one astronaut, 120 controls, 55 switches, and 35 control levers. These capsules entered space by riding on top of either an Atlas or Redstone rocket with the tip of the spacecraft's "cone" pointing upwards at launch. A small escape rocket attached to the top of the capsule could provide enough thrust to separate the capsule from the booster in the event of a malfunction during the ascent.

Figure 32: *First Steps into Space* ◄

Dimensions of Mercury: 2.90 meters (9.51 feet) long; 1.89 meters (6.20 feet) diameter

Weight of Mercury: 1,355 kilograms (2,987 lbs) in orbit on the *Friendship 7* mission

Crew: One astronaut for up to one day

Dimensions of Atlas: 29.0 meters (95.1 feet) high; 3.05 meters (10.0 feet) diameter

Weight of Atlas: 117,915 kilograms (259,958 lbs) on average

Redstone: Used for Mercury suborbital flights.

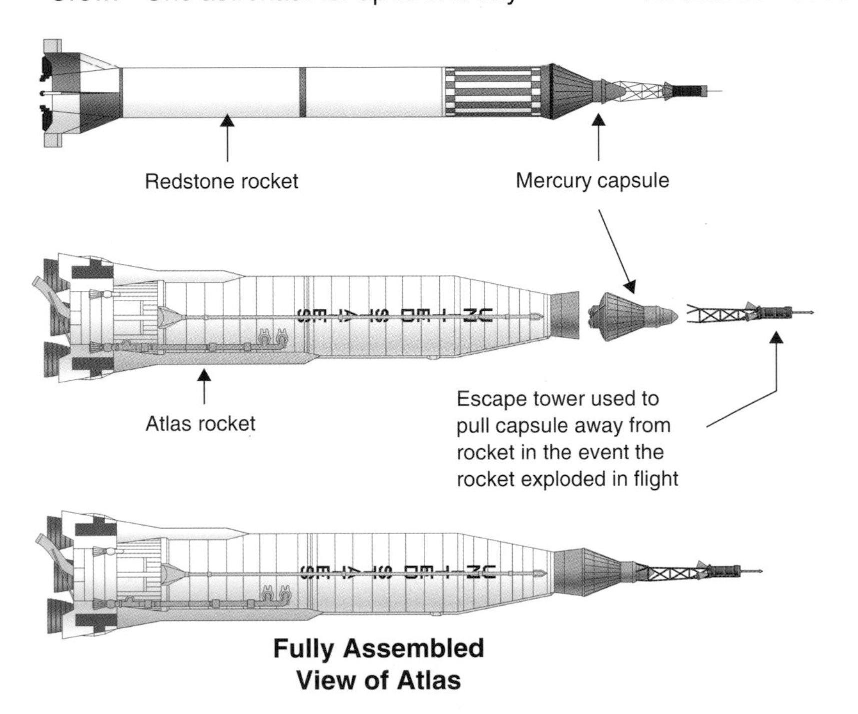

Redstone Launcher

This rocket was less powerful than the Atlas, and could only put the Mercury on a suborbital trajectory. It was used on the first two Mercury missions.

Atlas Launcher

NASA used the Atlas for the last four of the six Mercury missions. All of these four missions reached orbit. Atlas was a modified ICBM originally designed by the Air Force to drop nuclear weapons on the Soviet Union in the event of war. It was modified for astronaut use by NASA. Unfortunately, Atlas had a notorious reputation for exploding in flight.

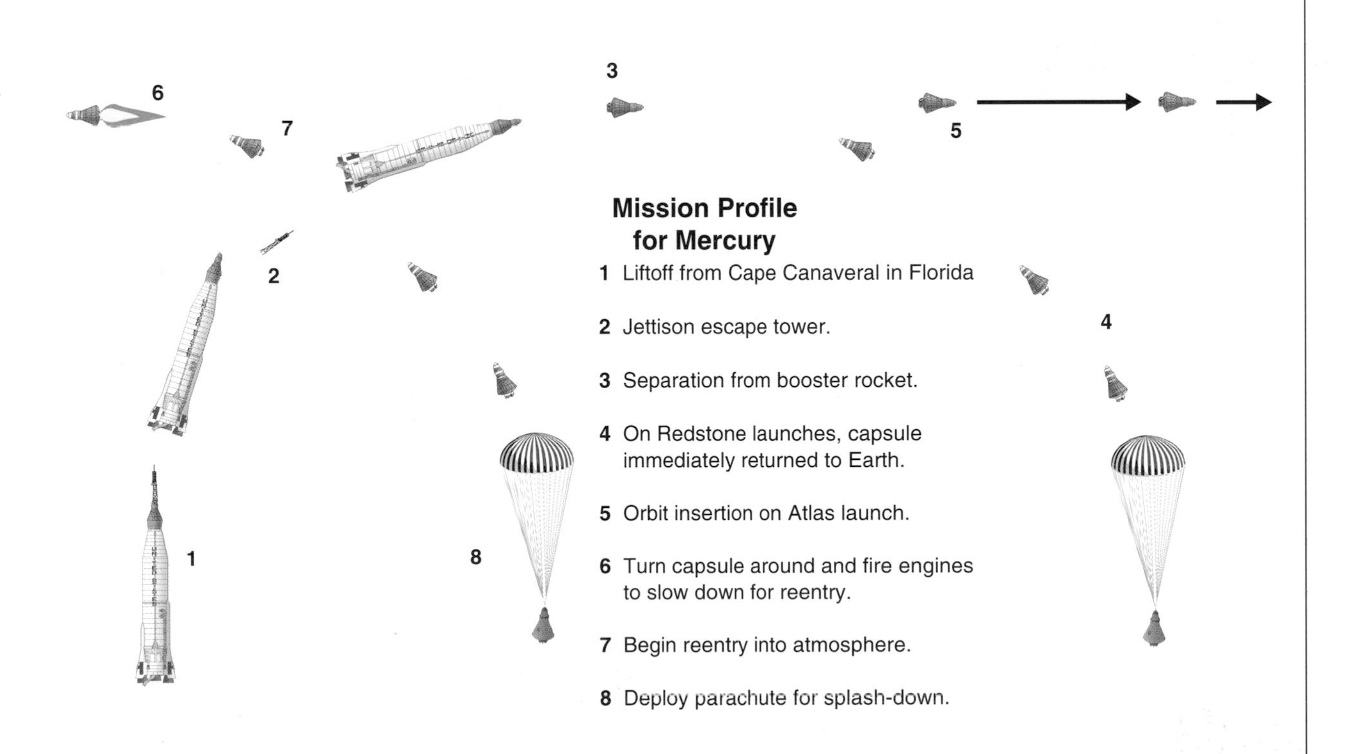

Above:
NASA selected the first seven American astronauts in April 1959 for the Mercury program. All of them were military test pilots. Front row, left to right, are Walter M. Schirra Jr., Donald K. Slayton, John H. Glenn, and M. Scott Carpenter. Back row, left to right, are Alan B. Shepard Jr., Virgil I. "Gus" Grissom, and Gordon Cooper Jr.

In space, Mercury astronauts essentially sat in the capsule and enjoyed the ride because they had no way to fly the spacecraft into a different orbit. The astronauts' lack of flying time did not bother NASA engineers because they were primarily interested in studying the effect of the space environment on humans and machines. However, a control stick similar to that found in airplane cockpits allowed the astronaut to rotate the capsule and to point it in any direction by activating tiny rocket thrusters. This type of maneuver played a crucial role in returning to Earth. At the end of a mission, the astronaut started the de-orbit process by rotating the spacecraft so that the base (blunt) end of the cone pointed forward in the direction of motion along the orbit. Several small rockets attached

to the base, called *retro-rockets*, then fired to slow the capsule and drop it into an orbit that lowered it into the atmosphere.

The process of falling through the upper atmosphere at near orbital speeds during the de-orbit process generated more than enough frictional heat to incinerate the capsule. Consequently, every Mercury capsule contained a heat shield attached to the base. This thick shield consisted of a silica-fiber type of resin highly resistant to heat. Mercury engineers designed the shield to slowly melt and char during re-entry into the atmosphere. The ashes from the shield floated off and carried excess heat away from the spacecraft. As the Mercury capsule fell into the lower atmosphere, a parachute packed into the top of the capsule deployed to slow the astronaut and capsule to a safe splashdown in the ocean.

NASA selected the first group of astronauts from a pool of 110 military test pilots chosen for their extensive flight test experience. They were also required to meet a rigorous set of physical standards. One of these requirements dictated an astronaut height of no greater than 178 centimeters (5 feet 10 inches) because of the small Mercury capsule size. Of the 110 test pilots, NASA ultimately chose Alan Shepard, Virgil "Gus" Grissom, John Glenn, Scott Carpenter, Wally Schirra, Gordon Cooper, and Deke Slayton to fly on Project Mercury. All seven trained hard, and the scientists and engineers worked long hours in an attempt to beat the Soviets into space. However, their target goal of a launch in March 1961 slipped to May 1961 because of technical problems in the spacecraft design. That schedule slip made all the difference in allowing Gagarin to become the first human to fly in space.

Above:
5 May 1961 – A Redstone rocket lifts off from its Cape Canaveral launchpad with a Mercury capsule containing Alan Shepard. On this flight, Shepard became the first American to enter space. The capsule is the black bell-shaped object at the top of the rocket, above the vertical black stripes painted on the side of the Redstone. The small tower-like structure above the capsule functions as an emergency escape rocket. It was designed to pull the capsule safely away from the rocket in the event of a malfunction during ascent.

A Flea's Jump

Nevertheless, on the morning of 5 May 1961, Alan Shepard climbed into his Mercury spacecraft atop the single-stage Redstone rocket for a ride into the pages of American history books. The Redstone's small size permitted NASA to place the launch control center less than 46 meters (50 yards) from the launchpad at Cape Canaveral. In comparison, today's Space Shuttle produces enough heat and exhaust to the point where observers must watch from at least 4.8 kilometers (3 miles) from the pad. However, the tiny Redstone provided just enough lifting power to put the United States back on track.

Millions of Americans watched on television and hoped for success as the countdown went through several delays that morning. NASA rewarded their patience by launching Shepard at 9:34 A.M. His capsule, nicknamed *Freedom 7*, received a ride from the Redstone to a peak altitude of 187 kilometers (116 miles) before falling back to Earth. During the short 15-minute suborbital flight, Shepard used the hand controllers to demonstrate the ability to orient the Mercury in different directions. He reported no discomfort during the five minutes of weightlessness that he experienced, and also looked out the

Above:

5 May 1961 - A U.S. Marine helicopter plucks astronaut Alan Shepard out of the Atlantic Ocean after the splashdown of his successful 15 minute suborbital flight. Shepard's *Freedom 7* capsule, seen floating in the ocean beneath the helicopter, was also rescued. From 1961 until the Space Shuttle days in 1981, every American who flew in space returned home by landing in the ocean.

Historical Fact

During Glenn's Mercury mission, he saw "firefly" like objects in space around his capsule. Later on, an astronaut named Scott Carpenter discovered that the "bugs" were really ice particles that had built up on the Mercury capsule and fallen off in orbit.

spacecraft to make observations of the Earth from space. Shepard's profound comments from orbit amounted to, "What a beautiful view!"

NASA never planned for America's first spaceflight to reach orbit. The mission amounted to a cannonball type of ride because the capsule received a shot upward and eastward, courtesy of the Redstone, before falling back to the Earth several hundred kilometers east of the launchpad. Shepard covered less than one-eighteenth the distance of Gagarin's flight, but still returned home a hero to the American public, which considered the flight as an unqualified success. Khrushchev casually dismissed the first Mercury mission as nothing more than a "flea's jump" as compared to the Russian flight. Unfortunately for NASA, Khrushchev made a realistic assessment of Shepard's mission. However, President Kennedy felt a little more optimistic about America's first spaceflight. Several weeks after the mission, he delivered a speech to Congress in which he announced to the world that the United States would land humans on the Moon. That speech set the tone for the American space program for the rest of the decade.

Gus Grissom followed Alan Shepard two months later to became the third human to reach space. On 21 July 1961, Grissom flew a mission that essentially duplicated Shepard's 15-minute suborbital flight. The only significant glitch occurred at the end of the flight as the capsule, with Grissom inside it, lay floating on the ocean awaiting the recovery helicopter. During Grissom's wait, something accidently triggered the escape hatch and blew the door right off the capsule. Navy divers rescued Grissom, but the capsule sank in deep water. Ironically, an engineer had painted a crack on the side of the capsule because Grissom named his spacecraft *Liberty Bell 7*. This crack became the source of a joke around Cape Canaveral that said Grissom's capsule was the last NASA would ever launch with a crack in it.

The Soviets upstaged NASA again only sixteen days after Grissom's short flight. This time, Soviet Air Force major Gherman Titov became the second human to reach orbit. He spent more than one day in space and circled the Earth 17 times. America still had not yet sent anybody into orbit. Khrushchev took this opportunity to further snub NASA by saying, "Titov's feat has shown once again what Soviet man, educated by the Communist Party, is able to do." The American space program seemed to be falling further behind, but NASA stayed the course and made steady progress towards its next launch half a year away.

Redemption

The morning of 20 February 1962 promised redemption for America. On that day, John Glenn climbed into his Mercury capsule, named *Friendship 7*, for the first American attempt at an orbital flight. Redstones lacked the thrust necessary to send a Mercury capsule into

orbit. Consequently, NASA switched to the more powerful Atlas rocket. Unfortunately, the early 1960s version of the Atlas gained a notorious reputation for exploding in flight. NASA's fears concerning an Atlas explosion never materialized, and Glenn became the first American to orbit the Earth.

The concern of Glenn reaching orbit safely paled in comparison with what happened in orbit. The most nervous part of the mission occurred when the computers on the ground received telemetry from the capsule in orbit about the possibility of a loose heat shield. If correct, Glenn was doomed to a fiery death during reentry into the atmosphere at the end of the mission. No methods existed at that time for a rescue in space, or for Glenn to venture outside to fix the problem. Ground control had no option other than to ask Glenn to perform the de-orbit maneuver and hope for the best. Fortunately,

Above:

Marine Lt. Colonel John Glenn squeezes into his Mercury capsule named *Friendship 7*. All Mercury capsules had the number "seven" appended to their names because the astronauts who flew them were the first seven ever chosen by NASA. Glenn is climbing in sideways because astronauts always entered space sitting with their backs against the bottom of the capsule, and face looking toward the top. This photograph was taken on 20 February 1962, shortly before Glenn's historic flight.

Above:

Astronaut Wally Schirra and his *Sigma 7* Mercury capsule blast off from Cape Canaveral on 3 October 1962 aboard an Atlas rocket. Schirra was the third American to orbit the Earth and completed six orbits during a near-perfect nine-hour mission.

Historical Fact

Mercury astronaut Gordon Cooper recorded a space "first" by releasing the first satellite from a spacecraft. It was a 152-millimeter sphere that contained a beacon. The goal was to test an astronaut's ability to visually track objects in space.

the heat shield held in place. Later on, Glenn described the reentry as "a real fireball." He returned home a hero after his three-orbit flight in *Friendship 7*. A jubilant nation, previously at a low point in national morale, treated the new hero to one of the largest ticker-tape parades in history.

Three more Mercury flights followed Glenn's during the next fourteen months. Scott Carpenter flew three orbits in May 1962, Wally Schirra followed with six orbits in October, and Gordon Cooper closed out the Mercury program with twenty-two orbits on a thirty-four-hour mission the following May. One of Project Mercury's objectives involved evaluating what happened to humans and astronaut performance during exposure to a weightless condition. Today, astronauts venture safely into space and experience weightlessness on a regular basis. However, NASA did not know the effects back in the early 1960s. Carpenter, Schirra, and Gordon tried various tasks, such as space photography, using stars for navigation, eating under a weightless condition, releasing small objects from the capsule to practice tracking objects in space, and spotting landmarks on Earth from orbit. Cooper's flight proved so successful that NASA canceled a risky two-day Mercury flight by Alan Shepard in order begin work on a new project called Gemini.

The Soviet Union, in the meantime, still scored impressive firsts that seemed to make the Mercury program look primitive. They launched two humans in two different capsules into space at the same time in August of 1962. Then, one month after Cooper's mission, they sent the world's first woman into space. Junior Lieutenant Valentina Tereshkova of the Soviet Air Force completed a spaceflight that lasted seventy-one hours. She spent more time in space and completed more orbits than all of the Mercury astronauts combined.

4.3 The Intermediate Step

NASA spent two years after Cooper's last Mercury flight designing the next generation spacecraft called Gemini. The primary purpose of this new project involved perfecting critical but still untried techniques needed for a piloted Moon flight. These skills included changing orbits, rendezvousing two spacecraft in orbit, docking two spacecraft together, and performing activities outside the spacecraft while in orbit. Astronauts needed to know how to perform these maneuvers because the Moon missions involved the use of two separate spacecraft.

Project Gemini

The ability to accurately maneuver a spacecraft in orbit represented the key goal of the Gemini project. Previously, astronauts on Mercury had only been able to change the orientation of the capsule in space. Engineers designed Gemini with rocket thrusters to allow it

to move forward, backward, and sideways in its orbital path in order to change orbits. The complexity of orbital change maneuvers, along with rendezvous and docking, required two people to fly the Gemini spacecraft. This fact combined with a planned schedule of ten Gemini flights in less than two years prompted NASA to chose another group of astronauts to complement the original seven. Many in the new group eventually flew to the Moon in the late 1960s and early 1970s. One of them, John Young, remained an astronaut into the early Space Shuttle days of the 1980s.

Gemini spacecraft looked like an enlarged version of the Mercury capsule (see Figure 33 on page 118). In fact, NASA originally named the new spacecraft design Mercury Mark II. The inside of a Gemini provided little room for comfort. Although the capsule weighed twice as much as a Mercury, it only contained 50% more cabin space for twice as many astronauts. John Young described the interior as comparable to, "sitting in a phone booth that was lying on its side." Some astronauts referred to the Gemini as the "Gus-mobile." This distinction arose because Gus Grissom was one of the few who could climb in and close the overhead hatch without banging the door on his flight helmet.

Mission planners scheduled some Gemini missions to last between one and two weeks to determine if humans could endure weightlessness for long periods of time. These long missions required a lot of electricity to run all of the flight instruments. While Mercury capsules employed batteries, Gemini employed a new gadget called the *fuel cell*. This device mixed liquid oxygen and hydrogen. The resulting chemical reaction produced electricity for the equipment, and drinking water for the astronauts as a byproduct.

Two astronauts combined with their life support systems, consumable supplies, and the propellant for space maneuvers added up to make Gemini weigh more than 3,700 kilograms (8,157 pounds). This weight amounted to twice as much as a Mercury capsule and too much for an Atlas rocket to lift. As a result, NASA turned to the Air Force's powerful Titan 2 intercontinental ballistic missile to launch Gemini capsules. The Air Force originally designed the Titan 2 to carry a nuclear weapon with 1,000 times more explosive than the Hiroshima atomic bomb to the Soviet Union in the event of a nuclear war. Fortunately, Titan boosters saw all of their action as space launch vehicles instead of as instruments of destruction.

Upstaged Again

Khrushchev also ordered his Soviet rocket scientists to work on their version of Project Gemini. Not surprisingly, they upstaged NASA again. The Soviet Union launched a Voshkod capsule into orbit with three humans onboard half a year before the first Gemini. Then, five days before Gemini's first flight, the Soviets sent another Voshkod spacecraft into orbit. The mission carried two crew members. One of them, Lieutenant Colonel Alexi Leonov,

Above:

This photograph shows the size and shape of the Gemini capsule as compared to adult humans. Notice that the Gemini looks like an enlarged version of the Mercury spacecraft. In the picture, Gemini astronauts David Scott and Neil Armstrong remain seated in their *Gemini 8* capsule with the hatches open and wait to be picked up by the Navy destroyer U.S.S. *Mason*. The men standing on the flotation raft around the capsule are Navy rescue divers. This photograph was taken on 16 March 1966, shortly after splashdown in the Pacific Ocean about 500 miles east of Okinawa.

▶ Figure 33: *On the Shoulders of Titans*

Dimensions of Gemini:	5.61 meters (18.4 feet) long 3.05 meters (10.0 feet) diameter	Dimensions of Titan 2:	33.2 meters (109.0 feet) high 3.05 meters (10.0 feet) diameter
Weight of Gemini:	3,760 kilograms (8,289 lbs) on the *Gemini 12* mission	Weight of Titan 2:	185,000 kilograms (407,855 lbs) on average
Crew:	Two astronauts for up to 14 days	Stages:	Titan 2 was a two-stage rocket

Titan 2 Launcher

The Titan 2 was designed as an ICBM capable of dropping nine-megaton hydrogen bombs on the Soviet Union in the event of nuclear war. NASA modified it for use in launching Gemini astronauts.

Gemini Spacecraft

Gemini was a two-section spacecraft. The astronauts sat in the front module, called the re-entry module. This section was the only part of the Gemini that returned to Earth. The rear section, named the adapter module, contained the propellant, water, oxygen tanks for the astronauts, and other supplies. It was jettisoned shortly before reentry.

Fully Assembled View of Gemini

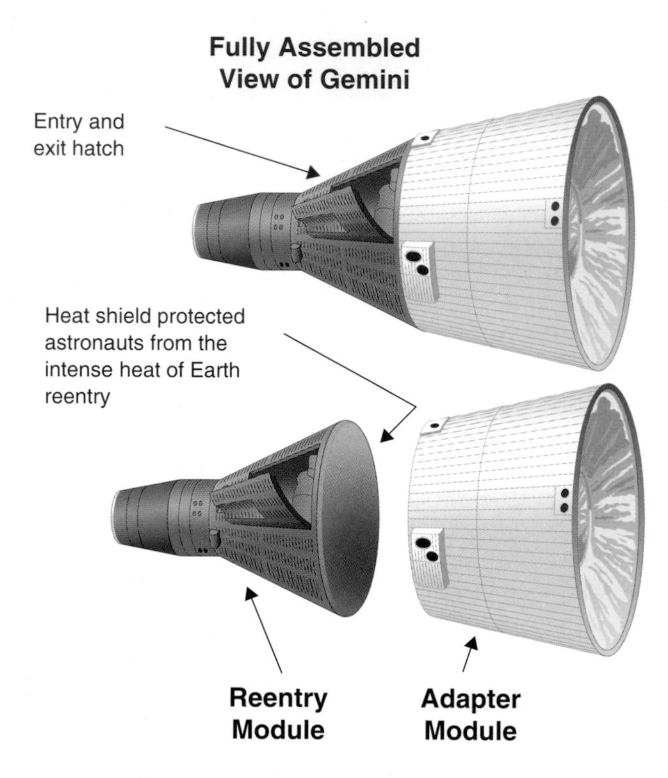

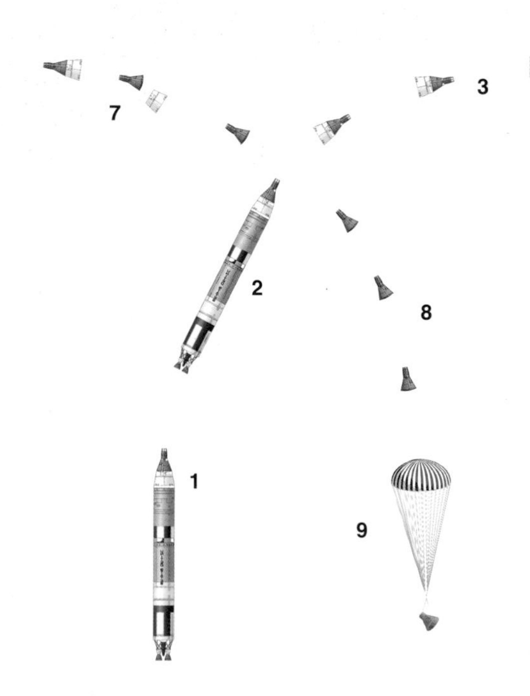

Typical Gemini Mission Profile

1 Liftoff from Cape Canaveral in Florida.

2 Ascend to orbit on Titan 2 rocket.

3 Separate from Titan 2 and insertion into orbit.

4 Rendezvous with Agena target vehicle.

5 Dock with Agena and practice spacewalks.

6 Turn vehicle blunt end first in preparation for reentry.

7 Jettison adapter module.

8 Reenter through atmosphere.

9 Deploy parachute for safe splash-down in ocean.

donned a special space suit and floated around outside the spacecraft for about 10 minutes while it orbited the Earth. This *space walk* added another impressive first to the list of Soviet accomplishments.

NASA kept pushing on despite the setbacks in political stature created by the Soviet accomplishments. On 23 March 1965, Gus Grissom and new astronaut John Young rode the first piloted Gemini capsule, called *Gemini 3,* into space from a Cape Canaveral launchpad half a mile north of the Mercury-Atlas pad. The mission's primary goal involved testing the maneuvering ability of the new spacecraft design. They fired thrusters to change the shape of their orbit, performed a slight inclination change, and dropped to a lower altitude. In the process, Grissom and Young became the first astronauts in history to perform an orbit change other than the de-orbit maneuver, and Grissom became the first human to fly in space twice.

Above:

A Titan 2 rocket carrying two astronauts in a Gemini capsule lifts off from Cape Canaveral's Launch Complex 19. The capsule is the black-and-white object (a black section sitting on top of a white section) at the top of the Titan. Notice that the Gemini had no launch escape rocket system like the Mercury. In the event the Titan malfunctioned before reaching space, the astronauts could eject out of the capsule and parachute to safety.

Above:

Ed White becomes the first American to walk in space on 3 June 1965 during the *Gemini 4* mission. He is holding a device called a *zip gun.* It acted as a miniature hand-held rocket and allowed White to maneuver outside the capsule. An extremely distorted reflection of the Gemini capsule can be seen in White's space suit visor. The photograph was taken by James McDivitt, who remained inside the spacecraft.

One of the highlights of the five-hour mission occurred when Young presented Grissom with half of a corned beef sandwich that he smuggled aboard. Normally, astronauts of the 1960s ate bland meals out of tubes to keep food from floating around in the weightless environment. NASA engineers worried about what the sandwich crumbs might do to the electronic instruments in the capsule. Consequently, mission control expressed extreme disapproval of the unauthorized snack.

The Gemini design employed a heat shield for reentry heat protection similar to the shield found on the Mercury capsules. Gemini capsules, like the Mercury ones, also used a parachute to provide for a safe splashdown in the ocean after a mission. After *Gemini 3* landed, Grissom refused to open the hatch until after the helicopter secured the recovery cable to the capsule because he did not want to take any chances of sinking a second spacecraft. Grissom also jokingly nicknamed his capsule *Molly Brown* in reference to the Broadway hit *The Unsinkable Molly Brown*. He hoped that the name would

bring good luck in keeping the capsule afloat after landing. The name worked because today, the *Gemini 3* capsule now rests on a display stand in Grissom's hometown of Mitchell, Indiana.

Vaulting into the Lead

James McDivitt and Ed White entered space in their *Gemini 4* capsule about two months after the Grissom and Young flight. The highlight of this mission came on the third orbit when White opened the hatch, floated into space, and performed the first American space walk. A tether cable attached to the Gemini, called an *umbilical cord*, kept him from floating away permanently. It also provided oxygen to the astronaut from the capsule's supply tanks. He tested the ability to use a pressurized gas gun to maneuver while on a space walk, and also conducted an initial experiment to find out whether an astronaut could perform useful tasks while floating weightless outside a spacecraft. White's short stroll in orbit lasted for about 20 minutes. Altogether, the mission lasted for 62 orbits over a 98-hour span, and set the world's record for the longest spaceflight. America had finally caught up with the Soviet Union.

On 3 June 1965, Gordon Cooper and Pete Conrad rode into orbit aboard *Gemini 5*. The crew's goal involved determining whether humans could function in a weightless environment for a length of time equal to the planned length of the future Moon missions. NASA scheduled this flight to last for twice as long as the *Gemini 4* mission. Consequently, Cooper wanted to use the phrase "eight days or bust" as their mission motto. High-ranking NASA officials did not approve because they did not want the public to think of the mission as a "bust" if an emergency warranted an early return to Earth. They also felt that the motto might be translated into other languages in a manner embarrassing to the United States government.

Cooper and Conrad succeeded in staying in space for the planned duration despite the lack of a mission motto, and Cooper also became the first human to orbit the Earth on two separate occasions. During the eight days in orbit, the two astronauts conducted scientific experiments and practiced orbital change maneuvers necessary for rendezvous attempts on future missions. However, a fuel cell malfunction prompted mission control to cancel most of the experiments other than finding out whether the crew could last for eight days. Conrad jokingly described the mission as comparable to spending "eight days in a garbage can" because he and Cooper spent most of their time in orbit without anything to do. After the end of the mission, Conrad further joked, "I wish I had brought a book."

Rendezvous in Orbit

NASA wanted the next Gemini crew to rendezvous and dock with an Agena target vehicle in orbit. An Agena was essentially the second stage of a modified Atlas booster. The original mission plan

Gemini Flight Log

Gus Grissom and John Young
Launched on 23 March, 1965
Gemini 3 Mission, 3 orbits

James McDivitt and Edward White
Launched on 3 June 1965
Gemini 4 Mission, 62 orbits

Gordon Cooper and Pete Conrad
Launched on 21 August 1965
Gemini 5 Mission, 120 orbits

Frank Borman and James Lovell
Launched on 4 December 1965
Gemini 7 Mission, 206 orbits

Walter Schirra and Thomas Stafford
Launched on 15 December 1965
Gemini 6 Mission, 16 orbits

Neil Armstrong and David Scott
Launched on 16 March 1966
Gemini 8 Mission, 7 orbits

Thomas Stafford and Gene Cernan
Launched on 3 June 1966
Gemini 9A Mission, 44 orbits

John Young and Michael Collins
Launched on 18 July 1966
Gemini 10 Mission, 43 orbits

Pete Conrad and Richard Gordon
Launched on 12 September 1966
Gemini 11 Mission, 44 Orbits

James Lovell and Edwin "Buzz" Aldrin
Launched on 11 November 1966
Gemini 12 Mission, 59 orbits

Of the two astronauts listed for each mission, the mission commander is listed first.

Historical Fact

Conrad and Cooper started a NASA tradition by designing the first mission patch in spaceflight history. A mission patch usually contains artwork that symbolizes the goals of the space mission. It is sewn onto the space suits of all the crew members who fly on that particular mission. Every NASA space crew since Conrad and Cooper's *Gemini 5* has designed its own mission patch.

Above:
Gemini 6, flown by Wally Schirra and Tom Stafford, maneuvers to within a meter (a few feet) of *Gemini 7*, flown by Frank Borman and Jim Lovell. It was the first orbital rendezvous in spaceflight history. This photograph of the rear of the *Gemini 7* capsule was taken on 15 December 1965.

called for an Atlas to boost an Agena into orbit shortly before the launch of *Gemini 6*. However, the Agena launch failed and NASA temporarily canceled the *Gemini 6* mission. Some engineers then proposed a brilliant alternative. *Gemini 7's* crew was supposed to begin a mission in December 1965 to find out if astronauts could survive in space for 14 days. The engineers proposed to send *Gemini 6* into orbit after the *Gemini 7* launch, and then rendezvous the two capsules in space.

With the modified flight plans of *Gemini 6* and *7* in mind, Frank Borman and James Lovell climbed into their *Gemini 7* capsule on 4 December 1965 to begin their 14-day endurance mission. NASA doctors knew from the Gemini 5 mission that humans could endure at least eight days of weightlessness. However, they also knew that some Moon missions might take between 12 and 14 days to complete. As a result, NASA flew the *Gemini 7* mission to determine if a two-week stay in space was feasible from a physiological standpoint. Borman and Lovell also tested a new lightweight space suit. They found that living in a tight capsule for two weeks and staying in the same clothes for the whole time was a miserable experience.

Wally Schirra and Thomas Stafford boarded the *Gemini 6* capsule eight days after the *Gemini 7* launch in an attempt to join their fellow astronauts in space. Something went drastically wrong that morning because the rocket engines on the Titan 2 shut off a split second after they ignited. If the Titan had risen even a few centimeters off of the launchpad, the shutdown would have caused the entire booster to fall back onto the pad and explode. Schirra trusted his senses, felt no motion from the rocket, and did not pull the emergency lever that would have immediately ejected him and Stafford out of the capsule and to safety. The crew's decision not to panic and eject preserved the physical integrity of the *Gemini 6* capsule and allowed NASA to attempt another launch.

Gemini 6 reached orbit three days after the initial botched launch attempt. The Titan 2 booster put Schirra and Stafford in orbit roughly 2,000 kilometers (1,243 miles) behind Borman and Lovell in *Gemini 7*. Schirra and Stafford then activated their radar to determine *Gemini 7's* exact location. Then, the flight computers in the capsule used the radar data to plot an intercept course. Altogether, *Gemini 6's* chase to catch *Gemini 7* took about seven hours (four orbits around the Earth).

After the rendezvous, both Geminis then flew around each other for the next seven hours. At times, the capsules came within one-third of a meter (1 foot) from each other. This two-capsule mission provided NASA with valuable rendezvous experience applicable to future Moon missions. *Gemini 7* remained in orbit for three uneventful and lonely days after the rendezvous to complete the fourteen-day endurance mission. Borman and Lovell heeded Conrad's previous advice and brought books onboard to read.

Historical Fact

Project Gemini was named after the constellation of stars in deep space that contains the twin stars Castor and Pollux. The twins symbolizes the fact that Gemini was a two-person spacecraft.

Docking with Disaster

America's next Gemini mission flew on 16 March 1966 with the objective of docking the capsule with an Agena rocket in orbit. NASA managed to launch an Atlas carrying the Agena and a Titan 2 rocket carrying the *Gemini 8* capsule within 40 minutes of each other that morning. Less than six hours after the launch, Neil Armstrong and David Scott rendezvoused their Gemini capsule to one meter from the Agena. Armstrong then slowly maneuvered the tip of the Gemini into the Agena's docking adapter and completed the first orbital docking between two spacecraft in history. What followed nearly turned into America's first disaster in orbit.

The two spacecraft, still locked together in the docking configuration, began to tumble slowly out of control. Armstrong thought that the Agena was at fault. He undocked and slowly backed the Gemini away. However, the rate of tumble increased and approached a dizzying 60 revolutions per minute. The crew reached the point of almost blacking out before Armstrong stabilized the capsule by using most of the capsule's maneuvering propellant. Consequently, the crew was forced to make an immediate, emergency splashdown only 10 hours after launch. Both astronauts were still dizzy after landing, and Scott was disappointed because he missed out on a planned space walk. However, he was happy to still be alive.

Gemini 9 began on an ominous note. One year before the planned launch, the original crew of Elliot See and Charles Bassett died when their training jet ironically crashed into the factory constructing their Gemini capsule. NASA sent astronauts Thomas Stafford and Gene Cernan into space on 3 June 1966 in place of See and Bassett. The replacement crew's agenda included several unfinished items left over from *Gemini 8,* such as a controlled docking with an Agena and a space walk by Cernan. Unfortunately, the Agena failed to jettison the nose cone protecting the docking adapter. That mishap prevented a successful docking attempt.

Above:
The crew of *Gemini 9*, Tom Stafford and Gene Cernan, approach the Agena for an attempted docking somewhere over South America. Unfortunately, the docking adapter that locked onto the Gemini's nose was located inside the nose cone of the Agena (the part that looks like the open mouth). Since the nose cone failed to separate as planned, the docking attempt was aborted. Cernan described the Agena as an "angry alligator."

Tying Up the Loose Ends

NASA finished the Gemini program with John Young and Michael Collins on *Gemini 10,* Pete Conrad and Richard Gordon on *Gemini 11,* and James Lovell and Buzz Aldrin on *Gemini 12.* The astronauts on these missions perfected all the tasks attempted on the previous missions. Young and Collins docked with one Agena, fired the Agena's rocket to propel the linked pair to a rendezvous with another Agena, and then docked with the second Agena. Conrad wanted NASA to send *Gemini 11* around the Moon, but settled for docking with an Agena and using its rocket to propel the coupled pair to a record high altitude of 1,368 kilometers (850 miles). *Gemini 12* also docked with an Agena. On that mission, Aldrin successfully completed three space walks, including one that lasted for two hours. During these three walks, he took pictures of the Earth, performed

Historical Fact

The laws of orbital mechanics combined with the amount of propellant on a Gemini constrained the *Gemini 11* launch to take place within a two-second time period in the morning. If they had missed this window of opportunity, they would not have been able to rendezvous with the Agena target vehicle.

Mercury & Gemini Report Card

Flight Dates:
Project Mercury, 1961 to 1963
Project Gemini, 1965 to 1966

Rockets Used:
Project Mercury, Redstone and Atlas
Project Gemini, Titan 2

Spacecraft Mass on Orbit:
Mercury Capsule, 1,355 kilograms
Gemini Capsule, 3,763 kilograms

Number of Flights:
Project Mercury, 6 total
Project Gemini, 10 total

Different Astronauts Flown:
Project Mercury, 6 total (1 per flight)
Project Gemini, 16 total (2 per flight)
3 Mercury astronauts also flew on Gemini

Project Mercury Highlights:
First Americans in space, demonstrated that humans and machines can survive and function in weightlessness, validated basic principles of spacecraft design and operations

Project Gemini Highlights:
Practiced critical skills for the Apollo Moon landings such as long-duration (up to 14 days) spaceflight, rendezvous of two spacecraft in orbit, and extra-vehicular activities

Historical Fact

In order to commemorate the end of a successful Gemini program, the crew of *Gemini 12* arrived at the launchpad with handwritten signs taped to the back of their space suits that said "The End."

simulated repairs on the Agena, and proved that humans could function effectively while floating outside a spacecraft.

Although Gemini missions never flew to the Moon, their ten missions provided NASA with the technical knowledge and experience necessary to successfully send astronauts on a journey to the Moon's surface and back. For example, *Geminis 5* and *7* showed that humans could endure a weightless environment for at least two weeks with no adverse physiological effects, and *Geminis 8* through *12* demonstrated the feasibility of maneuvering two separate spacecraft from different orbits to a safe docking. Furthermore, the tight schedule of ten flights in less than two years gave ground crews at Cape Canaveral much needed practice in launching rockets in a timely and efficient manner.

Project Gemini also provided a temporary sense of security to Americans by catapulting the United States ahead of the Soviet Union in the race to the Moon. By the end of *Gemini 12*, American astronauts had logged a total of 2,000 hours of valuable experience in space. The Russian total only amounted to 500 hours because not a single Russian flew into space during the duration of project Gemini. Many historians feel that this sudden lack of progress resulted from Khrushchev's insistence of performing space stunts, even at the expense of genuine technical progress and engineering development. However, at the time of Project Gemini's conclusion on 15 November 1966, the race to the Moon was still far from over.

4.4 Fulfilling the Dream

Kennedy made a risky decision in committing the United States to send astronauts to the Moon and back by 1970. At the time that he announced this goal in a speech to Congress on 25 May 1961, America's experience in human spaceflight amounted to a single 15-minute suborbital flight by Alan Shepard. NASA had not yet sent an astronaut into Earth orbit, and Kennedy was asking them to shoot for the Moon. Many experts within NASA did not know whether a Moon landing was possible by the end of the decade, or even possible at all. However, they also realized that Kennedy's challenge represented the chance of a lifetime to make their wildest dreams come true.

The Mode Decision

NASA named the new project Apollo and immediately began work on figuring out how to place astronauts on the surface of the Moon and then bring them back safely. Engineers working on solving the problem arrived at three possible scenarios. They called each one of the different ideas a "mode of flight." The resulting debate over which scenario to use sparked a fierce debate within NASA for a year.

Mode number one, called direct ascent, involved launching a single rocket from the Earth directly to the Moon. This scenario called for landing the rocket on the Moon by turning the rocket around, pointing its engines toward the surface, and then using the engines to brake the fall in a process opposite that of the launch. The same rocket would take the astronauts back to the Earth after the completion of lunar exploration activities. With further analysis, NASA found that a direct ascent would require a monster three-stage rocket (tentatively called Nova) weighing close to 5,400,000 kilograms (about 12,000,000 pounds). Initial studies also revealed that the third stage that landed on the Moon, blasted off, and then landed back on Earth would measure 27 meters (90 feet) tall. Engineers shuddered at the task of "backing" something the size and shape of an Atlas rocket down to a safe landing in a vertical orientation. They were already having enough problems trying to get the Atlas to go the other way.

Many within NASA favored a second flight mode called *Earth orbit rendezvous* (EOR). This scheme involved launching several rockets into Earth orbit. Each flight would carry either a segment of the vehicle that was supposed to ferry the astronauts to the lunar surface and then back to Earth, or the propellant for the vehicle. Astronauts or ground controllers would then somehow rendezvous all the parts in orbit, and then assemble them into one functional unit. Using EOR required rockets only half the size of the direct-ascent monster. However, in the early 1960s, nobody knew exactly how to perform a rendezvous in orbit, how to construct a spacecraft out of separate segments in orbit, or how to transfer dangerous and volatile propellants in the vacuum and weightlessness of space.

NASA called the third mode *lunar orbit rendezvous* (LOR). This scenario involved launching a main "command" spacecraft and a lunar surface "ferry," both docked together, into orbit around the Moon. After achieving lunar orbit, astronauts would transfer from the main craft to the ferry, and then ride the ferry to the surface. This ferry would also take them back to the main spacecraft in orbit around the Moon for the trip back to Earth. LOR utilized the concept of disposing of as much equipment as possible after use because sending something to the Moon takes a lot of energy, but sending something there and bringing it back takes even more energy. By not bringing the part of the spacecraft that landed on the Moon back to Earth, NASA could dramatically reduce the total amount of energy necessary to accomplish the mission.

Proponents of LOR tried to convince high-level managers within NASA that their idea represented the best mode. They said that LOR worked analogously to anchoring a large ship in a harbor and using a rowboat to reach land. This method, they argued, would require less effort than landing the entire ship. Opponents countered by arguing that rendezvousing two spacecraft together in Earth orbit presented enough of a challenge, and LOR supporters were recklessly proposing to rendezvous the ferry and main spacecraft in lunar orbit 400,000 kilometers from home. However, initial calculations

Above:

President John F. Kennedy appears in front of a joint session of Congress on 25 May 1961 to challenge America to land an astronaut on the Moon within nine years. This historical speech set the tone for the American space program during the 1960s.

Historical Quote

"For we cannot guarantee that we shall one day be first, we can guarantee that any failure to make this effort will make us last. We take the additional risk of making it in full view of the world."

President John F. Kennedy on 25 May 1961 regarding America's quest to land astronauts on the Moon.

Above:

Just how tight was it inside the lunar module? This photograph shows *Apollo 9* astronauts James McDivitt (left) and Rusty Schweickart (right) in a full-scale model of the lunar module during a December 1968 training simulation. As shown here, the inside of the lunar module contained no more space than two phone booths placed side by side. These two astronauts were the first to fly the lunar module. On their mission, they tested the spacecraft in Earth orbit to prove its "space worthiness."

showed that LOR required only one rocket about two-thirds the size of the booster proposed for the direct-ascent option. This economical savings convinced NASA management to adopt LOR as the flight mode for Apollo.

A Plan for Reaching the Moon

NASA's next step after making the mode decision involved designing the spacecraft for the astronauts to fly to the Moon, and the rocket to boost the two spacecraft to the Moon. Lunar orbit rendezvous-mission plans called for three astronauts and two spacecraft to fly into space on each mission. The three astronauts rode to the Moon and back aboard the Apollo spacecraft. The second craft, called the *lunar module* (LM), remained mechanically docked to the Apollo during the trip from the Earth to the Moon. There, its job involved ferrying two of the three astronauts from the Apollo spacecraft to the lunar surface and back. Figure 34 on page 127 shows what the two spacecraft looked like.

The Apollo spacecraft consisted of two distinct sections attached to each other. Astronauts occupied the cone-shaped capsule at the front end. This capsule, called the *command module*, provided much more room for each astronaut than the Gemini and Mercury spacecrafts. Crew members now enjoyed the previously unavailable luxury of taking off their bulky space suits inside the capsule. However, accommodations still remained tight. Imagine spending up to two weeks with three of your friends inside a car about the size of a mini-van, not taking a shower for the entire time, and using a bag while floating in midair to go to the bathroom. This description summarizes life inside the command module during a lunar voyage.

A cylindrical service module attached to the bottom of the command module and provided electricity, oxygen, and other life-support functions for the astronauts. The service module also provided the rocket propulsion necessary to place the Apollo spacecraft into a lunar orbit after the three- to four-day trip from Earth, and the propulsion required to break out of lunar orbit for the return to Earth. At the end of the mission, the command module separated from the service module, entered Earth's atmosphere blunt (base) end first with the protection of a heat shield, deployed a parachute to slow down, and splashed to a landing in the ocean.

The second spacecraft, called the lunar module (LM), appeared completely different from the traditional streamlined, aerodynamic cone-shaped capsule design. Instead, the LM looked like an odd metallic bug with two eyes and four legs. Two astronauts rode in the "head of the bug" and looked outside the LM through the "eyes," which were actually glass windows. The body of the bug, called the descent stage, attached below the head and contained a single rocket engine to slow the spacecraft from lunar orbit to a controlled landing on the Moon's surface. Slowing down to a safe landing required a rocket because devices such as parachutes, aircraft

Figure 34: *Tools for Apollo* ◄

Apollo Parts:	Command module (CM) and service module (SM), combined pair was called the CSM.
Dimensions of the CM:	3.23 meters (10.6 feet) high 3.90 meters (12.8 feet) width at base
Weight of the CM:	5,900 kilograms (13,007 lbs) fully loaded with propellant and crew.
Dimensions of the SM:	7.41 meters (24.3 feet) high 3.90 meters (12.8 feet) diameter
Weight of the SM:	24,500 kilograms (54,013 lbs) fully loaded with propellant and supplies.
Lunar Module Parts:	Ascent stage and descent stage, combined pair was called lunar module (LM).
Dimensions LM Overall:	6.98 meters (22.9 feet) high 9.45 meters (31.0 feet) width when measured diagonally
Weight of the LM:	16,440 kilograms (36,244 lbs) fully loaded for the *Apollo 17* mission. Mass varied with each mission depending on amount of supplies.
Who Made Them?	CSM - Rockwell International LM - Grumman

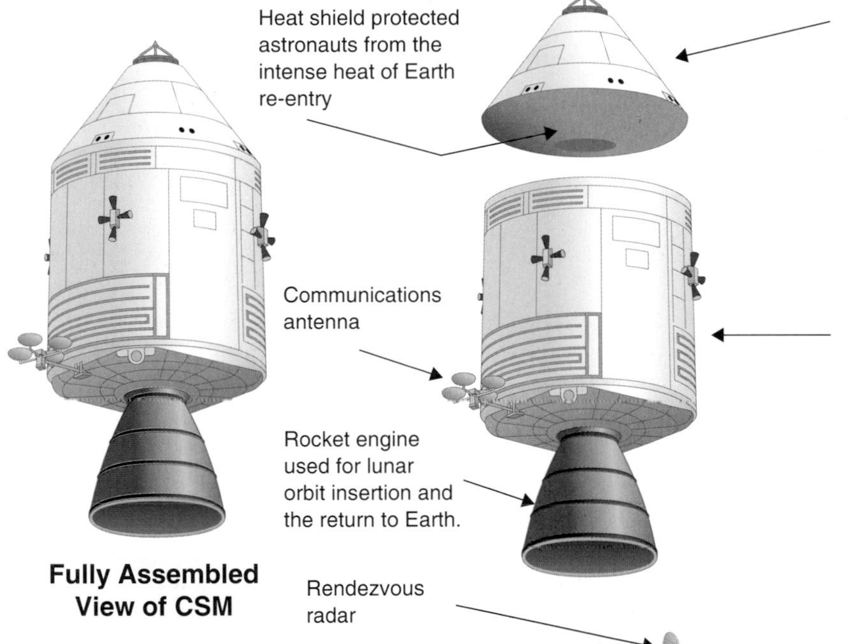

Command Module

All three Apollo astronauts sat in here for the ride from the Earth to the Moon and back. One of the three remained in the CM with the Apollo in lunar orbit while the other two explored the Moon. This module was the only part that returned to Earth.

Service Module

This module stored the propellant for the Apollo's rocket engine (SPS), housed oxygen tanks for the astronauts in the CM, and provided electricity to run the electronics and computers in the CM. The SM was jettisoned just hours before returning to Earth.

Ascent Stage

Two of the three Apollo astronauts sat in here during the descent from lunar orbit to the surface. It was also used to blast off from the Moon to rendezvous with the Apollo for the trip home.

Descent Stage

This stage contained the rocket engine that provided for a soft landing on the Moon, and served as the launch pad for the ascent stage for lunar lift-off.

▶ Figure 35: *Parts of the Saturn 5 Moon Rocket*

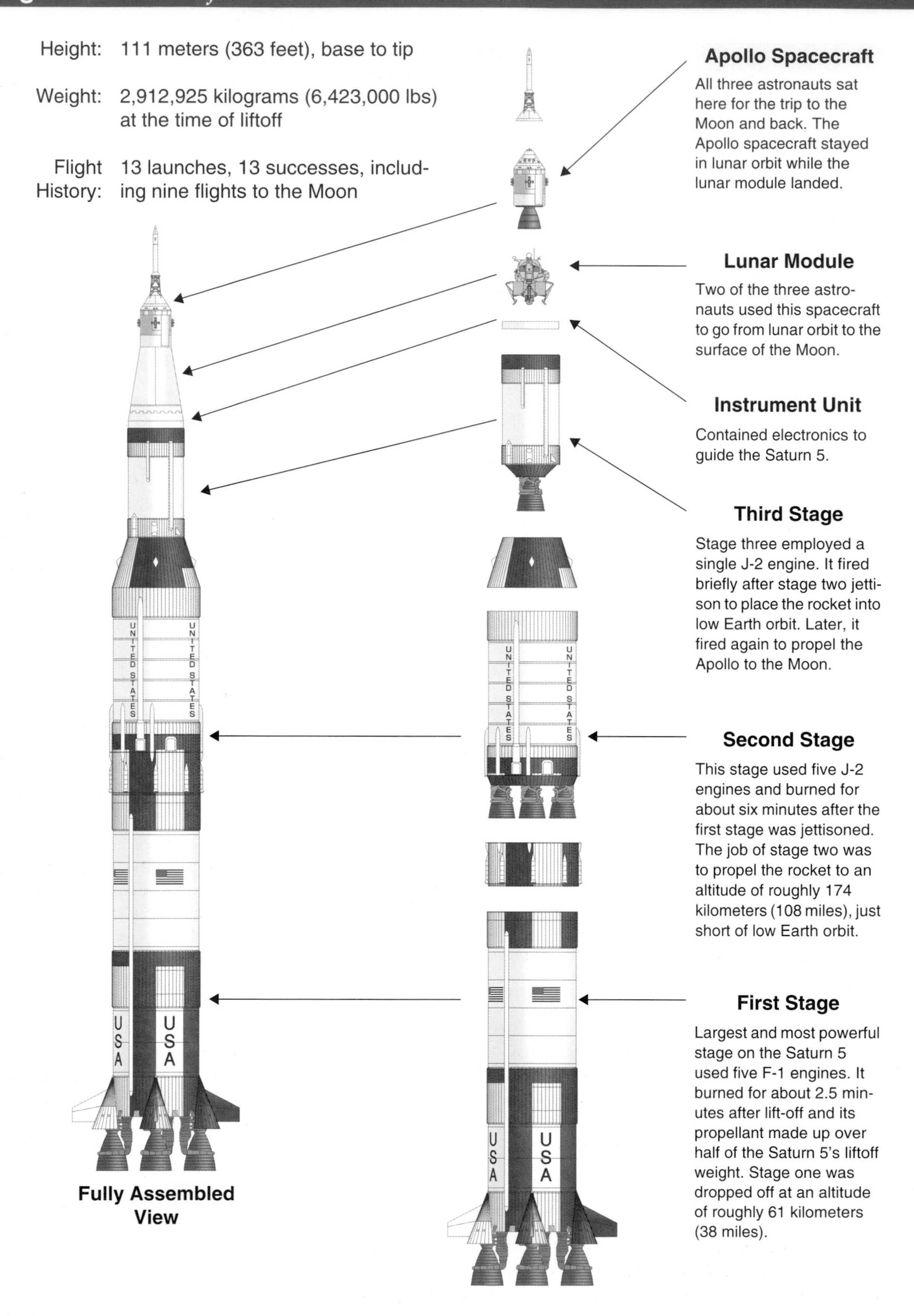

wings, and helicopter rotors did not work on the airless Moon. Four metallic legs that attached to the descent stage cushioned the landing and allowed the LEM to stand upright while on the Moon.

NASA rocket genius Wernher von Braun and his team of scientists at the Marshall Space Flight Center in Huntsville, Alabama, designed the gigantic rocket that boosted the Apollo and LM into space and to the Moon. Von Braun always named his designs after large planets. Consequently, he called the Moon rocket Saturn because he had already built a "Jupiter" that sent America's first satellite into orbit. Saturn rocket designs came in several different sizes and configurations. NASA chose to use von Braun's fifth variation, called the Saturn 5, for lunar orbit rendezvous. This gigantic three-stage rocket stood over 111 meters (363 feet) tall when assembled and weighed over 2,913,423 kilograms (6,423,000 pounds) when fully loaded with propellant and payload. In order to visualize the size of a Saturn 5, imagine a rocket two-thirds as high as the Washington Monument in the capitol and 3,000 times heavier than a ton of bricks.

Over 90% of the 2,913,423 kilograms of a fully loaded Saturn 5 consisted of propellant. Stage one (bottom stage) alone carried most of the rocket's total supply and its five engines consumed every last drop in less than three minutes after liftoff. The sheer power of this rocket sometimes disguised the delicacy of its construction. One of the secrets that allowed the Saturn 5 to fly was the super-thin, but extremely strong, aluminum alloy that the Alcoa and the Reynolds corporations jointly developed for the rocket's structure. This alloy kept the Saturn's external skin thickness to no more than 0.64 centimeters (one-quarter of an inch) at any location. Altogether, the payload plus the Saturn structure weighed 15 times less than the propellant needed for the trip into space.

The Saturn 5's first stage boosted the rocket to an altitude of 61 kilometers (38 miles) before consuming all of its propellant and dropping off about three minutes after liftoff. Stage two then took over the job of pushing the rocket. It benefited from the fact that all the propellant consumed by stage one's engines made the rocket four times lighter than at launch. Consequently, stage two managed to push and accelerate the rocket using engines only one-third as powerful as those on the first stage. Nine minutes after liftoff and about six minutes after separation of stage one, the second stage ran out of propellant and dropped of at the altitude of 174 kilometers (108 miles). Stage three then fired briefly to insert itself, the LM, and the Apollo spacecraft into a temporary low Earth orbit at an altitude of 185 kilometers (115 miles). The Saturn's three stages consumed enough propellant during the journey to Earth orbit to drop the rocket's weight to less than one-sixteenth the weight at the time of liftoff.

After the astronauts checked out all the major systems in low Earth orbit, they restarted the engine on stage three. This six-minute rocket burn, called the *trans-lunar injection* (TLI) maneuver, placed

Top Photograph:
With astronaut Alfred Worden at the helm, the Apollo spacecraft *Endeavour* orbits the Moon on the *Apollo 15* mission in August 1971. During this fourth lunar landing mission, David Scott and James Irwin explored the Moon's surface while Worden remained in lunar orbit to fly the Apollo.

Bottom Photograph:
Apollo 11 astronaut Michael Collins snapped this photograph of the lunar module (LM) *Eagle* from the Apollo spacecraft shortly before Neil Armstrong and Buzz Aldrin (inside the LM) began the descent to the lunar surface in July 1969. This mission marked the first landing of humans on the Moon.

Above:

A Saturn 5 Moon rocket ignites its first stage engines as it prepares to lift off from Launch Complex 39 at the Kennedy Space Center in Florida. Notice the tiny Apollo spacecraft at the top of the 111-meter-tall rocket. From 1967 to 1973, NASA launched 13 of these gigantic rockets. The Saturn 5 was, and still is, the only rocket design in the world that did not explode at least once during use. Even the Space Shuttle cannot claim this distinction.

what was left of the Saturn rocket on a trajectory toward the Moon. In reality, the constant motion of the Moon orbiting the Earth required the astronauts to aim for a point ahead of the Moon's position at the time of TLI. The hard part involved timing the burn so that they arrived at the same point in space at the same time as the Moon. An important concept to remember is that the Saturn rocket did not thrust all the way to the Moon. Instead, the astronauts coasted all the way there after the TLI maneuver. This burn worked analogously to using one great burst of energy to throw the Apollo and LM high into the air with enough force to reach the Moon.

TLI also used the last drops of propellant in stage three's tanks. The next step involved jettisoning the now useless third stage of the Saturn. However, the astronauts needed to first remove the LM from its protective canister between the third stage and the Apollo spacecraft. This task involved separating the Apollo from the Saturn, turning the Apollo around so that the tip of the capsule pointed back at the top of the LM in the protective canister, and then docking the two spacecraft by mechanically attaching the tip of the Apollo capsule to a docking adapter at the top of the LM's "head" portion. If all went well with the docking maneuver, the astronauts threw a switch to jettison the Saturn's third stage. Now, all that remained of the once mighty Saturn rocket was the Apollo and LM.

Both spacecraft, still docked together, approached the vicinity of the Moon about three full days after liftoff from the Earth. The next critical maneuver involved slowing the two spacecraft enough to allow the Moon's gravity to capture the two out of the trajectory from the Earth and into an orbit around the Moon. Some Apollo missions utilized what spaceflight engineers call a free return trajectory. This type of path allowed the astronauts to utilize the Moon's gravity to automatically sling the spacecraft back to Earth in case the slowdown maneuver failed. However, the lives of most of the Apollo astronauts depended on the success of the slowdown. Without it, they were doomed to fly past the Moon and away from the Earth forever. The slowdown maneuver, called the *lunar orbit insertion* (LOI) burn, typically occurred over the side of the Moon facing away from the Earth. Since the astronauts were behind the Moon, they could not depend on radio signals from the mission control center's giant computers back on Earth for help.

A successful LOI placed the two spacecraft in an elliptical orbit around the Moon, which the astronauts subsequently circularized at an altitude of about 96.6 kilometers (60 miles). They spent most of the fourth day of the mission in orbit around the Moon preparing for undocking the LM and descending to the surface. Sometime during day five of the mission, two of the three astronauts in the Apollo command module crawled into the LM through a built-in tunnel connecting the two docked spacecraft. The interior of the LM provided even less space than the cramped Apollo command module. Both astronauts stood, or more accurately, floated, in an upright position at their control consoles. From there, two windows that made up

Historical Fact

On the way to the Moon, the Apollo and LM were slowly spun like a rotisserie chicken. This action kept the sunny side of the spacecraft from overheating and the shaded side from freezing. The astronauts called this spinning "barbecue mode."

Figure 36: *Apollo Mission Schematic Diagram (Leaving Earth)*

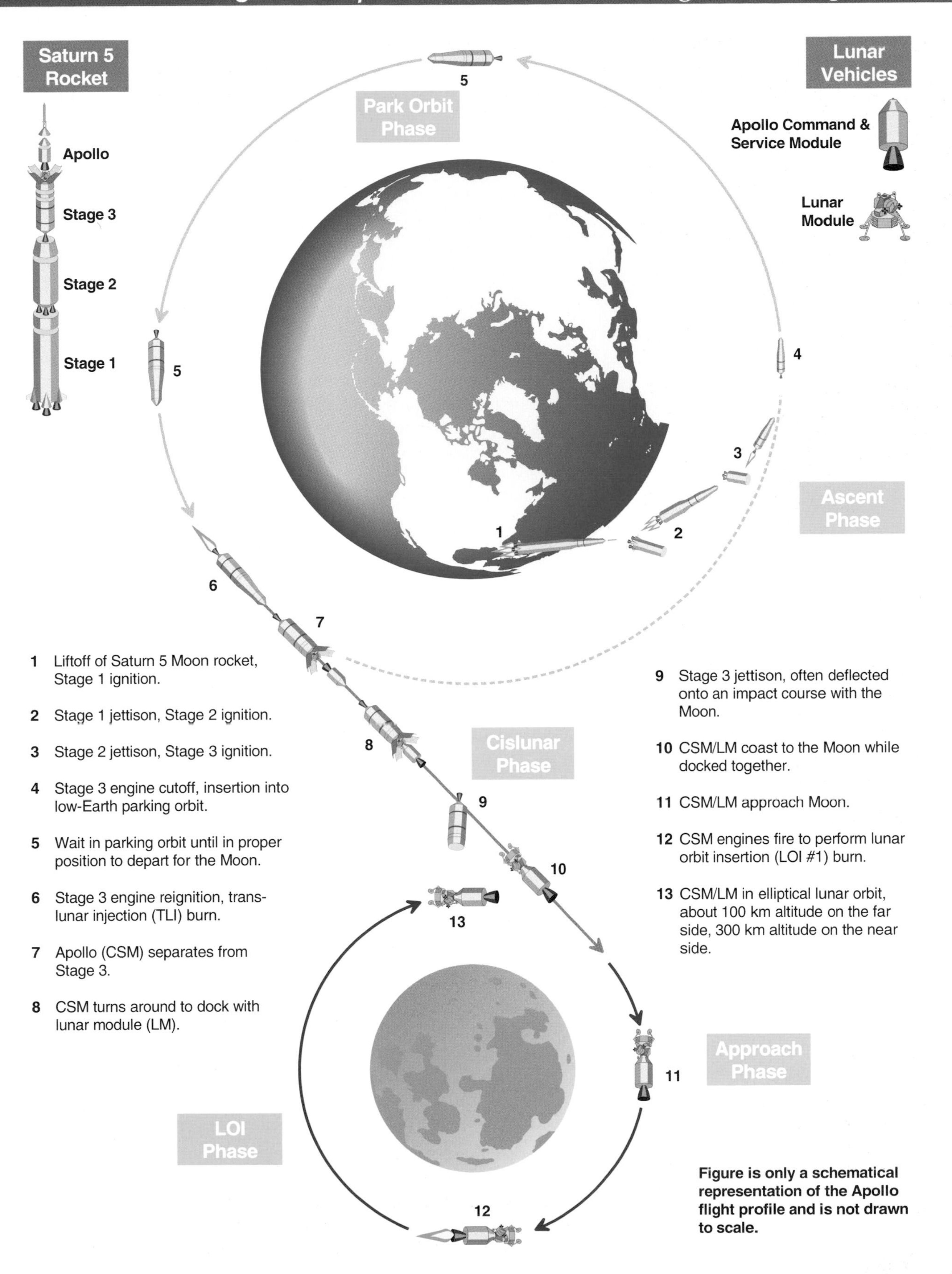

1 Liftoff of Saturn 5 Moon rocket, Stage 1 ignition.

2 Stage 1 jettison, Stage 2 ignition.

3 Stage 2 jettison, Stage 3 ignition.

4 Stage 3 engine cutoff, insertion into low-Earth parking orbit.

5 Wait in parking orbit until in proper position to depart for the Moon.

6 Stage 3 engine reignition, trans-lunar injection (TLI) burn.

7 Apollo (CSM) separates from Stage 3.

8 CSM turns around to dock with lunar module (LM).

9 Stage 3 jettison, often deflected onto an impact course with the Moon.

10 CSM/LM coast to the Moon while docked together.

11 CSM/LM approach Moon.

12 CSM engines fire to perform lunar orbit insertion (LOI #1) burn.

13 CSM/LM in elliptical lunar orbit, about 100 km altitude on the far side, 300 km altitude on the near side.

Figure is only a schematical representation of the Apollo flight profile and is not drawn to scale.

Historical Fact

Every Apollo crew consisted of three astronauts. The one in charge of the overall mission execution was called the *commander* (CDR). Both the commander and another astronaut, called the *lunar module pilot* (LMP), landed on the Moon's surface in the lunar module. In reality, the commander flew the LM and the lunar module pilot served as the copilot. The third astronaut, called the *command module pilot* (CMP), remained in lunar orbit to fly the Apollo while the other two explored the Moon.

RULE OF THUMB

The Apollo spacecraft was not designed to land on the Moon. If the LM crashed while landing, or if its ascent-stage engine failed to fire, the command module pilot would have been forced to return to Earth without the other two astronauts.

the "eyes" of the "bug" provided them with a view of their outside surroundings. The third astronaut remained behind in lunar orbit to keep an eye on the Apollo spacecraft.

A flip of a switch inside the LM mechanically disconnected it from the Apollo and allowed the two spacecraft to undock. After the lone astronaut remaining in the Apollo maneuvered away, the LM fired its descent engine briefly to put the odd-looking spacecraft into a lower orbit. This new elliptical path took the two lunar-surface-bound astronauts from a high point on their original 97-kilometer circular orbit and plunged them toward a low of 13.7 kilometers (8.5 miles). The astronauts in the LM then initiated another burn, called the *power descent insertion* (PDI), upon reaching the 13.7-kilometer altitude mark 45 minutes after the previous burn. PDI essentially dropped the LM out of its 97-kilometer-by-13.7-kilometer elliptical-descent orbit by nearly eliminating all of the spacecraft's forward velocity. A successful landing now depended on the two astronauts to skillfully use the LM's rocket engine to slow their fall to the lunar surface.

One of the most difficult landing tasks involved finding a safe touchdown site. NASA mission planners selected the general landing area before the mission, but it was up to the astronauts in the LM to find a safe spot to set down within the designated area. On many Apollo missions, the astronauts flew the LM to a desirable site only to realize at the last minute that large lunar boulders littered the ground. The problem was that no common reference objects such as trees or buildings existed on the Moon. Without familiar sights and familiar objects with which to gauge size, the astronauts had a hard time determining the size of the boulders on the lunar surface. There were several close calls with LMs landing next to boulders and craters. Searching for alternative landing sites at the last minute consumed a lot of propellant. Often, the final phases of landing occurred with less than one minute worth of descent propellant remaining in the tanks.

The "head" portion of the LM, called the ascent stage, provided the two lunar surface explorers with a way to return to their Apollo spacecraft in orbit around the Moon. After the astronauts finished the surface exploration portion of the mission, they climbed back inside the LM, fired the ascent stage's engine, and then flew the ascent stage back into lunar orbit to rendezvous with the Apollo spacecraft for the trip home. The astronauts essentially used the descent stage as a lunar launchpad for the ascent stage. Leaving the descent stage on the Moon reduced the total amount of propellant required to bring the astronauts back to the Apollo spacecraft.

Many Apollo astronauts felt helpless during the last few minutes before lunar liftoff. The reason was that the ascent stage contained only one rocket engine to blast the surface explorers back into lunar orbit for a rendezvous with the Apollo spacecraft. All the astronauts'

Figure 37: *Apollo Mission Schematic Diagram (Landing on the Moon)*

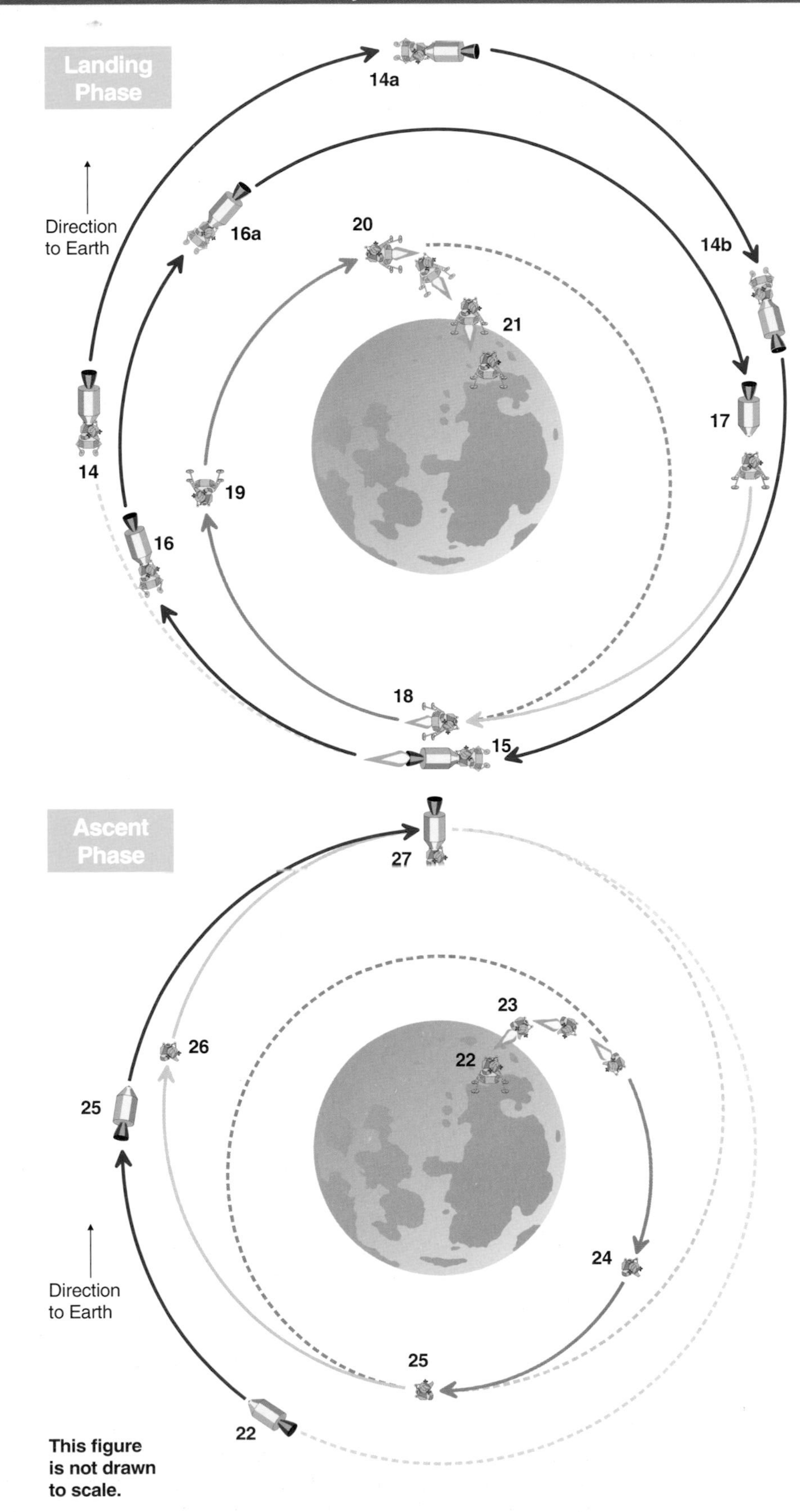

14 Apollo (CSM) and lunar module (LM) orbit the Moon 100 km over the far side, 300 km over the near side after Lunar Orbit Insertion burn (LOI #1).

15 Slow-down burn (LOI #2) puts CSM/LM into a circular orbit at about 100 km in altitude.

16 Astronauts wait in 100-km orbit and prepare for landing. Two of them transfer to the LM from the CSM.

17 LM undocks and separates from the CSM.

18 Descent orbit insertion burn (DOI) slows LM down and puts in an elliptical orbit with a low point of about 14 km.

19 LM plunges toward the lunar surface on the DOI orbit.

20 Powered Descent Insertion (PDI) burn eliminates LM's forward velocity and takes it out of the DOI orbit.

21 LM uses its rocket to slow to a gentle landing.

22 One astronaut remains with the CSM in a 100 km-altitude orbit around the Moon while the other two explore on the surface.

23 After surface exploration is complete, the two astronauts on the Moon blast off in the ascent stage of the LM, using the descent stage as a launchpad.

24 LM enters a low elliptical orbit with a 16-km low point on the near side, and 70-km high point on the far side. Since the LM is in a lower orbit than the CSM, it moves faster and begins to catch up.

25 At the right time, the LM uses tiny rockets to climb up to the CSM's altitude at 100 km.

26 LM approaches the CSM and performs final corrections to their rendezvous trajectory.

27 LM catches and docks with the CSM. The two astronauts in the LM transfer back to the CSM.

► Figure 38: *Apollo Mission Schematic Diagram (Returning Home)*

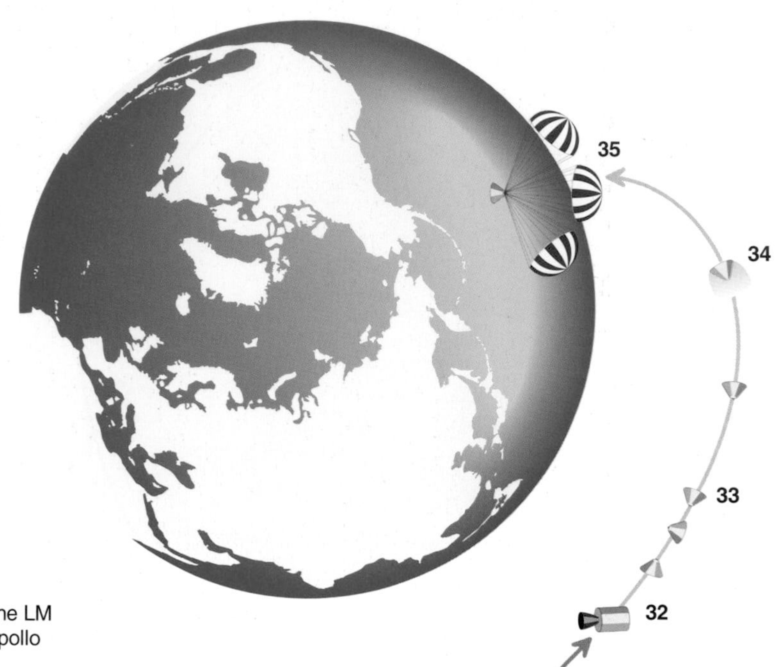

27 After liftoff from the Moon's surface, the ascent stage of the LM catches and docks with the Apollo spacecraft (CSM).

28 The two astronauts in the ascent stage of the LM transfer back to the CSM.

29 LM jettison

30 Main engine on CSM fires to perform the trans-Earth injection burn (TEI). This velocity speed-up allows the CSM to escape the Moon's gravity.

31 CSM falls toward the Earth on the return journey.

32 Service module (SM) jettison.

33 Command module (CM) turns around to orient blunt end of the capsule forward.

34 Heat shield on the blunt end of the CM protects the capsule and three astronauts during the descent through the atmosphere.

35 Parachute deploys for splashdown in the Pacific Ocean.

Lunar Departure Phase

31 27 28 29 30

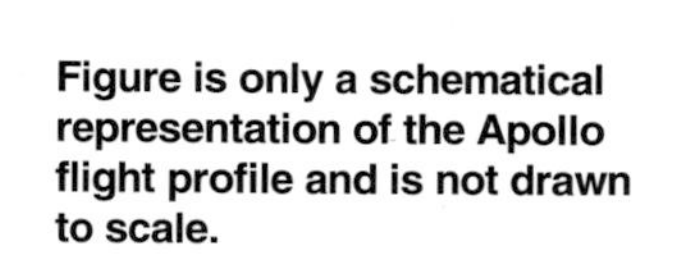

Figure is only a schematical representation of the Apollo flight profile and is not drawn to scale.

thousands of hours of training could not help them if the single engine failed to ignite because there was no backup and no second chance.

An engine burn that started at the time of lunar liftoff and lasted for seven minutes placed the LM's ascent stage back into lunar orbit. Typically, the two astronauts in the LM found themselves in a lower orbit than the Apollo and trailing it by several hundred kilometers. This lower orbit allowed the LM to orbit the Moon faster than the Apollo and eventually catch up after a three- to five-hour chase. The final rendezvous, approach, and docking required a sequence of complicated orbital maneuvers that required precise timing. After a successful docking, the two astronauts in the LM transferred back to the Apollo spacecraft for the trip home.

Returning back to Earth lacked the high drama of landing on the Moon, but required a perfectly functioning Apollo spacecraft and precise navigation. First, the three astronauts in the Apollo jettisoned the LM's ascent stage by undocking with it and leaving it in lunar orbit. Then, at the proper time, the Apollo service module's lone rocket engine performed the three-minute *trans-Earth-injection* (TEI) burn to break the Apollo out of lunar orbit and to send the spacecraft

Above:

Flight controllers in the Mission Control Room at the NASA Johnson Space Center (Houston, Texas) monitor the progress of an Apollo mission. The computers and their operators in this room kept watch on the Apollo spacecraft's electrical system, life support, communications, orbital trajectory, and more. Big screens at the front of the room provided television views of the mission in progress. Today, NASA uses this room and another one like it down the hall to control every Space Shuttle mission from liftoff to landing.

Above:

13 March 1969 - The *Apollo 9* crew awaits the arrival of a recovery helicopter from the U.S.S. *Guadalcanal*. Astronaut James McDivitt is standing in the hatch of his capsule while his crewmates Rusty Schweickart and David Scott have already climbed into the life raft. Visible in the photograph are three Navy divers who assisted with the recovery of the capsule and its crew.

on a trajectory back to Earth. A TEI miscue would have either stranded the astronauts in lunar orbit, or sent them into deep space. There was no room for error, and no backup engine in case the Apollo's rocket failed to ignite.

After the completion of TEI, the Apollo essentially fell "through the sky" for three days to cover the 384,400 kilometers (238,855 miles) from the Moon down to the Earth. About half an hour before arriving at the upper reaches of the Earth's atmosphere, the cone-shaped command module separated from the service module. Reentry into the atmosphere occurred at a velocity of more than 39,000 kilometers (24,233 miles) per hour. This extreme speed demanded precise navigational control. If the capsule descended through the atmosphere at too steep an angle, it would burn up even with the protection of the heat shield. At too shallow a descent angle, the capsule would skip off the atmosphere like a thrown rock skipping across water. Finally, parachutes opened after the capsule fell into the lower atmosphere and slowed the three astronauts to a safe splash-down in the Pacific Ocean.

A Fiery Beginning

NASA engineers made preparations for the Apollo flights all throughout the duration of Project Gemini. By the time of the last Gemini mission in November 1966, NASA managers had already penciled the first Apollo mission into the flight schedule for February of the next year. They selected Gus Grissom (second American in space and first to fly the Gemini), Ed White (first American space walker), and rookie astronaut Roger Chaffee for the mission with the goal of conducting extensive tests of the Apollo spacecraft in Earth orbit. Since the mission did not involve a flight to the Moon or a lunar module, NASA opted to use the Saturn 1B rocket. This booster represented a scaled-down, less powerful version of the Saturn 5 design.

On 27 January 1967, the three astronauts proceeded to the launchpad at the Kennedy Space Center, entered the Apollo spacecraft on top of the Saturn 1B rocket, and prepared to spend the day practicing for the launch with ground personnel by conducting a full-scale "dress rehearsal" of the pre-launch procedures. Several hours into the test, ground controllers heard a warning from the radio link to the inside of the Apollo. "We've got a fire in the cockpit," Chaffee calmly announced over the intercom. Moments later, the calm announcement turned into a panicked, "We're burning up!" It took the rescue crews five minutes to get the hatch open because of the intense heat from the fire. By that time, the astronauts were already dead. Most people incorrectly assumed that they burned to death. In reality, the space suits protected the astronauts from the flames. They died from the inhalation of toxic gasses caused by the fire. In any event, it was a national tragedy.

Historical Quote

"How are we going to get to the Moon if we can't talk between two buildings?"

Astronaut Gus Grissom commenting on technical difficulties with the Apollo's communications systems shortly before the *Apollo 1* fire on 27 January 1967.

Project Apollo seemed jinxed. President Kennedy, the spiritual inspiration of the American space program, had been assassi-

nated in 1963. Now, the death of Grissom, White, and Chaffee represented the second tragic setback to strike the Apollo program in four years. However, President Lyndon Johnson and NASA vowed to press on. A quote that fallen astronaut Grissom had said several weeks before the fire summarized the desire to keep the Apollo program moving forward. He had previously told reporters, "If we die, we want people to accept it. We're in a risky business and we hope if anything happens to us, it will not delay the program."

An intensive investigation that followed determined that an electrical short circuit caused the fire. Thousands of NASA engineers work long hours and spent the next twenty-one months redesigning the Apollo spacecraft. In total, they made 1,341 changes, including a redesign of the entire electrical system. Unfortunately, the deadline to the Kennedy challenge of landing astronauts on the Moon before 1970 was rapidly approaching.

Preparing to Meet the Kennedy Challenge

NASA restored American hopes by finally launching the first piloted Apollo on 11 October 1968. On that day, a Saturn 1B rocket boosted the crew of *Apollo 7* into low Earth orbit to conduct an extensive flight test of the new spacecraft. The crew consisted of Wally Schirra, Donn Eisele, and Walter Cunningham. Schirra, now forty-five years old, became the oldest person to fly into space and the only astronaut to fly Apollo, Gemini, and Mercury missions. Even though Apollo's larger cabin provided much more room than the Gemini, the 11-day mission turned out to be brutal because the food was bad, and all three developed colds. Nevertheless, the *Apollo 7* mission proved the soundness of the redesigned Apollo spacecraft.

The next logical step involved testing both the Apollo spacecraft and the still unflown lunar module in low Earth orbit. Unfortunately, the lunar module was not yet ready for flight. Another problem cropped up as well. Intelligence reports at the time indicated that the Soviet Union would attempt to upstage NASA by sending a crew to fly around the Moon in December 1968. Consequently, NASA engineers proposed to rearrange the flight schedule to send *Apollo 8* to the Moon instead of into low Earth orbit. The *Apollo 8* crew would not be able to land without a lunar module, but they could orbit the Moon and come back. In the process, NASA would learn valuable information about communications and navigation at distances far beyond low Earth orbit. High level officials within NASA approved the new plan.

On 21 December 1968, a Saturn 5 rocket launched Frank Borman, James Lovell, and William Anders in their *Apollo 8* spacecraft. The first two stages of the Saturn lifted the Apollo into low Earth orbit. Then, the third stage boosted the Apollo into a trajectory bound for the Moon. They reached the Moon and entered orbit around it on the morning of Christmas Eve. For the next twenty hours, they took photographs of the surface for the purpose of determining future

Rule of Thumb

Due to the large number of astronauts waiting for a chance to fly, NASA had an unwritten rule within the astronaut corps that essentially said, "He who has commanded an Apollo mission shall not command another." Only command module pilots (CMPs) were allowed to fly two Apollo missions. Typically, CMPs were offered command of an Apollo flight six missions later. For example, David Scott (CMP, *Apollo 9)* commanded *Apollo 15*, and John Young (CMP, *Apollo 10)* commanded *Apollo 16*. It was rumored that Michael Collins (CMP of the first lunar landing on *Apollo 11)* was offered command of *Apollo 17*. He declined and decided to retire from the astronaut corps to pursue a career in international diplomacy.

Historical Fact

Of the original seven Mercury astronauts, only the ones with last names beginning with "S" flew Apollo missions - Schirra *(Apollo 7)*, Shepard *(Apollo 14)*, and Slayton *(Apollo-Soyuz)*. Gus Grissom was scheduled to command the first Apollo mission, but died during the *Apollo 1* fire in 1967.

► Figure 39: *Apollo Mission Log*

Apollo 1 Mission Summary

Liftoff Date	Splashdown Date
Did not fly	N/A
Command Module Name	**Lunar Module Name**
Apollo 1	No LM on this mission

Crew Members
Mission Commander, Gus Grissom (middle)
Lunar Module Pilot, Roger Chaffee (right)
Command Module Pilot, Edward White (left)

Previous Space Experience
Grissom, flew on *Mercury 4* and *Gemini 3*
Chaffee, no previous missions
White, flew on *Gemini 4*

Significant Events on Mission
First planned flight of Apollo, crew died during a launch pad accident

Apollo 7 Mission Summary

Liftoff Date	Splashdown Date
11 October 1968	22 October 1968
Command Module Name	**Lunar Module Name**
Apollo 7	No LM on this mission

Crew Members
Mission Commander, Walter Schirra (middle)
Lunar Module Pilot, Walter Cunningham (right)
Command Module Pilot, Donn Eisele (left)

Previous Space Experience
Schirra, flew on *Mercury 8* and *Gemini 6*
Cunningham, no previous missions
Eisele, no previous missions

Significant Events on Mission
First Apollo flight, tested Apollo spacecraft in low Earth orbit

Apollo 8 Mission Summary

Liftoff Date	Splashdown Date
21 December 1968	27 December 1968
Command Module Name	**Lunar Module Name**
Apollo 8	No LM on this mission

Crew Members
Mission Commander, Frank Borman (right)
Lunar Module Pilot, William Anders (left)
Command Module Pilot, James Lovell (center)

Previous Space Experience
Borman, flew on *Gemini 7*
Anders, no previous missions
Lovell, flew on *Gemini 7* and *Gemini 12*

Significant Events on Mission
First humans to orbit Moon, first pictures of Earth from "deep" space

landing sites. After they finished the 10th orbit, the *Apollo 8* crew ignited the service module's engine to put the spacecraft on a trajectory back to Earth. On the way home after the burn, astronaut Jim Lovell announced to the world over the radio, "Pleased be informed that there is a Santa Claus."

In March 1969, the *Apollo 9* crew completed *Apollo 8's* original goal by testing both the Apollo spacecraft and the lunar module in low Earth orbit. The crew of James McDivitt, David Scott, and Rusty Schweickart put the two spacecraft through a rigorous set of tests by repeatedly docking and undocking the two spacecraft, firing the lunar module's engines, and separating the two craft in orbit by more than 100 kilometers and then re-rendezvousing them. Schweickart also performed a space walk to test the functionality of the new Apollo space suit designed for use on the lunar surface.

The goal of the next mission, flown in May 1969, involved repeating the *Apollo 9* mission, but in lunar orbit. In short, Thomas Stafford, John Young, and Gene Cernan of *Apollo 10* performed a full-scale dress rehearsal of the first lunar landing. While in orbit around the Moon, Stafford and Cernan transferred from the Apollo to the lunar module and then flew it to within 15.2 kilometers (50,000 feet) of the surface. Then, they rendezvoused with Young in the Apollo spacecraft for the trip home. This mission allowed NASA to gain valuable experience with the techniques necessary to carry out a successful Moon landing in the future. NASA would have scheduled *Apollo 10* to land on the Moon if the mission had occurred near the end of 1969. However, the unqualified success of *Apollos 7, 8,* and *9* put NASA ahead of schedule and allowed them to fly *Apollo 10* as a practice mission for "extra insurance."

Historical Fact

Although *Apollo 8* did not land on the Moon, many historians consider America's race to the Moon with the Soviets to have been won with the successful completion of this mission. Never before had humans left Earth orbit. Many NASA engineers who worked on Apollo considered *Apollo 8* as the highlight of their careers. On the other hand, the Soviet Union, in its unsuccessful attempt to land humans on the Moon, designed a rocket larger than the Saturn 5. This gigantic Soviet rocket employed over 30 engines on the first stage alone. However, it never flew successfully. All of them exploded within several minutes of leaving the launch pad.

A Giant Leap

NASA now stood ready to launch the *Apollo 11* mission to finally accomplish Kennedy's goal of landing astronauts on the Moon by the end of 1969. The crew for this historic flight consisted of Neil Armstrong, Michael Collins, and Buzz Aldrin. They named their Apollo spacecraft *Columbia* and the lunar module *Eagle*. Coincidentally, famous literary author Jules Verne wrote a book in 1865 called *From the Earth to the Moon,* about a fictional spacecraft called the *Columbiad*. In the book, he tells the tale of three astronauts aboard the *Columbiad* who were shot to the Moon from a gigantic cannon in Florida. Verne had no idea how faithfully twentieth-century history would eventually replicate his nineteenth-century prose.

Rule of Thumb

NASA always assigned a full backup crew to each Apollo mission in case one of the prime crew members became injured or ill. In general, the backup astronaut crew for a given Apollo mission was assigned as the prime crew three missions later. Astronauts called this rule, "back one, skip two, fly one."

Apollo 11 left the Earth for the Moon on the morning of 16 July 1969. A little more than three days later, the three astronauts and their two spacecraft arrived at the Moon and entered orbit. Armstrong and Aldrin then climbed into the *Eagle* for the trip to an area called the Sea of Tranquility, located near the lunar equator. That trip almost turned into a disaster. Midway through the descent, the *Eagle's* computer issued a 1201 executive overflow alarm. The 1201

Above and Opposite Page:

Dramatic four-photograph sequence taken from the launch tower captures the historic moment when the *Apollo 11* astronauts left the Earth for the Moon. Liftoff of the gigantic Saturn 5 rocket from Launch Complex 39-A at the Kennedy Space Center occurred on 16 July 1969 at 9:32 a.m. Eastern Daylight Time.

was the computer's cryptic way of informing the astronauts that it was being worked beyond maximum capacity. Then, as they approached the designated landing site, Armstrong found it covered by a field of large jagged boulders. Fortunately, Armstrong and Aldrin found an alternate site in time. Instruments showed that they landed with less than 30 seconds worth of propellant remaining in the *Eagle's* tanks. Upon landing, Armstrong announced to the world over the radio, "Tranquility base here, the *Eagle* has landed."

At 10:56 P.M. eastern time on 20 July 1969, roughly six hours after landing, Armstrong climbed out of the lunar module and stepped onto the surface of the Moon. He said upon stepping onto the lunar dirt, "That's one small step for man, one giant leap for mankind." Those words will remain as one of the most famous quotes throughout history. Aldrin followed Armstrong onto the surface a few minutes later. In the two hours that they spent outside the *Eagle*, the two astronauts planted the American flag, collected surface rocks, drilled into the surface to collect deeper samples for study by NASA geologists back on Earth, and took pictures of everything they saw.

An estimated 600 million people around the world watched this historic event unfold live on television. Michael Collins, the astronaut who remained with the Apollo spacecraft in lunar orbit, was probably the only person with an interest in the landing who did not get to see it live on television. He had to settle for listening to the events over the radio link.

In November of the same year, Pete Conrad, Dick Gordon, and Alan Bean of *Apollo 12* repeated *Apollo 11's* mission. However, *Apollo 12* landed in a different area called the Ocean of Storms. Conrad said upon stepping onto the Moon, "That may have been a small step for Neil, but it's a long one for me." He and Bean more than doubled the amount of time spent outside the lunar module by Armstrong and Aldrin. Altogether, the *Apollo 12* surface crew made two separate excursions separated by a period of rest inside the lunar module. During the first excursion, they set up experiments that measured the Moon's seismicity, and also deployed devices to measure the flux of atomic particles from the Sun. On the second excursion, they hiked to the landing site of a NASA spacecraft called *Surveyor*.

Historical Quote

"That's one small step for man, one giant leap for mankind."

Astronaut Neil Armstrong on 20 July 1969 after his first step on the Moon

Above:
20 November 1969 - *Apollo 12* astronaut Pete Conrad inspects the soil scoop on spacecraft *Surveyor 3* during NASA's second Moon landing mission. That spacecraft landed on the Moon in 1967. It carried no astronauts, but transmitted valuable scientific data back to Earth. *Apollo 12* landed near *Surveyor*, and the astronauts brought pieces of it back to Earth for analysis.

Opposite Page Top:
20 July 1969 - *Apollo 11* astronaut Buzz Aldrin climbs out of the lunar module to join Neil Armstrong on the surface of the Moon on the first lunar landing mission.

Opposite Page Bottom:
During the first lunar landing, astronaut Buzz Aldrin sets up equipment on the Moon to gather scientific data. An American flag stands proudly on the lunar surface in the center background of the picture (near the horizon) and next to the lunar module (upper right-hand corner of the picture). Neil Armstrong took the picture. In all the *Apollo 11* pictures from the Moon, none show Armstrong because he carried the only camera. In all the excitement, they forgot to take his picture.

This automated machine had previously landed on the Moon in 1966. Conrad and Bean removed parts of *Surveyor* to bring back to Earth for analysis. Scientists were interested to find out how the materials held up after exposure to the lunar vacuum for several years.

Unlucky 13

Apollo 13 flew in April of 1970 with James Lovell, Fred Haise, and Jack Swigert aboard. Two days after the launch and well on the way to the Moon, the oxygen tank in the Apollo's service module exploded. Swigert radioed back to mission control in Texas, "Okay, Houston, we've had a problem here." That statement turned out to be one of the greatest understatements of the century. Not only did the oxygen tank supply air for the astronauts, it was also mixed with liquid hydrogen to produce electricity for the spacecraft's instruments. The explosion left the Apollo spacecraft without power, and with a rapidly dwindling supply of air. To make matters worse, the laws of orbital mechanics precluded making an immediate U-turn for a return to Earth. The only way to return involved flying around the Moon and using a combination of its gravity and an engine burn to slingshot the crew back home.

The crew took refuge inside the lunar module that they had named *Aquarius*. Unfortunately, *Aquarius* only provided air for two astronauts for 48 hours while the return trip required 87 hours. The mission control engineers at Houston, Texas, helped to save the day by inventing a carbon dioxide-removing device that the astronauts could build using spare parts in the spacecraft such as plastic hoses, cardboard covers from notebooks, and electrical tape. Without it, the crew would have suffocated long before returning home. The astronauts later described the four-day return trip as cold, uncomfortable, miserable, and tense. Fortunately, they arrived without harm. *Apollo 13* proved the space program's ability to deal with a major unexpected crisis successfully.

America returned to the Moon in February of 1971 with the *Apollo 14* crew of Alan Shepard, Stuart Roosa, and Edgar Mitchell. Shepard, the first American in space, became the only Mercury astronaut to fly to the Moon. On this flight, Shepard and Roosa landed in the Fra Mauro region, a location only a handful of kilometers away from the *Apollo 12* landing site. They pulled a rickshaw-type cart along to carry their scientific equipment. The cart allowed them to gather more rocks for return to Earth than the *Apollo 11* and *12* missions combined. Unfortunately, a planned excursion to the 305-meter (1,000-foot)-wide Cone Crater was canceled when the astronauts lost their way on the lunar surface. Although later estimates showed that they had made it to within 30.5 meters (100 feet) of the crater's rim, the two had become disoriented in the alien, lunar terrain.

Above:

1 August 1971 – On the *Apollo 15* mission, astronaut James Irwin salutes the American flag. They kept the flag flying proudly on the airless Moon using a stiff wire. In the picture, the lunar module rests in the middle background in front of the mountain, and the lunar rover sits off to the right of the lunar module. The Hadley Delta Mountain rises in the background more than 4,000 meters (13,124 feet) in altitude above the *Apollo 15* landing site. About five kilometers separates the base of the mountain from the landing site. This photograph was taken by David Scott, the mission commander of *Apollo 15*.

The J Missions

Apollo 14 marked the last flight where the mission objectives primarily involved a safe landing on the Moon. Now, NASA turned to missions involving multiday stays on the Moon. Such stay times allowed the astronauts to accomplish much more scientific research and geological exploration than previously possible. These scientific flights represented what NASA code-named "J" missions. The "J" came from an alphabetical ordering scheme based on the increasing complexity of the mission objectives. For example, *Apollo 8* was code named "C prime," *Apollo 9* was "D," *Apollo 11* was "G," and *Apollos 12, 13,* and *14* were all "H" missions.

Apollo 15 marked the first mission of the "J" series. On this August 1971 flight, David Scott (from *Apollo 9*), James Irwin, and Alfred Worden flew to the Moon. Scott and Irwin landed at the Hadley Rille plains, a mountainous region higher than the Sierra Nevada in California. During the mission, they drove around using a lunar rover jeep designed specifically for use on the Moon. This lunar jeep allowed them to travel more than 27 kilometers during their geological expeditions. In total, Scott and Irwin explored more of the Moon's

Figure 40: *Apollo Mission Log*

Apollo 9 Mission Summary

Liftoff Date	Splashdown Date
3 March 1969	13 March 1969
Command Module Name	**Lunar Module Name**
Gumdrop	*Spider*

Crew Members
Mission Commander, James McDivitt (left)
Lunar Module Pilot, Rusty Schweickart (middle)
Command Module Pilot, David Scott (right)

Previous Space Experience
McDivitt, flew on *Gemini 4*
Schweickart, no previous missions
Scott, flew on *Gemini 8*

Significant Events on Mission
First flight of lunar module, tested lunar module in low Earth orbit

Apollo 10 Mission Summary

Liftoff Date	Splashdown Date
18 May 1969	26 May 1969
Command Module Name	**Lunar Module Name**
Charlie Brown	*Snoopy*

Crew Members
Mission Commander, Thomas Stafford (middle)
Lunar Module Pilot, Gene Cernan (left)
Command Module Pilot, John Young (right)

Previous Space Experience
Stafford, flew on *Gemini 6* and *Gemini 9*
Cernan, flew on *Gemini 9*
Young, flew on *Gemini 3* and *Gemini 10*

Significant Events on Mission
Tested lunar module in lunar orbit, first color TV from a space mission

Apollo 11 Mission Summary

Liftoff Date	Splashdown Date
16 July 1969	24 July 1969
Command Module Name	**Lunar Module Name**
Columbia	*Eagle*

Crew Members
Mission Commander, Neil Armstrong (left)
Lunar Module Pilot, Buzz Aldrin (right)
Command Module Pilot, Michael Collins (middle)

Previous Space Experience
Armstrong, flew on *Gemini 8*
Aldrin, flew on *Gemini 12*
Collins, flew on *Gemini 10*

Significant Events on Mission
First humans to land and walk on the Moon

Above:

11 December 1972 - On the last Apollo Moon mission *(Apollo 17)*, astronaut Gene Cernan takes a test drive in the lunar rover prior to loading it up with equipment. The rover, when fully loaded, carried a television camera, communications antennas, and scientific gear for geological exploration. At top speed, the rover moves along at a mere 18 kilometers (11 miles) an hour. This photograph was taken by scientist-astronaut Dr. Harrison Schmitt.

surface than the *Apollo 11, 12,* and *14* missions combined. They also brought back the greatest geological treasure of the Apollo program, a 4.5-billion-year-old sample of ancient lunar crust nicknamed the "Genesis Rock."

NASA closed out the Moon landings with two more missions. The *Apollo 16* crew of John Young (from *Apollo 10*), Thomas Mattingly, and Charlie Duke reached the Moon in April of 1972. At first, a malfunction in the Apollo spacecraft's engine control system almost caused mission control to cancel the landing. However, the problem was quickly solved, and Young and Duke landed at an area called the Descartes Highlands. Scientists chose this region because of suspected volcanic activity long ago in the Moon's history. In total, they spent a record 71 hours on the lunar surface and took three separate exploration excursions using the lunar rover. Although the test for previous volcanic activity yielded mixed results, the *Apollo 16* mission discovered the highlands to be extremely rich in aluminum. This discovery reaffirmed the beliefs of many scientists that humans may eventually mine the Moon to extract valuable natural resources and ores.

The crew of the final Moon flight, *Apollo 17*, consisted of Gene Cernan (from *Apollo 10*), Ronald Evans, and Harrison Schmitt. In the middle of December of 1972, Cernan and Schmitt spent close to three days conducting geological surveys in a region of the Moon called the Taurus Littrow Valley. Schmitt, a professional geologist, became the first American to fly into space who did not make a living by being a pilot. Altogether, the two on the Moon collected a record 109 kilograms (240 pounds) of lunar rock for return to the Earth, roamed for more than 33 kilometers (20.5 miles) on the lunar rover, and discovered orange-colored soil caused by a venting of oxidized iron ore in gaseous form. In the opinion of scientists back on Earth, this last mission to the Moon also proved to be the most productive. In total, the Apollo astronauts who went to the Moon brought back close to 385 kilograms (850 pounds) worth of lunar rocks, thousands of photographs, and miles of computer tapes containing geophysical data gathered from lunar orbit. These lunar samples and data allowed scientists to finally study the chemical composition of Earth's closest celestial neighbor.

Historical Fact

The first three Apollo crews that landed on the Moon were quarantined for two to three weeks after returning to Earth. NASA wanted to make sure that they did not bring back any unknown viruses or bacteria. This practice was dropped for the last three Moon landing missions.

Scientists found that the rocks on the Moon and the rocks on the Earth contain the same chemical elements, but in different proportions. For example, Moon rocks contain more titanium, aluminum and calcium. They also contain more elements with high melting points such as hafnium and zirconium. However, Earth rocks contain elements with lower melting points such as sodium and potassium. Consequently, some scientists speculate that billions of years ago, the Moon may have formed in a higher temperature environment than the Earth. Such a theory may explain the absence of water in the lunar samples. However, evidence suggests that water may exist in the soil near the lunar poles. Since the end of the Apollo flights, more than

Figure 41: *Apollo Mission Log*

Apollo 12 Mission Summary

Liftoff Date	**Splashdown Date**
14 November 1969	24 November 1969
Command Module Name	**Lunar Module Name**
Yankee Clipper	*Intrepid*

Crew Members
Mission Commander, Pete Conrad (left)
Lunar Module Pilot, Alan Bean (right)
Command Module Pilot, Richard Gordon (middle)

Previous Space Experience
Conrad, flew on *Gemini 5* and *Gemini 11*
Bean, no previous missions
Gordon, flew on *Gemini 11*

Significant Events on Mission
Explored lunar region Ocean of Storms, visited *Surveyor 3*

Apollo 13 Mission Summary

Liftoff Date	**Splashdown Date**
11 April 1970	17 April 1970
Command Module Name	**Lunar Module Name**
Odyssey	*Aquarius*

Crew Members
Mission Commander, James Lovell (left)
Lunar Module Pilot, Fred Haise (right)
Command Module Pilot, John Swigert (middle)

Previous Space Experience
Lovell, flew on *Gemini 7*, *Gemini 12*, and *Apollo 8*
Haise, no previous missions
Swigert, no previous missions

Significant Events on Mission
Explosion on Apollo spacecraft halfway to Moon cancelled landing

Apollo 14 Mission Summary

Liftoff Date	**Splashdown Date**
31 January 1971	9 February 1971
Command Module Name	**Lunar Module Name**
Kitty Hawk	*Antares*

Crew Members
Mission Commander, Alan Shepard (middle)
Lunar Module Pilot, Stuart Roosa (left)
Command Module Pilot, Edgar Mitchell (right)

Previous Space Experience
Shepard, flew on *Mercury 3*
Roosa, no previous missions
Mitchell, no previous missions

Significant Events on Mission
Explored Fra Mauro region on Moon, first lunar golf game

► Figure 42: *Apollo Mission Log (J Missions)*

Apollo 15 Mission Summary

Liftoff Date	**Splashdown Date**
26 July 1971	7 August 1971
Command Module Name	**Lunar Module Name**
Endeavour	*Falcon*

Crew Members
Mission Commander, David Scott (left)
Lunar Module Pilot, James Irwin (right)
Command Module Pilot, Alfred Worden (middle)

Previous Space Experience
Scott, flew on *Gemini 8* and *Apollo 9*
Irwin, no previous missions
Worden, no previous missions

Significant Events on Mission
Explored Hadley Rille region on Moon, first use of lunar rover

Apollo 16 Mission Summary

Liftoff Date	**Splashdown Date**
16 April 1972	27 April 1972
Command Module Name	**Lunar Module Name**
Casper	*Orion*

Crew Members
Mission Commander, John Young (middle)
Lunar Module Pilot, Charlie Duke (right)
Command Module Pilot, Thomas Mattingly (left)

Previous Space Experience
Young, flew on *Gemini 3*, *Gemini 10*, and *Apollo 10*
Duke, no previous missions
Mattingly, no previous missions

Significant Events on Mission
Explored Descartes highlands region on Moon, 71 hour stay on Moon

Apollo 17 Mission Summary

Liftoff Date	**Splashdown Date**
7 December 1972	19 December 1972
Command Module Name	**Lunar Module Name**
America	*Challenger*

Crew Members
Mission Commander, Gene Cernan (bottom)
Lunar Module Pilot, Harrison Schmitt (left)
Command Module Pilot, Ronald Evans (right)

Previous Space Experience
Cernan, flew on *Gemini 9* and *Apollo 10*
Schmitt, no previous missions
Evans, no previous missions

Significant Events on Mission
Last lunar landing, explored Taurus-Littrow Valley on the Moon

100 scientific teams from around the world have studied the lunar samples.

After the Moon

NASA had scheduled three more lunar landings after *Apollo 17*. Unfortunately, budget cuts forced them to cancel the flights of *Apollo 18, 19,* and *20*. Those missions would have been dedicated to lunar scientific research like the *Apollo 15, 16,* and *17* missions. Instead, NASA focused on Earth-orbit applications for the next batch of Apollo missions by launching the space station *Skylab* in 1973. This spacecraft was built from the third stage of the giant Saturn 5 rocket and served as an orbital laboratory for astronauts. *Skylab*'s interior contained more open space than a small one-bedroom house. However, as compared to the Apollo capsule, *Skylab*'s accommodations were luxurious. This space station contained a dining room table, three separate sleeping areas, a bathroom, and a "zero-gravity" shower.

Above:

The *Skylab 3* crew of Alan Bean, Owen Garriott, and Jack Lousma inspect the outside of the *Skylab* space station during a "fly around" before docking. The white appendages at the top (object that looks like blades from a windmill) and left edge of the photograph are solar panels that provide electricity to the space station. During this mission, the three astronauts spent fifty-nine days in Earth orbit conducting scientific experiments. This photograph was taken on 28 July 1973.

In 1973 and 1974, NASA launched three Apollo spacecraft to dock with *Skylab* on three separate occasions. Each mission involved a crew of three astronauts. The first crew stayed for twenty-eight days, the second stayed for fifty-nine days, and the third stayed for eighty-four days. Each mission showed that astronauts could function in the weightless environment of microgravity for long periods of time. The previous space endurance record of fourteen days was set by the crew of *Gemini 7* in 1965. While on *Skylab*, the astronauts performed medical experiments, studied the Sun, conducted astronomical observations, and used various types of sensors to image the Earth for the purpose of determining the locations of hidden natural resources.

NASA finished the Apollo program in 1975 with a historic mission. On 15 July, they launched an Apollo spacecraft carrying Tom Stafford, Vance Brand, and Deke Slayton. They rendezvoused and docked with a Soviet spacecraft carrying Alexi Leonov and Valeri Kubasov. The two spacecraft remained docked for close to two days. During that time, the two crews exchanged flags and gifts, visited each other's spacecraft, and conducted joint scientific experiments. Project Apollo, initiated in the middle of the cold war, ended with a mission of goodwill and international cooperation. ❑

USA
NASA
Atlantis

On Twin Pillars of Fire

Chapter 5: How the Space Shuttle Works

Throughout the 1960s, President John F. Kennedy's bold vision of tomorrow for America, the safe landing of astronauts on the Moon, set the major theme for the space program. In 1969, tomorrow suddenly arrived with the successful completion of the Apollo 11 mission. Five more triumphant landings followed over the next three years. During that time, NASA officials debated the future direction of American exploration of the final frontier. The space agency proposed a reusable space plane to shuttle astronauts and satellites to and from orbit, a large space station in orbit dedicated to science, and humans on Mars by 1980. Unfortunately, the late 1960s ushered in an era of social unrest in America. Burning issues on the home front such as the country's involvement in a protracted guerrilla war in Vietnam stifled national interest in space science. The only portion of NASA's post-Apollo plan that survived national apathy and congressional budget cutting was the reusable space plane, more commonly known as the Space Shuttle.

Opposite Page:

3 October 1985 – Space Shuttle *Atlantis* rockets into space from NASA's Kennedy Space Center on its maiden voyage. This mission was the 21st in Space Shuttle history.

5.1 What Is the Space Shuttle

NASA engineers and rocket scientists of the 1960s found that the cheapest and easiest method to send astronauts into space involved strapping them tightly into small metal capsules, hurling the capsules into orbit, and bringing the so-called spacecraft back to Earth on a flaming descent path through the atmosphere that nearly melted the capsule. The astronauts survived, but the capsule did not. Consequently, every mission required a new spacecraft. The rockets used to hurl the capsule into orbit also failed to survive. Most of them fell back into the ocean and sank. Eventually, the engineers improved capsule design and sophistication to the point where they sent astronauts to the surface of the Moon and brought them back safely. Unfortunately, these new designs represented the peak evolutionary point of an old idea. NASA saw that a new method of spaceflight was needed to replace the antiquated and uneconomical method of the throwaway capsule.

The Space Shuttle represents NASA's first attempt at making space easier and more affordable to reach. Unlike any other piloted spacecraft in history, a Space Shuttle soars into space vertically like a rocket, returns to Earth and lands on a runway like an ordinary airplane, and returns to fly into space another day. Just as important, NASA also recycles the rockets that carry the shuttle into space. Never before in the history of spaceflight have spacecraft, rockets, and rocket engines found their way onto more than one flight in their operational lifetime.

Above:

Not only do Space Shuttle orbiters look like airplanes, they also land on runways like airplanes. In this photograph, Space Shuttle *Columbia* looks a little worse for wear after its November 1982 landing at Edwards Air Force Base, California, on its fifth mission. The orbiter's reflection at the bottom of the photograph was caused by a puddle formed by rainstorms a few days before the landing. Many of the runways at Edwards lie on the hardened clay surface of a large, dry lake bed. Unfortunately, the runways tend to collect water during storms. Today, NASA prefers to land orbiters at the Kennedy Space Center in Florida.

RULE OF THUMB

The main rocket engines on the orbiter are designed only to propel the spacecraft into orbit. The orbiter does not have jet engines to fly within the atmosphere like a plane. When the orbiter lands, it glides to the runway unpowered.

While in low Earth orbit, shuttle astronauts perform various tasks. They can deploy satellites and auxiliary rockets to carry the satellites to geosynchronous altitude, deploy payloads in low Earth orbit that weigh too much for other rockets to carry, rendezvous with and repair damaged satellites, retrieve satellites from orbit for repair on Earth, or conduct scientific experiments. Currently, astronauts are constructing the world's largest space station in order to conduct scientific research in orbit on a permanent basis. Clearly, the shuttle is the most versatile spacecraft ever to circle the Earth. However, development costs soared higher than predicted, operational costs exceeded projected values, and the number of missions flown fell below the planned schedule. NASA also discovered that using the shuttle to launch satellites often costs more than using the "throwaway" rockets of the past. For these reasons, NASA restricts shuttle payloads to those too heavy for launch on other rockets, or those that require a human presence in space.

The orbiter is the part of the Space Shuttle system that looks like an airplane (see Figure 43 on page 153). Its size is similar to that of a standard medium-size commercial passenger jet such as the DC-9, MD-80, or Boeing 727. However, an orbiter is smaller than the large jumbo-jet airliners like the DC-10 or Boeing 747 that serve international flights or crowded domestic routes between airline hub cities. The orbiter's similarities to medium-size passenger jets end with its size. On a standard commercial jet, the passengers, crew, and payload (the luggage) occupy the entire length of the aircraft. Passengers occupy the upper deck, and the luggage rides in the belly of the plane beneath the passenger deck. Astronauts on the orbiter only occupy an area in the front end of the spacecraft called the *crew cabin*. During a mission, the crew cabin is the only area of the orbiter that contains pressurized air for the astronauts to breathe. The current cabin design supports a crew size of up to seven astronauts for missions lasting between ten to fifteen days. Payload storage space takes up the rest of the orbiter's body.

To date, six orbiters have been constructed. The first, *Enterprise* (code number OV-101), never flew in space. Instead, *Enterprise* rode on top of a 747 jumbo jet to an altitude of about seven kilometers (23,000 feet), and was dropped in midair. Test pilots onboard *Enterprise* then guided it to an unpowered runway landing to test the aerodynamic properties of the orbiter design. The *Enterprise* was named after Captain James T. Kirk's starship on the television series *Star Trek* after thousands of the show's fans wrote to NASA. Much to their disappointment, the fans did not realize that NASA never intended to send the *Enterprise* into space.

Five operational orbiters followed the *Enterprise*. The first one, *Columbia* (OV-102), flew into space for the first time in 1981. *Challenger* (OV-99) followed in 1983, *Discovery* (OV-103) in 1984, and *Atlantis* (OV-104) in 1985. NASA named these orbiters after famous exploring ships in American history. After *Challenger* was destroyed during a mission in 1986, Congress authorized the appropriation of a

Figure 43: *What the Orbiter Looks Like*

What:	Currently, it is America's only piloted spacecraft. The shuttle orbiter provides routine access to space for a variety of missions. Also the world's first reusable spacecraft	NASA's Fleet:	Four in operation (*Atlantis*, *Columbia*, *Discovery*, *Endeavour*), one test vehicle never flown in space (*Enterprise*), one lost in the line of duty (*Challenger*)
Crew:	Between two to seven astronauts	Orbit:	28° to 57°, low Earth altitudes only
Mission Duration:	Standard lengths between seven to ten days. Longer missions possible	Cost:	Each orbiter costs about two billion dollars to manufacture
Payload Capacity:	23,500 kg (51,800 lbs) to 28° orbit, 17,000 kg (37,800 lbs) to 57° orbit	Launch Site:	Kennedy Space Center (KSC) in Florida, on the Atlantic coast

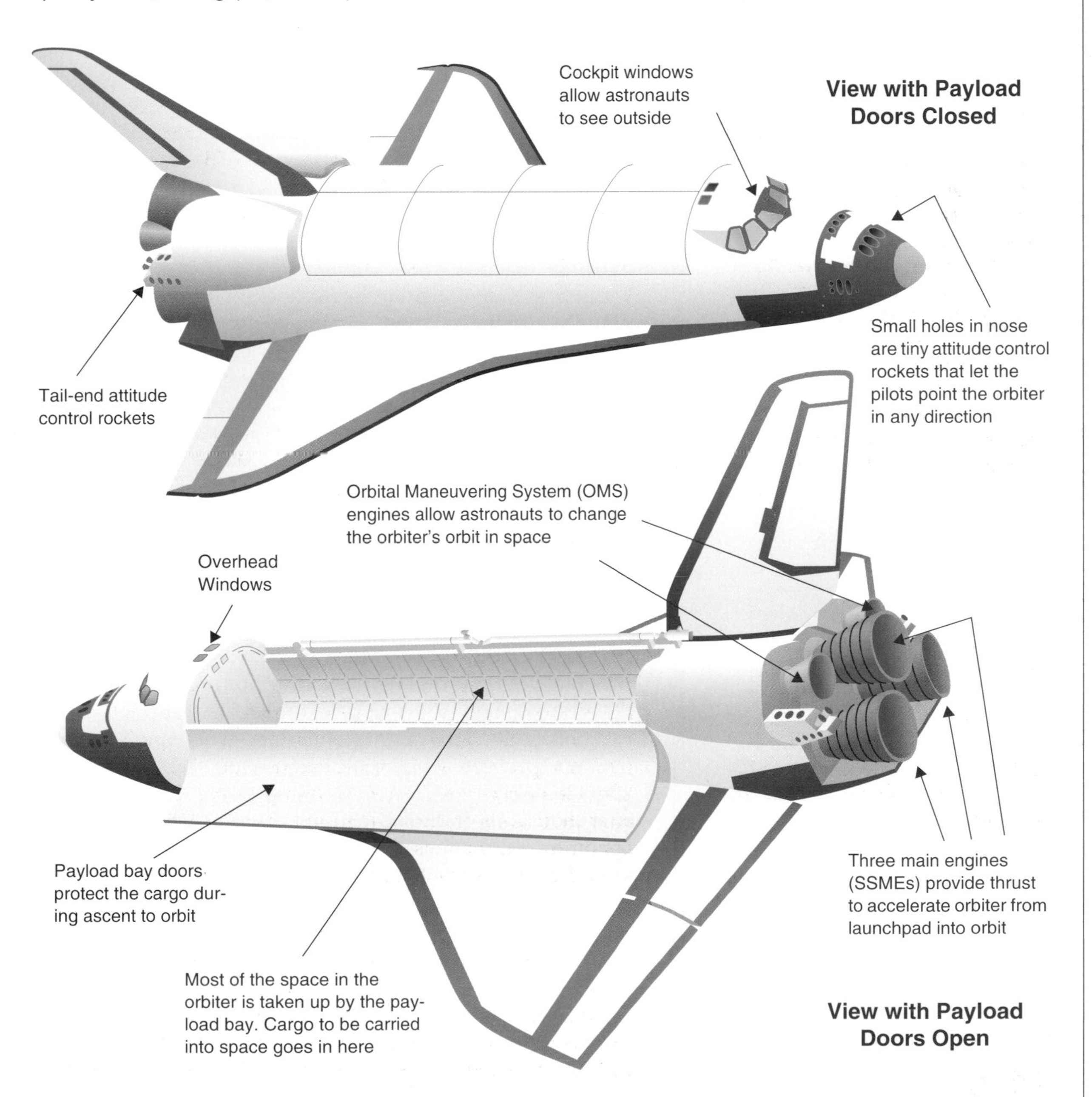

RULE OF THUMB

If you think flight numbers for commercial airlines seem confusing, try figuring out NASA's scheme for numbering shuttle missions. NASA assigns all shuttle flights an STS (short for Space Transportation System) number. The first nine missions were called STS-1 through STS-9 and flew in sequential order.

Starting with the 10th mission, NASA switched to a different system to try and convey more information. Flight number 10 was renamed STS-41B. The first number was the last digit in the fiscal year in which the launch was originally scheduled. For example, the "4" in 41B stood for fiscal 1984. The second number designated the launch site. In this case, the "1" in 41B indicated that the shuttle launched from the Kennedy Space Center in Florida. A "2" stood for Vandenberg Air Force Base in California. However, no shuttles were ever launched from there. Finally, the letter at the end indicated the order of launch within the fiscal year with "B" meaning the second launch of the year.

After the *Challenger* accident (flight number 25, but called STS-51L), NASA returned to its original system. The next flight was numbered STS-26. In general, flight numbers are assigned about 19 months ahead of time and do not change even when the mission is delayed. For example, STS-51L flew in 1986, and STS-41 flew before STS-38.

SSME stands for Space Shuttle main engine. These rocket engines are the most sophisticated ever built. Each one is designed to last for up to 55 launches.

sixth and final orbiter. That orbiter, *Endeavour* (OV-105), first flew in May 1992. Currently, NASA does not plan to build any new orbiters.

Each orbiter received an identification number. "OV" stands for *orbiter*, and a higher number indicates a newer orbiter. Originally, numbers less than 100 indicated test orbiters not destined for spaceflight, and numbers greater than 100 indicated operational orbiters. Indeed, *Challenger* (OV-99), not *Enterprise*, was originally built as a test vehicle. However, sometime during construction, NASA decided to make *Challenger* into an operational orbiter.

One of the orbiter's many functions includes carrying large-size payloads into space. The orbiter's payload bay (located behind the crew cabin) measures 18.3 meters (60 feet) long, 5.2 meters (17 feet) wide, and 4 meters (13 feet) deep. Items as large as a Greyhound passenger bus or several Indian elephants can easily fit in the bay with room to spare. Most of the time, the bay stores satellites that the orbiter delivers to space, or a small laboratory module that gives astronauts room to conduct scientific experiments. Two doors that stretch the entire length of the payload bay protect the contents of the bay from aerodynamic forces during the ascent to orbit and descent from orbit. Even if there is no payload to protect, it is absolutely essential that the doors remain closed during ascent and descent because the aerodynamic forces present during these phases of the flight are strong enough to rip the doors off the hinges.

A cluster of three rocket engines, called the Space Shuttle main engines (SSMEs), provides some of the propulsion required to reach orbit. Figure 43 on page 153 shows the location of the three main engines on the rear end of the orbiter. These engines represent the most sophisticated ever built and they burn a cryogenic combination of liquid hydrogen fuel and liquid oxygen oxidizer. During normal operations, a single SSME turbo-pump sends liquid hydrogen to the engine at a rate of 171,396 liters (45,283 gallons) per minute. This rate is fast enough to fill a midsize backyard swimming pool in less than one minute.

Each main engine is rated to produce 1,668,000 Newtons (375,000 pounds) of thrust at sea level. This thrust level means that each one of the three main engines has the power to cause the equivalent mass of 188 small cars (approximately 166,800 kilograms or 375,000 pounds) to hover in midair. The main engines are also designed to be throttled. Most liquid-fueled rocket engines developed before the Space Shuttle had two modes of operation. They were either completely shut off or thrusting at 100% capacity. A Space Shuttle main engine possesses the capability to thrust at as little as 65% of the rated thrust, or up to 109% of the rated thrust for emergency purposes.

An orbiter designed with internal tanks to contain all the propellant necessary to achieve Earth orbit would be much too large for practical purposes. Instead, an orbiter stores all of its propellant used

Figure 44: *A Look at the Complete Space Transportation System*

Purpose:	Boosts the orbiter from the launchpad to low Earth orbit
Key Parts:	One orbiter (OV), one external tank (ET), and two solid rocket boosters (SRBs)
Fate of Parts after Use:	Orbiter is reused, SRBs are recovered and its parts are refurbished; ET burns up in the atmosphere
First Flew:	12 April 1981; more than 100 successful launches as of June 1999; only one launch failure in history
Length:	56 meters (184 feet) tall
Width:	24 meters (79 feet) wide as viewed from the front
Depth:	23 meters (75 feet) wide as viewed from the side
Weight:	About two million kilograms (4.4 million pounds) when fully loaded with propellant on the launchpad
Launch Rate:	Currently, NASA plans call for about six to seven launches per year

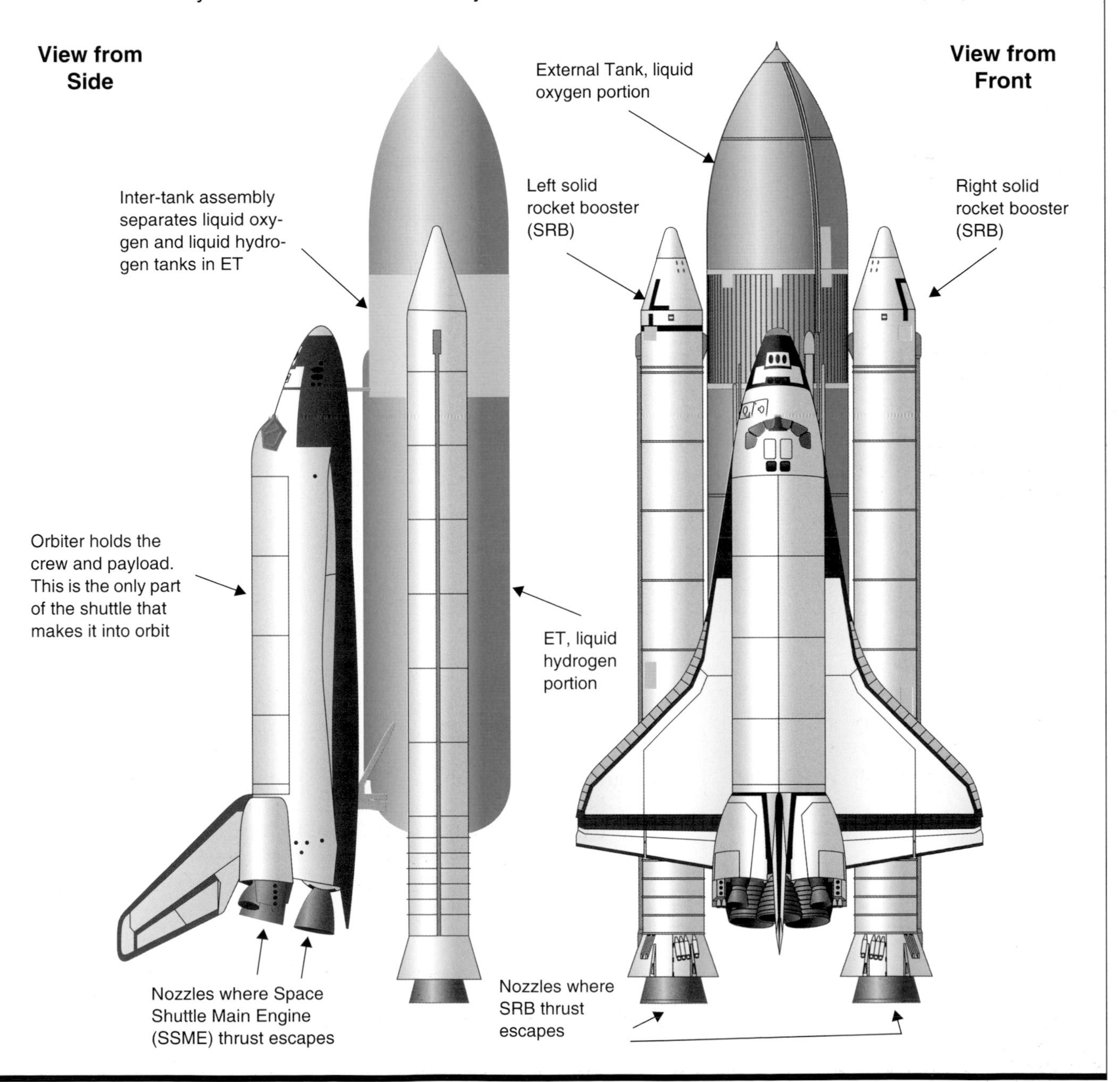

▶ Figure 45: *Why the Shuttle Needs Solid Rockets*

Key Fact:
A rocket needs to produce an amount of thrust in excess of its weight in order to rise off of the launchpad. Without the solid rocket boosters (SRBs), the shuttle would be unable to reach orbit.

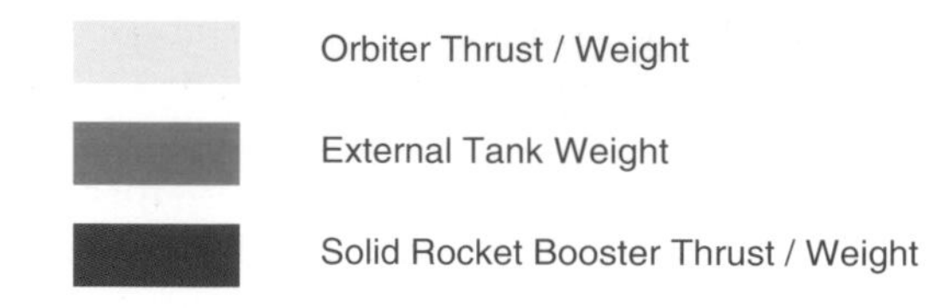

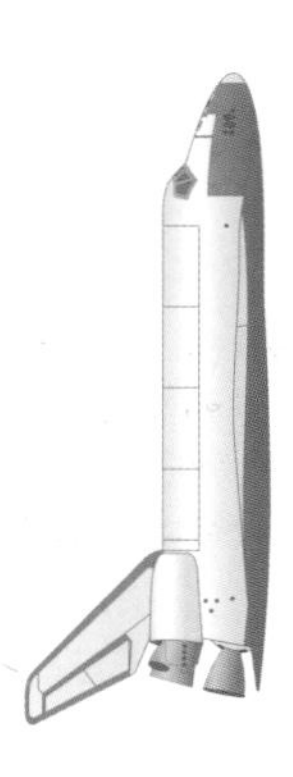

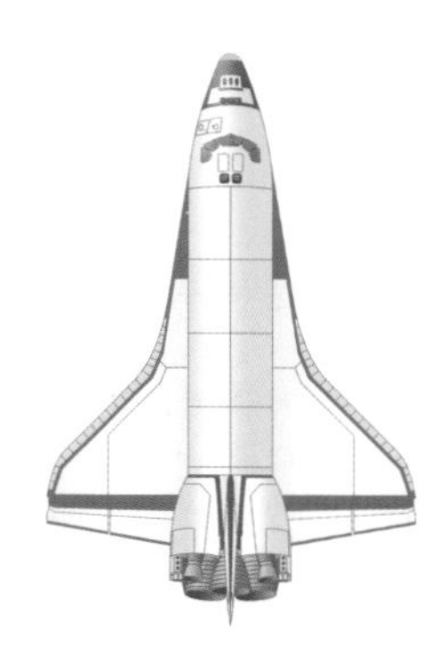

The orbiter's three main engines (SSMEs) produce enough thrust to lift the orbiter. However, the orbiter by itself does not carry enough propellant to reach orbit.

Thrust

Weight

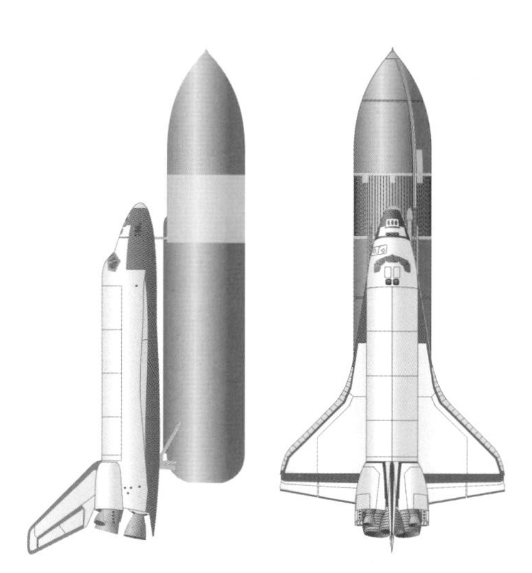

Once the extra propellant is added, the weight of the propellant and the external tank (ET) exceeds the thrust capacity of the three main engines.

Thrust

Weight

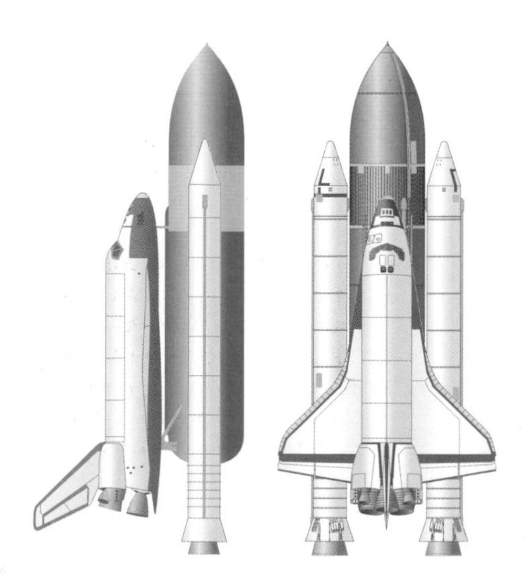

Two solid rocket boosters (SRBs) provide the additional thrust. They are powerful enough to lift their own weight and have enough thrust left over to lift the orbiter and the ET. Alone, the two SRBs could lift the entire Space Shuttle without the help of the SSMEs. However, the SSMEs must also help thrust because the SRBs do not contain enough propellant to send the shuttle into orbit.

Thrust

Weight

during ascent in a large brownish orange–colored tank, carries the tank piggyback style during ascent, and then jettisons the tank after consuming all of the propellant. This tank, called the *external tank* (ET), measures 48 meters (154.2 feet) long with a diameter of 8.4 meters (27.5 feet). In reality, an external tank consists of two tanks connected by a middle structure called the *inter-tank assembly* (see Figure 44 on page 155). The top tank contains 541,428 liters (143,060 gallons) of liquid oxygen and the bottom tank holds 1,449,905 liters (383,066 gallons) of liquid hydrogen.

ET stands for *external tank*. It looks like a giant orange bullet and contains the propellant for the SSMEs. The tank is the only major part of the Space Shuttle system that is not reused.

Although the hydrogen tank holds 2½ times the volume of the oxygen tank, it weighs only ¼ as much when both are filled to capacity because liquid oxygen weighs 16 times more than liquid hydrogen. A large, 43.2-centimeter (17-inch) pipe carries liquid oxygen to the orbiter. This pipe emanates from the inter-tank and runs down the side of the hydrogen tank to the bottom of the ET. Another pipe of the same size protrudes from the bottom of the hydrogen tank to deliver liquid hydrogen to the orbiter. Both of these pipes attach to the orbiter's bottom side near the main engines at a point called the *disconnect*.

SRB stands for *solid rocket booster*. Two of these attach to the external tank and help the SSMEs lift the shuttle for the first two minutes of flight. They are the largest solid propellant rockets ever built.

Boil-off presents a major problem when loading the cryogenic propellants into the ET before launch. The ambient temperature at the Space Shuttle launchpad typically exceeds 27° C (80° F), while the boiling point of liquid oxygen and hydrogen is roughly 200° below zero. To keep the boil-off rate at an acceptable level, a thin layer of spray-on polyisocyanurate foam covers the exterior of the ET. This foam gives the ET its brownish orange color and keeps ice from forming on the tank's structure. On the first two Space Shuttle launches, the foam was painted white to make the external tank match the orbiter for aesthetic purposes. However, this practice was discontinued because the several thousand pounds of paint needed to cover the tank serve no useful purpose from an engineering perspective.

The mass of the orbiter, payload in the orbiter, external tank, and all of the propellant in the external tank adds up to roughly 861,825 kilograms (1,900,000 pounds), which is the equivalent mass of 950 small cars. This amount of mass requires a minimum thrust of roughly 8,617,250 Newtons (1,900,000 pounds) just to lift it off of the launchpad and to have it hover in midair. In addition, enough additional thrust must be present to accelerate the vehicle from rest to orbital velocity. The three main engines on the orbiter, as powerful as they are, combine to produce only 5,004,000 Newtons (1,125,000 pounds) of thrust, enough only to cause an equivalent mass of 564 small cars to hover in midair. All three main engines on the orbiter, throttling at full power, could not begin to budge the orbiter and external tank off the launchpad, not to mention accelerating the vehicle to orbital velocity. As a result, two large solid rocket boosters (SRBs) augment the thrust from the main engines. Figure 45 on page 156 illustrates this concept.

Historical Fact

No rocket launches involving astronauts before the Space Shuttle had ever used solid rockets. Although solid rockets are simple, they cannot be shut off in the event of an emergency because they thrust until all of the propellant burns out. NASA had originally wanted to use liquid propellant boosters instead of solid ones, but Congress cut the Space Shuttle development budget in the early 1970s. NASA was then forced to introduce the SRBs into the design as a cost-saving measure.

RULE OF THUMB

Gimbaling involves rotating a rocket engine's nozzle to point in different directions. Such an action allows for some steering control during the ascent to orbit. Both the Space Shuttle main engines and the solid rocket boosters have the ability to gimbal.

STS stands for *Space Transportation System.* It is NASA's official name for the Space Shuttle.

Historical Fact

KSC stands for *Kennedy Space Center*, the launch site of all Space Shuttle missions. KSC lies on the Cape Canaveral National Seashore in Florida and used to be called "Cape Kennedy."

The space center is an hour's drive east of Orlando, Florida, and about an hour and a half south of Jacksonville. Although access to the center is restricted, NASA offers escorted tours that allow the general public to see most of the interesting facilities. The area around the center is also an excellent area to observe wildlife.

These SRBs, the largest solid rockets ever than designed, measure 45.5 meters (149.2 feet) in height. Each booster produces 11,790,000 Newtons (2,650,000 pounds) of thrust at liftoff, which constitutes enough force to cause an entire parking lot of more than 1,200 small cars to hover stationary in midair. A SRB consists of four individually manufactured segments containing propellant and a nose cone that stack vertically on top of each other to form the complete booster. The nose cone of the booster contains the propellant igniter, the booster's vital electronic devices, and the parachutes. A nozzle on the bottom segment directs the thrust and can gimbal by up to 7° to provide steering for the Space Shuttle during ascent.

The reason for manufacturing the booster in parts lies in the solid propellant fabrication process. Solid rocket propellant looks and feels similar to a hardened putty-like material or a rubber typewriter eraser. However, the propellant originates from a dense viscous plaster-like liquid that is poured inside the solid rocket body, where it is cured (dried) over several days. During the pouring and curing process, it is absolutely essential to keep the mixture uniform. For this reason, it is easier to fabricate solid rocket boosters in four different segments than as one long segment.

The chemical mixture that forms the solid propellant is identical in each segment. By weight, this mixture is 16% atomized aluminum powder fuel, 70% ammonium perchlorate oxidizer, 12% polybutadiene acrylic acid acrylonitrile that acts as a binding agent to hold the mixture together, and 2% epoxy, which helps during the curing process. Traces of iron oxide within the mixture act as a reaction catalyst to control the burn rate during flight. Each solid rocket booster contains 502,125 kilograms (1,107,000 pounds) of solid propellant. Despite this enormous weight, both SRBs thrusting together produce enough thrust to lift themselves off the launchpad, and still have enough thrust left over to help the main engines propel the orbiter and external tank towards orbital velocity.

Together, the solid rocket boosters, external tank, and orbiter form what NASA calls the National Space Transportation System or STS for short. The terms *orbiter* and *Space Shuttle* often confuse people because the press and other information sources often interchange the terms. Technically, an orbiter represents just one of the three major components of the Space Shuttle system. The full term *Space Shuttle* refers to the entire STS, which includes all three components.

5.2 Preparing for Flight

Every Space Shuttle journey to orbit begins in the middle of an alligator-infested Florida swamp at a place called the Kennedy Space Center (KSC). This NASA facility lies on the Atlantic coastline immediately adjacent to the United States Air Force's Cape Canaveral launch facility. To a bureaucrat, the distinction between the two facilities is clear. NASA, a civilian agency, runs the Kennedy Space Cen-

ter, and the military runs Cape Canaveral. However, to most observers, both launch centers seem to operate as one entity, as some NASA offices and spacecraft preparation facilities lie on the Air Force's side of the fence. Entry into either of the facilities is restricted to employees with valid security badges, but workers at KSC or the Cape may travel in between the two facilities without passing through a guard gate. For these reasons, members of the press often mistakenly refers to the entire place as "Cape Kennedy." No such facility exists in reality.

Most of the land at Kennedy Space Center and Cape Canaveral remains undeveloped because it is one of the largest remaining areas of wetlands along the Atlantic coast. Many species of wildlife, including many endangered species, make their homes and flourish on the space center's grounds. NASA sets extremely stringent regulations to protect the wildlife. For every acre of habitat destroyed to make room for Space Shuttle facilities, NASA policy calls for a good faith re-creation of two new acres of habitat somewhere else on the space center's grounds. Normal government policy only requires one-quarter of a new acre created for every acre destroyed for construction.

America's Space Port

#1 - Orbiter Processing Facility (OPF)
Set of three "high-tech" hangers where orbiters are readied for flight

#2 - Vehicle Assembly Building (VAB)
Gigantic building where orbiters are mounted onto the external tank and solid rockets

#3 - Launch Control Center (LCC)
Building used to control the launch countdown

#4 - Crawl-Way
5.5-kilometer road from VAB to launchpad

#5 - Launch Complex 39A
Shuttle launchpad near Atlantic beach

#6 - Launch Complex 39B
Shuttle launchpad near Atlantic beach

The road at the bottom of the photo leads to the Space Shuttle runway, several kilometers from the left edge of the picture.

Above:

At Edwards Air Force Base in California, the ground processing crew prepares to tow Space Shuttle *Columbia* off the runway in preparation for its return to Florida. The hoses that connect the rear of the orbiter carry coolant fluid and electrical power into the shuttle. This photograph was taken on 14 April 1981, shortly after the conclusion of the first flight in shuttle history (STS-1).

Opposite Page Top:

The Orbiter Processing Facility (OPF) serves as a "high-tech" maintenance hangar for orbiters. Here, a complex array of platforms surrounds the orbiter to allow engineers and technicians easy access to every part of the spacecraft. The platform array almost completely obscures the view of the orbiter. However, the nose end can be seen sticking out on the right edge of the photograph. This photograph of *Columbia* was taken on 25 September 1979, about 17 months before the first Space Shuttle mission.

Opposite Page Bottom:

After performing "routine maintenance" activities on the orbiter, engineers and technicians at the Kennedy Space Center begin to push Space Shuttle *Discovery* out of the OPF. Next, they will move the orbiter next door into the gigantic Vehicle Assembly Building and attach *Discovery* to its solid rockets and external tank. This photograph was taken on 21 June 1988, several months before the STS-26 mission.

Ground Processing

Launching Space Shuttles into orbit involves much more than strapping the astronauts into the orbiter and pushing the mystical "red firing button" in the launch control center. Every launch requires years of planning and months of "ground processing" activities to ready the orbiter, solid rockets (SRBs), and external tank (ET) for flight. The processing activities to prepare for a mission begin almost immediately after an orbiter lands from the previous mission. Several hours after landing, a tractor tows the orbiter off the runway (not shown in the Kennedy Space Center photograph). This tractor looks and functions like the standard tractors used to push passenger jets back from the gate at a commercial airport. The orbiter's specific destination is a shop facility called the Orbiter Processing Facility (OPF). This facility is a conglomeration of three nearly identical 31-meter- (100-foot-) tall buildings that serves as a combined hangar and maintenance shop for the orbiters. Each hangar, called an OPF high bay, houses one orbiter at a time.

After towing the orbiter into the OPF, engineers jack the entire vehicle up off its landing gears. The orbiter is then surrounded with a bewildering array of platforms and scaffolding that holds the spacecraft in place in midair. In addition, the scaffolding provides access to every conceivable location on the orbiter's surface. The complexity and density of the platform network obscures the view of the orbiter in many places. Large overhead cranes that run on rails mounted to the high bay ceiling lift heavy equipment on and off the platforms as well as large cargo items in and out of the orbiter's payload bay. One of the first tasks involves removing items from the payload bay used by astronauts on the previous mission.

Orbiters spend an average of four to six weeks in the OPF in between missions. During that time, every system on the orbiter goes through a thorough and meticulous inspection, followed by replacement or repair if necessary. Most of this work occurs in the OPF. However, some orbiter components may be sent to other KSC areas or other NASA centers for refurbishment. Such a process typically takes longer than six weeks. In these cases, engineers install refurbished components flown on previous missions on other orbiters.

Some of the specific activities that occur in the OPF include inspecting every one of the 32,000 tiles and replacing of those that are loose, damaged, or missing. These tiles are about the size of small paperback books. Together, they cover the outer surface of the orbiter for the purpose of preventing it from melting during reentry into the Earth's atmosphere from orbit. Other engineers remove the main engines as well as the associated fuel pumps and send them to other KSC areas for refurbishment. They may choose to reinstall the same engines, new engines, or refurbished engines from previous missions. Crews of electricians scour the inside of the orbiter to verify and repair electrical circuits, maintain the operational status of the onboard computers, and install replacement components for the

Above:

After the completion of ground processing activities on an orbiter, it is towed on its landing-gear wheels into the Vehicle Assembly Building (VAB) where it is joined with the solid rockets and external tank. In the picture, engineers and technicians closely monitor the late stages of *Discovery's* move from the OPF to the VAB. Notice that protective end pieces cover the mouth of the main engine nozzles. This picture was taken on 21 June 1988, about three months before the 26th Space Shuttle mission. It was the first mission after the *Challenger* accident in January 1986.

OPF stands for *Orbiter Processing Facility*. NASA uses these three buildings to store and maintain the orbiters.

VAB stands for *Vehicle Assembly Building*. It is extremely difficult to appreciate the sheer size of this building by looking at photographs. Since Kennedy Space Center lies in the middle of a swamp, there are no other large buildings with which to make a comparison. The inside of the building is almost entirely hollow. From inside, it is possible to look up and see the roof 50 stories above.

communications, guidance, and navigation systems. Other crews serve as state-of-the-art housekeepers and clean the inside of the orbiter. The task list is endless; an entire book could be written on all the activities that occur in the OPF. After all of the processing activities finish, engineers lower the orbiter down to the ground so that the spacecraft rests on its landing-gear wheels.

The next step involves towing the orbiter next door to the gigantic Vehicle Assembly Building (VAB). This large white monument to America's space program towers more than 160 meters (525 feet) above the Florida marshlands, equivalent to the height of the Washington Monument in Washington, D.C. or half the height of the World Trade Center in New York. Measured by volume, the VAB ranks as the largest building in the world under a single roof. In addition, the building contains enough open floor space inside to place an entire Major League Baseball stadium with room to spare, and enough volume to hold the equivalent space of more than two Empire State Buildings. In the VAB, engineers will eventually attach the orbiter to the two solid rocket boosters and the external tank.

Processing of the solid rocket boosters and external tank occur in parallel with activities in the OPF. Each solid rocket arrives at Kennedy split into four different segments. Most of the time, these segments have been used six months prior on a previous flight. During a mission, the solid rockets jettison two minutes after launch and fall into the Atlantic Ocean with the aid of parachutes. Two ships staffed with professional divers recover the boosters and tow them back to the Cape. Engineers then disassemble the boosters and send the segments back to the manufacturer in Utah by rail. There, the booster segments are reloaded with solid propellant and sent back to Florida on a train. Although the SRB segments will fit on an airplane, NASA safety engineers consider solid propellants too dangerous for shipment by air.

New external tanks travel to Kennedy from the manufacturing plant in New Orleans lying horizontally on a barge because the tanks are too large to travel to Kennedy on trucks or on an airplane. Every mission requires a new tank because the orbiter jettisons the tank shortly before reaching orbit, about eight minutes after launch. The used, empty tank then proceeds to burn up in the upper atmosphere over the Indian Ocean. Reloaded solid rocket segments and the new external tank typically arrive at KSC about half a year before the scheduled launch date. Technicians move these components into the VAB immediately upon arrival at Kennedy.

Assembly of all the Space Shuttle components into their launch configuration occurs on the side of the VAB closest to the ocean. The interior arrangement allows engineers to assemble two separate Space Shuttles for flight at any one time. This process, called *stacking* by those who work in the VAB, takes anywhere between four to six weeks. Stacking occurs on a mobile launch platform or MLP for short. This platform measures two stories tall, weighs 3.73

million kilograms (8.23 million pounds), and looks like a rectangular slab of metal with pipes and rails decorating the sides. Four large posts, one under each corner, hold the bottom of the MLP about two stories above the floor of the VAB.

Stacking begins with the transfer of the bottom segments of the SRBs to the mobile launch platform. Each bottom segment rests on several support posts on the MLP. These posts hold the segment upright so that the bottom of the segment lies slightly above the surface of the platform. Engineers then transfer the second segment of each booster and stack one on top of each bottom segment. This process continues until all four segments of each booster are stacked together, one on top of another. Now, the entire solid rocket booster mass rests on the support posts.

Assembling solid rocket booster segments on top of each other takes a lot of time and involves much painstaking work despite the use of gigantic cranes to move the segments. Together, all the seg-

Above:

May 1988 - Solid rocket boosters (SRBs) arrive from their manufacturer in four segments that must be joined together to form the complete booster. Here, technicians inside the Vehicle Assembly Building (VAB) prepare the bottom segment of the right-side SRB for the STS-26 mission. Later, the crane holding the booster segment will carry it to another part of the VAB where the three other segments will be stacked on top of this one.

Above:

Inside the cavernous Vehicle Assembly Building, workers complete the meticulous process of lifting Space Shuttle *Discovery* off the ground and into a vertical orientation. Once in this position, the orbiter will be lifted high off the ground by a gigantic crane and attached to its external tank.

Opposite Page:

A gigantic crane in the VAB lowers a Space Shuttle orbiter into place next to the external tank (dark object in the middle surrounded by the two solid rocket boosters that are painted white). The two bumps near the bottom of the tank are where the propellant pipes from the external tank attach to the orbiter. The pipe that runs down the length of the tank (on the right side) carries liquid oxygen from the top of the tank to the orbiter's main engines. To get an idea of the size of components involved here, look at the person standing on a platform in front of the left solid rocket booster near the middle of the picture.

ments must function as a single rigid body during flight. Connecting bolts must keep each segment in precise alignment with the segments below and above. Engineers literally spend days checking the alignment of the entire booster. Each boundary between two segments, called a *field joint*, requires a complete seal to prevent hot flames from the combustion of the propellant from escaping through the joints during flight. Three O-rings analogous to washers found in pipes line the inner side of the metal segment casing and help to seal each joint. Much time is sunk into verifying the tightness of the seals. A failure of one of the O-rings contributed to the destruction of the *Challenger* in January 1986.

The next step in the stacking process involves joining the external tank to the two solid rocket boosters. A gigantic crane capable of lifting 227,000 kilograms (499,400 pounds) suspends the tank in midair in between the two solid rockets while VAB engineers bolt the boosters to the tank. The engineers often refer to the process by which two components are bolted together as *mating*. Similar to the solid-rocket assembly process, the mating of the external tank to the solid rockets involves several days of painstaking work to verify a proper alignment of the tank between the boosters.

After mating the external tank to the solid rockets, the VAB engineers turn their attention to mating the orbiter with the external tank. Two large metallic slings fit over the orbiter like a harness to allow the gigantic 227,000-kilogram crane to lift the spacecraft. The crane initially lifts the shuttle orbiter high off the ground and rotates it from a horizontal to a vertical orientation where the nose points toward the ceiling and the main engine nozzles point toward the ground.

By this time, many sets of platforms at different levels have already been installed around the solid rockets and external tank to provide easy access to the entire surface area of the semi-complete Space Shuttle stack. In order to bypass the maze of platforms, the gigantic crane must first lift the orbiter over 100 meters (330 feet) above the floor of the VAB, slide the orbiter over the top of the platforms, then lower it into place next to the external tank. The slings that suspend the orbiter from the crane hold the spacecraft in place while engineers bolt it to the external tank. After verification of a secure mate, the crane lifts the slings away from the Space Shuttle to complete the stacking process. In the launch configuration, the orbiter attaches to the external tank. The tank attaches to the two solid rocket boosters. As a consequence, the entire mass of the Space Shuttle launch stack rests on the mobile launch platform support posts that hold the solid rockets in place.

The next major step involves transporting the entire mobile launch platform with the Space Shuttle stack on top to the launchpad at Launch Complex 39-A or 39-B. This is not an easy task. A gigantic flatbed vehicle called the *crawler* accomplishes this gargantuan feat. This machine looks like a one-story-tall rectangular slab of metal with

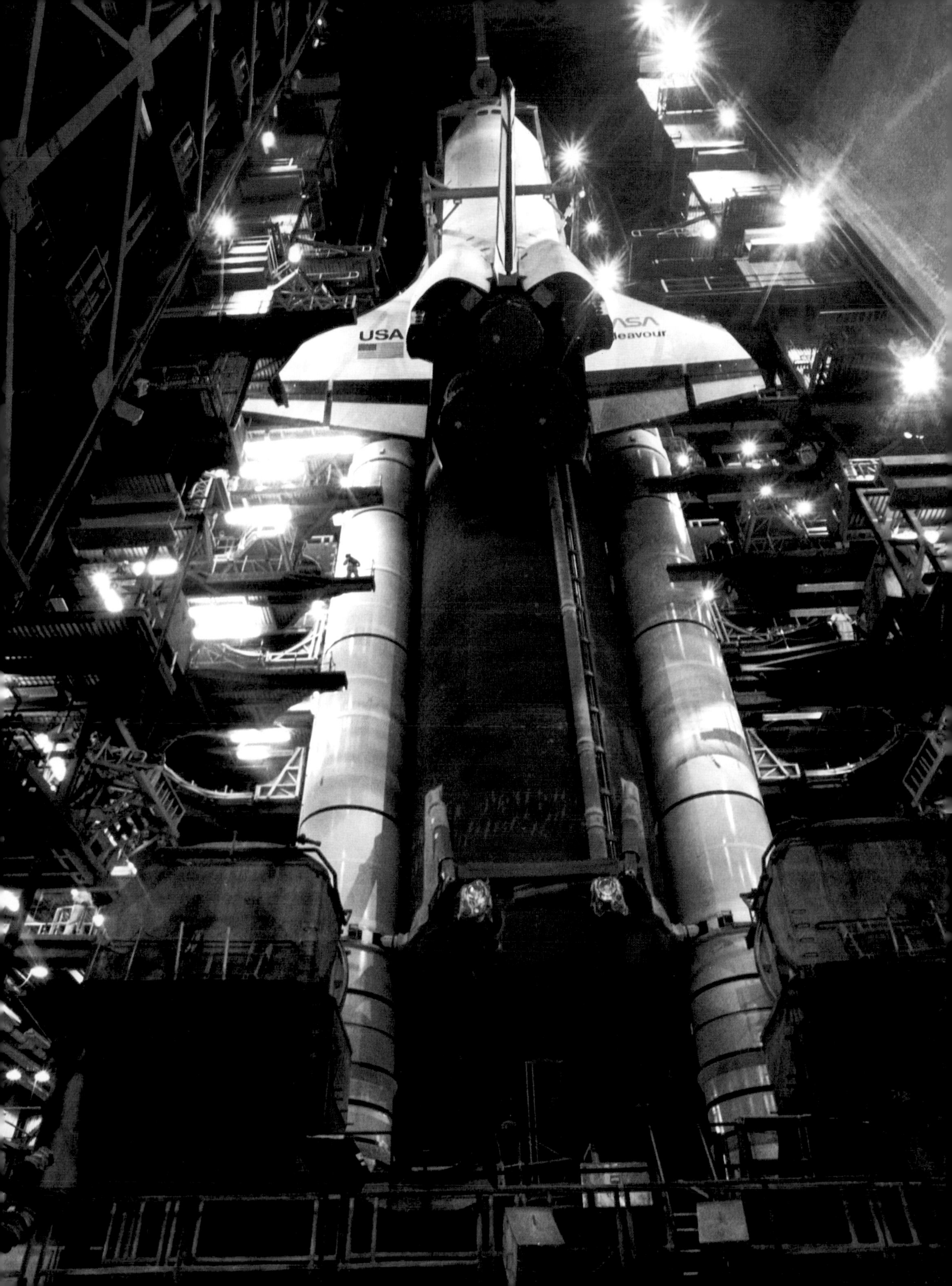
USA

two metallic treadmill-type tracks (similar to those found on military armored vehicles) located under each corner of the slab. These treadmill tracks are as large as a medium-size fire engine and serve as the crawler's wheels. In total, the crawler measures 40 meters (131 feet) long, 34.75 meters (114 feet) wide, and weighs 2,721,000 kilograms (6 million pounds) when unloaded. NASA owns two of these machines.

In order to pick up the mobile launch platform and the Space Shuttle stack, the crawler must first drive into the VAB through a tall vertical door that stretches from the floor to the ceiling. The VAB possesses four of these doors. Two of them are visible in the photograph of the VAB shown earlier in this chapter. Posts under each corner of the mobile launch platform hold the platform about two stories above the floor of the VAB. This position allows the crawler to drive into the VAB and under the MLP. Hydraulic jacks inside the crawler lift the surface of the crawler's flatbed and push the bottom of the MLP so that the platform elevates slightly above the posts. Crawler operators then slowly drive the machine out of the VAB with the MLP containing the Space Shuttle stack on top.

Engineers at NASA use the term *roll out* to describe the transport of the Space Shuttle from the VAB to the launchpad. A large two-lane gravel roadway, called the *crawlway*, runs from the VAB to both of the launchpads. This crawlway contains a 15.2-meter (50-foot) wide grass median that separates each 12.2-meter (40-foot) wide lane. Many observers typically guess that the two lanes allow a crawler to travel in each direction (to and from the launchpad) at any one time. However, the crawler's extensive width forces it to take up both lanes at the same time. The 5.5-kilometer (3.4-mile) journey to the pad typically takes about six hours because a fully loaded crawler creeps along the crawlway no faster than 1.6 kilometers (1 mile) an hour. At the launchpad, the crawler deposits the MLP on support posts in a procedure opposite that used to lift the MLP off the support posts in the VAB. Crawler operators then drive the gigantic vehicle away from the launchpad.

Both launch complexes look alike. Each one occupies about 0.7 square kilometers (165 acres) of land and takes the shape of an octagon surrounded by a barbed wire fence for security. The physical cement launchpad that the Space Shuttle lifts off from, sits in the middle of the complex surrounded by a large grass field that takes up most of the space within the complex grounds. Essentially, the grass field serves as a buffer zone between the exhaust that the Space Shuttle's engines generate and the tanks that contain the propellant to fill the tanks on the Space Shuttle.

On the pad, the shuttle sits next to a large steel structure that looks like a tower of twisted metal. This open frame tower, called the *fixed service structure* (FSS), looks like the skeleton of a recently constructed building. The FSS contains twelve decks. Together, they allow launchpad technicians to work on parts of the Space Shuttle stack from the ground to the top of the external tank. Several steel

Above:

One of the two gigantic crawler transporters inches back toward the Vehicle Assembly Building after depositing a fully assembled Space Shuttle stack at the launchpad. Notice the size of the crawler as compared to the van on the left edge of the photograph.

Opposite Page:

25 January 1990 - Space Shuttle *Atlantis*, complete with solid rockets and external tank, rolls out from the Vehicle Assembly Building (middle of the photograph) to Launch Complex 39-A. The rectangular slab beneath the shuttle is the mobile launch platform (MLP), not the crawler. Although most of the crawler is hidden in this photograph, its front tracks can be seen sticking out from beneath the MLP. Notice that the crawler takes up both lanes of the gravel road that leads to the launchpad. NASA calls this road the crawlway. In order to visualize the true size of the objects in this photograph, consider the fact that the crawlway measures close to half a football field in length from edge to edge.

Above:

This aerial photograph shows the physical layout of Launch Complex 39-A. Notice that the complex looks octagonal and that the physical launchpad and launch tower sit in the middle. The two spherical tanks at the right edge of the complex contain the liquid oxygen and hydrogen propellant used by the shuttle. Complex 39-B looks identical to this one.

Opposite Page:

April 1991 - At the pad, Space Shuttle *Discovery* sits on the mobile launch platform, located next to the tower. In turn, the platform rests over a deep trench, and is held up by posts. At launch, the rocket exhaust flows through square holes in the platform, into the trench, and away from the pad. The crawler is wide enough to straddle this trench when it drives up to place the shuttle next to the launch tower. In the photograph, the cap at the end of the oxygen vent access arm (steel arm leading to the external tank from the tower) is being deployed to cover the top of the tank.

walkways, called *access arms*, extend from the tower to various parts of the Space Shuttle. One of the arms, called the *crew access arm* (CAA), allows launchpad technicians and the astronauts to enter the orbiter's crew cabin from the tower's seventh deck. The famous clean room or "white room" with white painted walls lies at the end of the CAA and makes a firm seal over the side hatch leading into the crew cabin. Anybody entering the white room area must wear special clean suits that minimize the amount of dust and debris entering the orbiter.

The arm at the top, called the *oxygen vent access arm*, extends from between the ninth and tenth decks of the tower. A cap at the end of the arm forms a tight seal over the top of the external tank. This "beanie cap" seal provides the external tank with protection from the rain and warms the top of the tank, which contains super cold liquid oxygen. Warming the exterior of the top prevents ice crystals from forming, falling off, and damaging the orbiter upon launch.

Countdown to Launch

Launches do not arbitrarily occur when the pad engineers happen to finish preparing and loading propellant onto the Space Shuttle. Instead, launches must occur during a specified time period

USA
NASA
Discovery

USA

during a day called the *launch window*. This window defines a period of time when a rocket can launch into space and meet all of the mission objectives and mission safety criteria. Most windows for Space Shuttle missions last between one and several hours. However, some launch windows may last only several minutes. Countdown plans ensure that the thousands of events necessary to launch a rocket occur in the right order and at the precise time to guarantee a liftoff during the launch window.

Countdown terminology expressed in the language of rocket science sometimes leads to confusion. The phrase *t minus* (written as "T-") denotes the time left in the countdown. Liftoff occurs at T- 0 hours, 0 seconds. In reality, the countdown time rarely equals the amount of time left until liftoff. For example, T- 12 hours indicates that 12 hours remain in the countdown. However, due to unscheduled problems in preparing the vehicle, the actual time to launch may exceed 12 hours. During these unscheduled difficulties, the clock stops until somebody fixes the problem. NASA calls these periods *holds* and includes them as part of the normal Space Shuttle countdown.

Preplanned holds allow engineers to catch up on any work that falls behind schedule, and allows for variability in the amount of time required to complete a launch preparation task. For example, suppose that loading fuel into the shuttle takes anywhere between seven and eleven hours to complete. In this hypothetical case, the clock counts down during the first seven hours of fueling. Then, it enters a four-hour preplanned hold to allow for the possibility that fueling may take up to four additional hours. Keep in mind that it is impossible to start the clock before a preplanned hold expires even if the catch-up activities finish early. Starting the clock before a preplanned hold expires may result in T- 0 seconds arriving before the beginning of the launch window. In contrast, during an unplanned hold, it is advantageous to restart the clock as soon as possible to prevent T- 0 from occurring after the end of the launch window.

A team of engineers under the direction of the launch director runs Space Shuttle countdowns from a building called the Launch Control Center (LCC) located next to the Vehicle Assembly Building. Countdown schedules vary slightly for every mission, depending on the specific mission parameters. Most of them start with the clock at the T- 43 hour mark, roughly 72 hours from liftoff with preplanned holds factored in.

Most of the excitement begins once the clock has counted down to the T- 6 hour mark. At that time, the count enters a one-hour preplanned hold during which pad engineers complete preparations for loading the external tank with liquid hydrogen and oxygen. The launch director must make sure that the weather forecast for the launch window looks favorable before instructing the launch team to fill the external tank. For a typical morning launch, the T- 6 hour hold occurs sometime between 1:00 A.M. and 3:00 A.M.

RULE OF THUMB

The launch window defines a period of time during the day when a rocket can be launched into space and still meet all the mission and safety constraints. Some of the factors that influence the timing of the window include the laws of orbital mechanics, the desired lighting conditions at the time of liftoff and landing, the position of other spacecraft that the orbiter might rendezvous with, and the position of the Sun in space relative to the orbiter during times when critical activities need to be performed. The reason for the last factor is that many of the satellites that the orbiter carries into space, and the orbiter itself, use devices that track the stars to aid in navigation. If the Sun is in the wrong position, the sensors could be "blinded."

Opposite Page:

This rare view, photographed on 5 September 1990, shows both launch complexes (39A in the foreground, 39B in the background) occupied. Several kilometers separate the two nearly identical launch complexes. Space Shuttle *Discovery* (at 39B) was launched on 6 October. *Columbia* (at 39A) was scheduled for launch on 6 September, but a problem discovered the day of the launch delayed the mission until December. What is rare about the photograph is not that both launchpads are occupied, but that both shuttles are clearly visible. Most of the time, a metal structure, located to the left of the tower, covers the orbiter for protection against the harsh salt air and humidity of the Florida coast. That structure, called the *revolving service structure*, swings around to cover the orbiter by means of a large hinge attached to the tower. This photograph was taken over the Atlantic Ocean.

Historical Fact

The VAB, crawler, and launchpads at 39A and 39B are all left over from the Apollo Moon exploration days. NASA modified them for Space Shuttle use in order to save money.

Above:
29 September 1988 – The crew of STS-26 depart their pre-launch crew quarters at the Operations and Checkout Building on their way to Launch Complex 39B. From front to back are Rick Hauck, Richard Covey, Mike Lounge, Dave Hilmers, and George "Pinky" Nelson.

Opposite Page Left:
At the launchpad, the crew access arm provides a bridge between the Space Shuttle and the launch tower. The "white room," located at the end of the access arm next to the orbiter, provides a clean environment for technicians to make final preparations for astronauts about the board the orbiter.

Opposite Page Right:
Several hours before launch, STS-47 astronauts walk across the crew access arm toward the "white room" before climbing onboard Space Shuttle *Endeavour.*

Chilldown of the pipes that carry the cryogenic propellants to the ET begins as soon as the count leaves the T- 6 hour hold. This action is necessary because cryogenic liquids only remain in liquid form at temperatures several hundred degrees below zero. The chilling of the pipes minimizes the amount of propellant that boils off during the transfer from the storage tanks on the ground to the external tank.

Both the hydrogen and oxygen tanks in the external tank are filled and topped off by the time the clock reaches the T- 3 hour mark. However, propellant tends to boil inside the external tank despite the fact that a thick layer of insulation covers the outside of the tank to keep the inside cold. From now until several minutes before launch, the supply tanks on the ground constantly replenish the propellant that boils off inside the external tank. The gas resulting from the boil-off passes out of the external tank through vents near the top of the tank.

The clock stops counting for a preplanned two-hour hold at the T- 3 hour mark. At this time, a team of inspectors at the pad verifies that no ice has built up on the outer surface of the external tank. If they find ice, the launch director extends the two-hour hold to allow the ice to melt. The reason for this action is that ice chunks that fall off of the external tank during launch may damage the orbiter.

During the T- 3 hour hold, a team of Kennedy Space Center physicians wake the astronauts up. Even if the launch is scheduled for the evening, the astronauts are always sound asleep in the hours leading up to the T- 3 hour hold. The reason involves the fact that mission rules call for them to gradually adjust their circadian rhythms so that the launch occurs in the "morning" according to their biological clocks. After the astronauts wake up, they undergo a final physical exam, eat breakfast, then put on heavy orange pressure suits for protection against a sudden loss of air in the orbiter during ascent.

The astronauts leave their crew quarters in the Operations and Checkout Building (located several kilometers south of the VAB) at the T- 2 hours, 55 minutes mark. Television news stories often show the astronauts exiting the famous double doors of the Operations and Checkout Building and entering the van for the drive to Launch Complex 39, complete with the flashing lights of a police escort. Two decades ago, the Moon-bound Apollo astronauts left for the launchpad from the same double doors. The drive to the launchpad takes about 20 minutes.

White-room technicians, dressed in lint-free clean suits similar to those worn by surgeons, await the arrival of the astronauts at the end of the crew access arm, near the top of the launch tower. The technicians' job involves helping the astronauts board the orbiter and strapping the space-bound crew securely into their pre-assigned seats. After a final traditional thumbs up hand signal for good luck, the white-room staff exits the orbiter seals the hatch behind them,

Columbia
United States
NASA

Above:

Tension in the Launch Control Center mounts as the launch director conducts the "go" or "no go" poll of all the members of the launch control team minutes before the scheduled liftoff. The launch controllers in front of the computer consoles monitor the hundreds of systems on the shuttle during the countdown. They must make sure that all systems are "go" for a successful launch. The display screens on the left wall show critical countdown information, and the windows on the right wall face the launchpad.

takes the elevator to the bottom of the tower, and then drives back to the VAB area so that they can view the launch from a safe distance. The astronauts and more than one million kilograms of explosive propellant are now alone on the launchpad.

During the next two hours, the launch team in the Launch Control Center constantly monitors the weather, pressure in the external tank, and sensor instruments around the orbiter that detect pressure changes and leaks in the fuel lines. Inside the orbiter, the commander and pilot verify the operational readiness of the communications systems, calibrate the navigational units with the exact longitude and latitude of the launchpad, and prepare the internal primary flight-control systems for launch. Another astronaut, not assigned to fly on the current mission, flies a plane called the shuttle training aircraft (STA) around the Kennedy Space Center's shuttle landing runway. He or she verifies that clear weather exists for the orbiter to return to KSC after launch in the event of a major emergency.

The clock counts down to another preplanned 10-minute hold at T- 20 minutes to allow the launch director to conduct a final

briefing for the launch control team. After the clock restarts, the primary and backup flight computers on the orbiter transition from standby to active mode. The count enters the last preplanned hold at the T- 9 minute mark. At this time, the launch director completes the final polls for launch. Console by console, the launch controllers in the launch control center responsible for each major system on the shuttle answer the roll call with a "go" (which means my system is ready for launch) or "no-go" (which means my system is not ready for launch). This rapid-fire poll continues until all the stations answer that they all respond "go for launch."

Normally, the T- 9 minute hold lasts for 10 minutes. However, the weather around the Kennedy Space Center plays a big role in determining whether to continue the count at the end of 10 minutes. Several major weather rules constrain Space Shuttle launches. One of the rules deals with the location and density of clouds near the launchpad. All clouds must lie above 2,440 meters (8,000 feet) and the thickness must be less than 50% (a partly cloudy day). This rule seems to imply that the Space Shuttle needs 2,440 meters to accelerate to a velocity that allows it to punch through the clouds. In reality, if something goes drastically wrong during the first few minutes of flight, the capability exists for the orbiter to ditch the solid rockets and external tank, turn around, and glide back to KSC for a landing on a runway near the launchpad. If this contingency occurs while the clouds cover KSC, the pilots will not have a clear view of the runway to make a safe landing.

Another major constraint restricts launches when the winds at low altitudes are too strong. Remember that the solid rocket booster nozzles can only gimbal about 7° to steer the Space Shuttle. This limited steering ability might not be enough to overcome sudden gusts of wind. The last major constraint of interest restricts launches when any clouds capable of producing lightning exist within 80.5 kilometers (50 miles) of the launch pad. Many times, the launch director extends the T- 9 minute hold to await better weather. Keep in mind that liftoff must occur by the end of the launch window. As a result, the T- 9 minute hold must end with at least nine minutes remaining in the window.

If the weather looks unfavorable, the launch director may decide to scrub (cancel) the launch during the T- 9 minute hold. In the event of a launch scrub, the astronauts leave the orbiter, and pad engineers drain the external tank propellants. Most of the time, it is not necessary to restart the entire 43-hour countdown. Typically, the next launch attempt usually occurs at the same time the next day. NASA engineers call this schedule a "24-hour scrub turn-around." However, if the launch attempt on the second day fails, the earliest possible third attempt must wait for the fourth day. This "day off" rule exists primarily to allow fatigued launch crews to rest.

Once the launch director gives the official "go" to resume the count at T- 9 minutes, the automatic ground launch sequencer that

Final Countdown Events

T- 9 minutes
Clock is stopped for at least 10 minutes to conduct final "go/no-go" poll and to assess the weather situation

T- 7 minutes, 30 seconds
Crew access arm leading from the tower to the orbiter is retracted

T- 6 minutes
Auxiliary power unit pre-start

T - 5 minutes
Auxiliary power units transition from warm-up to ready mode

T- 4 minutes, 30 seconds
Orbiter switches to internal power

T- 3 minutes
Main engines gimbal to launch position

T- 1 minute, 57 seconds
Oxygen vent access arm covering the top of the external tank retracts

T- 31 seconds
Ground launch sequencer hands control of countdown over to computer on the orbiter

T- 11 seconds
Sound suppression system activates and floods the base of the launchpad with water

T- 6 seconds
Main engine "ignition sequence start" and all three main engines fire up 0.12 seconds apart

T- 0 seconds
Solid rocket booster ignition and liftoff

RULE OF THUMB

The rules that govern a shuttle countdown are extremely complicated. In fact, they occupy several books containing several thousand pages each. These books form the basis for the shuttle's "Launch Commit Criteria."

Opposite Page, Top Left:

At T- 6 seconds, the shuttle's main engines ignite (bright glow under the orbiter). All three engines must build up to full thrust by T- 0, the time of liftoff. The cloud of smoke emerging from in front of the shuttle is simply steam from water in the flame trench (under the launchpad) being vaporized by the exhaust flames of the engines.

Opposite Page, Top Right:

At T- 0, the two solid rocket boosters ignite. This photograph captures that exact moment. The bright glow at the bottom of the picture comes from the super-hot rocket exhaust of the orbiter's three main engines and the two solid rockets. Prior to this point in time, the computer can shut the main engines off and abort the countdown in the event of a detected malfunction. However, once the solids ignite, there is no turning back.

Opposite Page, Bottom Left

This photograph captures the shuttle inching off the launchpad split seconds after liftoff. The bright exhaust trail under the shuttle comes from the twin solid rocket boosters. In contrast, the exhaust of the shuttle's three main engines consists of mainly water vapor and is nearly invisible.

Opposite Page, Bottom Right:

The shuttle takes about three seconds to clear the height of the launch tower. At this time, the shuttle is moving slower than a car on a local road, but will accelerate to faster than the speed of sound in about 30 seconds. After clearing the tower, control of the mission is handed over from the Launch Control Center in Florida to Mission Control at the Johnson Space Center in Houston, Texas.

controls the countdown activates. At T- 7 minutes, 30 seconds, the crew access arm retracts from the orbiter and swings into a position where it will not interfere with the launch. In the event of an emergency, the arm can redeploy in less than 20 seconds to allow the astronauts to evacuate the orbiter.

At T- 6 minutes, the three auxiliary power units (APUs) begin to warm up in an event called the APU pre-start. These devices provide the power to the hydraulic motors that provide the force to gimbal the main engines during ascent. Hydraulic motors force their parts to move by forcing a dense incompressible liquid through relatively small tubes. All three APUs transition from warm-up mode to ready mode at the T- 5 minute mark.

At T- 4 minutes, 30 seconds, the orbiter switches from using electrical power generated at the launchpad to power generated by the internal fuel cells. These cells mix liquid oxygen and hydrogen in a chemical reaction that produces electricity for the orbiter's electronics. As a byproduct of the oxygen-hydrogen reaction, the fuel cells also produce drinking water for the astronauts.

Between T- 4 and T- 3 minutes, the computer automatically gimbals the main engines through various positions to test the gimbaling ability and to condition the hydraulic system. This activity is analogous to allowing oil to thoroughly lubricate the inside of a car motor before driving the car. At T- 3 minutes, the orbiter's engines gimbal to their launch positions.

The propellant tanks in the external tank reach the proper pressure required for flight during the next minute. At T- 1 minute, 57 seconds, the computer commands the tower to retract the oxygen vent access arm with the "beanie cap." Retraction takes approximately 30 seconds. At T- 31 seconds, the ground launch sequencer hands control of the countdown over to the orbiter's onboard computers. From this point on, the orbiter controls its own countdown. Although few visible events happen between the countdown handoff and rocket ignition, the orbiter's computers check and recheck all of the systems and sensors thousands of times every second. These computers can detect possible abnormalities and halt the countdown faster than the quickest human reaction time.

Roughly 11 seconds before liftoff, a water tower located near the pad begins to supply freshwater to pipes that empty into a trench below the mobile launch platform. In addition, these pipes also supply water to the "rainbird" sprinklers that spray water on top of the mobile launch platform behind the solid rockets. The water helps to reduce the incredible amount of noise that the two solid rockets and three main engines on the orbiter produce at the time of liftoff. Without water in the trench, the powerful sound waves that emanate from the rocket engines would echo off the cement pad with enough energy to bounce back, strike, and damage the Space Shuttle. Instead,

Atlantis
United States

USA
United States
NASA

the sound waves lose energy upon hitting the water. Consequently, the sound echoes off the cement launchpad with less energy.

Sparks begin to appear beneath the main engines at T- 8 seconds. This scene often appears on television news stories without a proper explanation. Reporters sometimes give the impression that the sparks ignite the main engines similar to a person igniting a furnace pilot flame with a match. In reality, the onboard computers begin the flow of liquid hydrogen to the main engines several seconds before the flow of liquid oxygen. This sequence produces a hydrogen rich environment in the engines' combustion chamber upon ignition. Some of this hydrogen boils off and escapes out the nozzle. A device beneath the main engines burns off the escaped hydrogen gas to prevent an extremely explosive cloud of hydrogen from forming under the engines. This device causes the sparks.

The excitement begins to peak at T- 6.6 seconds as the computers command the Space Shuttle main engines to start. The first engine starts at T- 6.6 seconds, the second at T- 6.48 seconds, and the third at T- 6.36 seconds. Starting each main engine 0.12 seconds apart reduces the suddenness of the tilt and the overall stress on the system. When the main engines start, the sudden force tilts the entire Space Shuttle stack in the direction of the external tank by up to a meter before recoiling back into the original position. Astronauts often compare this effect to sitting on top of a thin, tall tree swaying in the breeze.

As the few remaining seconds until T- 0 count down, the super-hot exhaust from the main engines flows through an opening in the mobile launch platform, vaporizes the water in the trench, and forms a large cloud of steam to the south of the launchpad. Moments before T- 0, the computer looks at the main engine performance one last time to verify that thrust in each main engine exceeds 90% of its rated value. If not, the computer assumes that one or more of the engines has malfunctioned, immediately stops the countdown, and shuts down all three engines.

At precisely T- 0, the computer commands the igniter in both of the solid rocket boosters, located inside the boosters at the top, to shoot a 45-meter pillar of fire down inside the rocket to ignite the solid propellant. Both solid rockets must ignite simultaneously to assure a safe liftoff from the launchpad. Simultaneously, the computer orders the explosion of the giant bolts that hold the Space Shuttle to the mobile launch platform, allowing liftoff to occur. All systems must function properly at the time the solid rockets ignite because once they ignite, they must thrust until all the propellant burns out. Nothing short of a total catastrophe can stop the flight at this time.

As the Space Shuttle slowly inches its way off the launchpad and into the air, the exhaust from the two solid rockets meets the water in the trench and creates another cloud of steam in addition to

RULE OF THUMB

Want to Watch a Launch on Television?

As of June 1995, Cable News Network (CNN) is the only station that still shows shuttle launches live. Incidently, CNN has two stations. The one that carries launches live is the normal station, not *Headline News*. CNN usually picks up coverage from KSC about two minutes before the opening of the launch window. Finding out when the launches occur is another story. Many newspapers do not bother to write articles about upcoming Space Shuttle launches. Fortunately, NASA has provided a solution. You can call the Kennedy Space Center at (407) 867-INFO for a recorded message on the status of the next scheduled launch.

Opposite Page:

4 March 1994 - On a clear morning, Space Shuttle *Columbia* lights up the morning sky as it lifts off from Complex 39B on the start of a 14-day mission dedicated entirely to scientific research. This photograph, taken less than 10 seconds after launch, shows the shuttle already at an altitude of more than 400 meters (1,312 feet). Notice the reflection in the water of the extremely bright solid rocket booster exhaust. The clouds of smoke formed from the super-hot exhaust that completely obscure the launchpad will take several minutes to clear. By then, the shuttle will be halfway to orbit.

RULE OF THUMB

Want to Watch a Launch at KSC?

Entrance to KSC is usually restricted to authorized employees with a valid security badge. However, NASA distributes free passes that allow the general public to enter the space center's grounds to watch a shuttle launch. Write to:

Public Affairs, Mail Code PA-PASS
Kennedy Space Center, FL 32899

Since NASA only hands out a limited number of these passes for each launch, it is best to write at least several months in advance. Each pass is good for a specific mission. If you know what mission you want, request that one. Otherwise, tell them the approximate dates that you plan to be in the Kennedy Space Center area and they will attempt to accommodate you. Without a pass, the best place to view a launch is along U.S. Highway 1 on the Atlantic coast near Titusville, Florida.

Opposite Page:

22 October 1992 - Space Shuttle *Columbia* leaves Launch Complex 39B on its way toward a 10-day mission with a crew of five NASA astronauts and one Canadian. This photograph was taken by another astronaut flying in a chase plane. At this time, less than 30 seconds after launch, *Columbia* has already reached an altitude of close to 3 kilometers (slightly more than a mile). Notice that the exhaust from the twin solid rocket boosters forms a pillar of fire and smoke that reaches all the way down to the launchpad. The background in the photograph comes, not from the sky, but from the blue water of the Atlantic Ocean.

the cloud created by the main engines on the orbiter. However, this new cloud flows in the opposite direction, north of the pad. The Space Shuttle takes approximately three seconds to clear the height of the launch tower. Until now, the personnel in the Launch Control Center at KSC controlled the mission. After the tower is cleared, control of the mission transfers to Johnson Space Center (JSC) in Houston, Texas.

From the closest observation sites three miles from the launchpads, all appears silent as the Space Shuttle climbs off the launchpad. The only clue that marks the beginning of the launch is the glow from the solid rocket exhaust and two clouds of steam flowing in opposite directions from the launchpad. However, the sound that will eventually reach listeners miles away takes time to emanate from the launchpad. At first, the launch energy produces a deep rumbling noise like a giant train approaching in the distance. The deep noise eventually gives way to a sharp crackle that snaps through the air like the crackle from twigs burning at a campfire. However, this crackle is deep enough to be felt like the sound from a giant bass speaker, and sharp enough to vibrate metal structures like a small earthquake.

The first minute of flight after the launch provides a spectacular display of great quantities of energy harnessed for human use. As the Space Shuttle accelerates upwards, two streams of bright fire and orange exhaust emanate from the solid rocket nozzles and eventually form a single column of gray smoke that extends all the way down to the base of the launchpad. This column of smoke allows spectators to easily follow the vehicle's trajectory. At the same time, only a bright glow at the bottom of the orbiter announces the fact that the main engines are also thrusting. The reason is that the product of the combustion of liquid hydrogen and liquid oxygen is an exhaust trail of super-hot, invisible water vapor.

NASA likes to portray Space Shuttle launches as routine. However, keep in mind that launches are extremely dangerous and are anything but routine. The infamous *Hindenburg* zeppelin exploded because it carried hydrogen gas to keep it afloat. The Space Shuttle external tank carries close to one million kilograms of explosive hydrogen and oxygen. Essentially, the astronauts ride a flying bomb into orbit.

5.3 How Do Shuttles Get into Orbit?

At the time of launch, both of the solid rocket boosters and all three main engines thrust in unison as the Space Shuttle ascends straight up into the air and off the launchpad. Before liftoff, the orbiter sits on the pad, attached to the external tank, with the payload bay doors facing south down the Florida coastline. After the Space Shuttle clears the launch tower, it starts to spin around its vertical axis (while climbing) so that the payload bay doors face eastward

▶ Figure 46: *How the Shuttle Gets into Orbit (Part 1)*

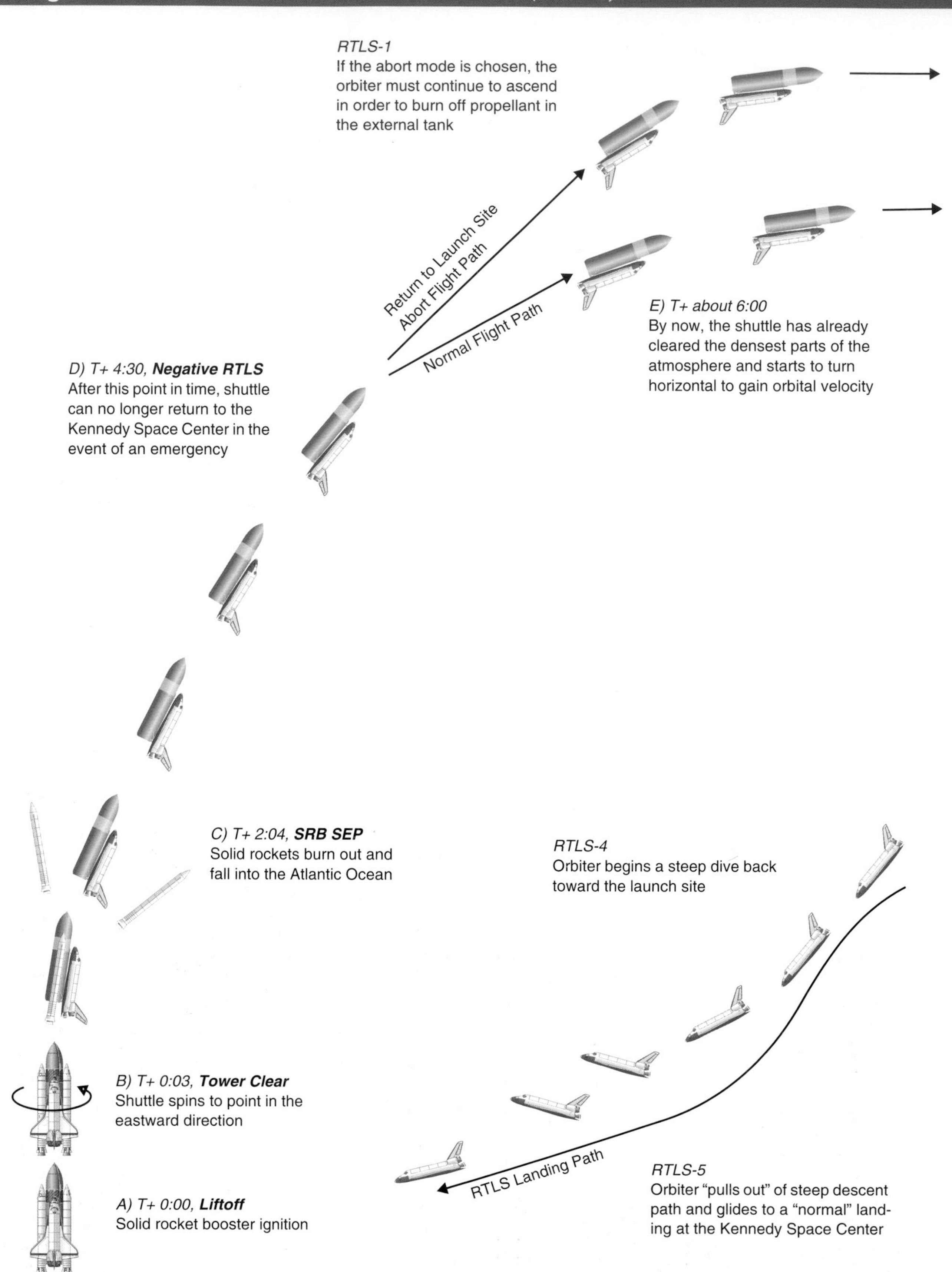

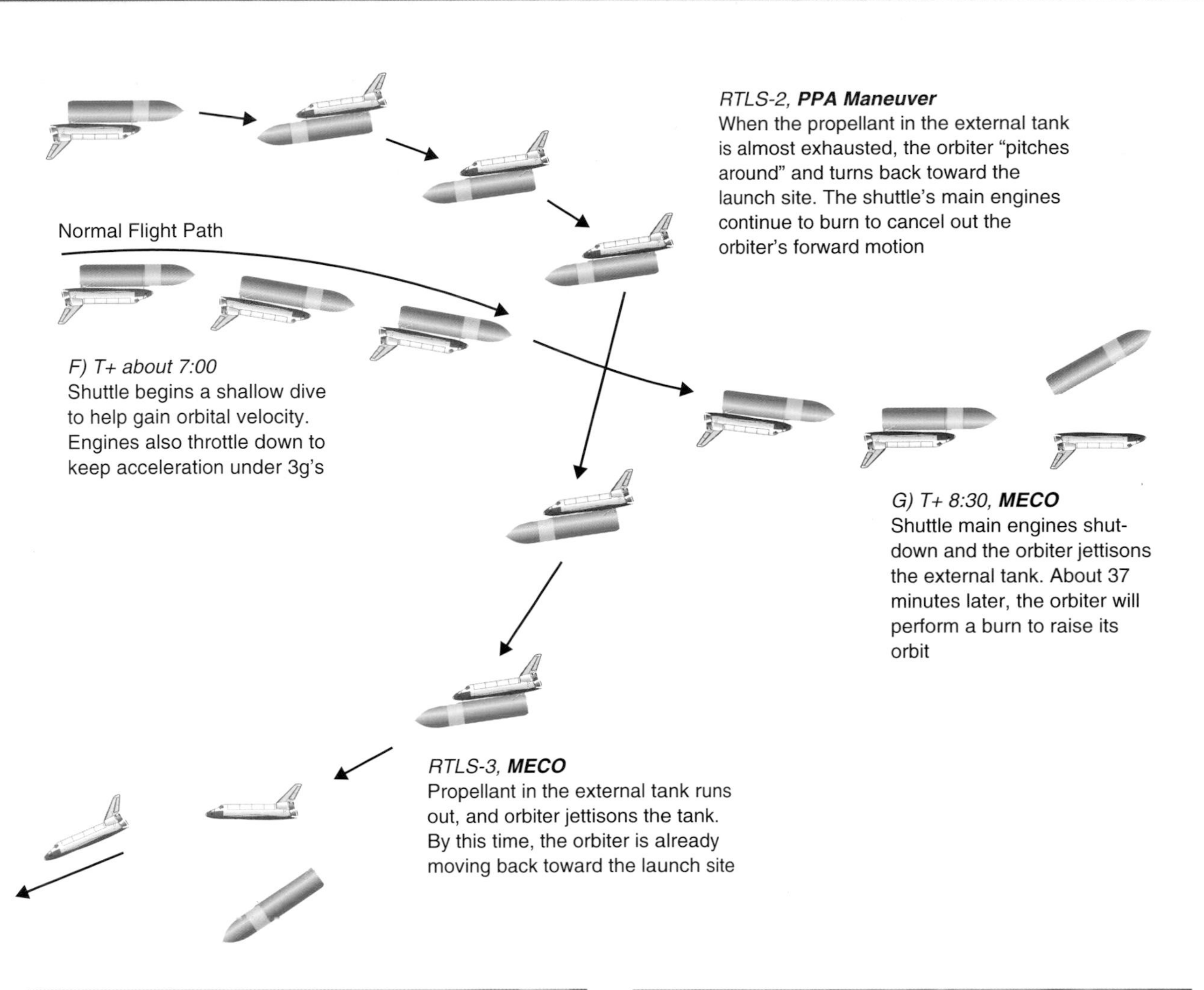
RTLS-2, **PPA Maneuver**
When the propellant in the external tank is almost exhausted, the orbiter "pitches around" and turns back toward the launch site. The shuttle's main engines continue to burn to cancel out the orbiter's forward motion
Normal Flight Path
F) T+ about 7:00
Shuttle begins a shallow dive to help gain orbital velocity. Engines also throttle down to keep acceleration under 3g's
G) T+ 8:30, **MECO**
Shuttle main engines shutdown and the orbiter jettisons the external tank. About 37 minutes later, the orbiter will perform a burn to raise its orbit
RTLS-3, **MECO**
Propellant in the external tank runs out, and orbiter jettisons the tank. By this time, the orbiter is already moving back toward the launch site

Above:

12 April 1981 - A ground-based tracking camera at the Kennedy Space Center took this photograph of *Columbia* and its external tank (right side of the picture) separating from their twin solid rocket boosters (left side of the picture) roughly two minutes and five seconds after liftoff. Notice that the glow from *Columbia's* three main engines indicate that all are still thrusting. Although the two solid rockets are still thrusting in the picture, they burned out completely several seconds after the picture was taken. The shuttle is about 46 kilometers (29 miles) above the ground in this picture. In order to take this amazing shot, the camera used a gigantic lens with a focal length of 10.2 meters (400 inches).

toward the Atlantic Ocean. At the same time, the rocket engines on the solid rockets gimbal so that the direction of climb changes from straight up to almost straight up, but slightly off vertical (see Figure 46 on page 182). This maneuver, called the roll and pitch-over, allows the Earth's gravity to slowly tilt the Space Shuttle so that it will eventually fly eastward in the horizontal direction instead of upward.

Fifty seconds after launch, the Space Shuttle has climbed to an altitude of 9.1 kilometers (5.7 miles), and has accelerated past the speed of sound (about 1,190 kilometers per hour or 740 miles per hour). About this time, a combination of the shuttle's velocity and the density of air maximizes the amount of pressure and force that the shuttle experiences from the atmosphere. NASA engineers call this region of maximum dynamic pressure "max Q." To compensate for the maximization in pressure, the onboard computers throttle the main engine thrust from 100% down to 67%. This action minimizes the chance of damage from aerodynamic forces.

With every second that passes after "max Q," the shuttle climbs higher, the atmosphere thins, and the aerodynamic pressure decreases dramatically. Between 60 and 70 seconds after launch, the dynamic pressure from the atmosphere decreases to the point where the computers can safely command the main engines to throttle back up to full thrust. Keep in mind that decreasing the thrust does not

decrease the velocity. Instead, the rate at which the Space Shuttle increases its velocity decreases.

Both solid rockets begin to burn out two minutes after launch, with the shuttle at an altitude of 46.3 kilometers (29 miles) and at a downrange (east of the launchpad) distance of 38.1 kilometers (23.7 miles). At two minutes and four seconds, small explosives disintegrate the bolts that hold the solid rockets to the external tanks. This action allows tiny rockets at the top of both solid rocket boosters to push the nearly spent but still thrusting boosters away from the rest of the Space Shuttle. After the boosters burn out, they fall into the Atlantic Ocean on giant parachutes. Then, two ships drag them back to KSC for refurbishment.

The single ribbon of smoke marking the Space Shuttle's trajectory splits into two columns after solid rocket separation (called "SRB sep" in NASA jargon) because the boosters jettison in opposite directions. On an extremely clear day, some viewers may catch a glimpse of the separation without the aid of binoculars. Also, viewers on the ground typically lose sight of the Space Shuttle at this time because the glow from the main engines tends to blend with the bright sky, and the engines do not leave a visible cloud of smoke to mark the flight path. However, during a night launch, viewers may see the shuttle long after solid rocket separation.

The thrust from the solid rocket boosters gives the astronauts a violent ride for the first two minutes of the flight. Some astronauts describe the vibrations from the boosters as comparable to driving an old pickup truck with bad shock absorbers over railroad ties at freeway speeds. The main engines continue thrusting after solid rocket separation to propel the orbiter and external tank towards orbital velocity. In contrast, they provide a relatively smooth ride. At the time of launch, the Space Shuttle climbs almost straight up into the sky. By this time, its angle of climb has decreased to 30° off vertical. The shuttle's velocity, already close to four times the speed of sound (4,632 kilometers or 2,880 miles per hour), will triple two times during the next six minutes.

At four minutes and thirty seconds after launch, the space shuttle reaches "the point of no return." Until now, the orbiter had the capability to turn around, jettison the external tank, and glide back to the runway at KSC in the event of a major emergency. The combination of the Space Shuttle's velocity, altitude, and downrange distance from the launchpad contribute toward making a return to KSC impossible after this point in time. NASA engineers refer to this threshold as "negative return to launch site" (negative RTLS).

During the next two minutes, the Space Shuttle ascends from its "negative RTLS" altitude of about 103 kilometers (64 miles) to an altitude of between 113 and 121 kilometers (70 and 75 miles). A 10-kilometer gain in two minutes amounts to little progress in the vertical direction. However, in those two minutes, the shuttle's horizontal

Above:
After the shuttle jettisons the solid rocket boosters two minutes after launch, they fall into the Atlantic Ocean. There, the SRBs are recovered by two boats, called the *Liberty Star* and *Freedom Star*, and towed back to land for use on a future flight. NASA recycles the SRBs as a way to reduce the cost of the shuttle program.

Historical Fact

The first time that NASA flew the Space Shuttle back in 1981, they found that the SRBs were much more powerful than anticipated. At the time of SRB separation, *Columbia* was close to two miles (three kilometers) higher than the flight engineers had predicted based on the original design specifications.

RULE OF THUMB

The ascent from the launchpad to low Earth orbit takes about eight and one-half minutes. During the first six and one-half minutes, the shuttle expends most of its energy climbing in order to clear the dense parts of the lower atmosphere. The last two minutes are spent accelerating the spacecraft horizontally to reach orbital velocity.

MECO stands for *main engine cutoff*. This event occurs at 8 minutes, 30 seconds after launch and marks the time when the main engines shut down. They are not used at any time during the remainder of the mission. The orbiter jettisons the external tank shortly after MECO. During the next half hour, the tank plunges back into the atmosphere and eventually burns up over the Pacific Ocean due to re-entering the atmosphere at a near-orbital velocity.

Historical Fact

Some NASA engineers suggested a plan in the early 1980s that would have involved leaving the external tanks in orbit rather than dumping them to a fiery death over the Pacific Ocean. Astronauts could then proceed to build a space station out of several empty propellant tanks. After a study of this scenario, NASA concluded that building a space station from scratch was a more feasible idea.

speed increases by close to 8,000 kilometers (5,000 miles) per hour, a velocity gain comparable to the gain during the first four and one-half minutes of flight. The reason for the large gain during a short period of time is that after solid rocket booster separation, the Space Shuttle directs an increasing amount of its energy output towards gaining the horizontal velocity required to achieve orbit. Between "negative RTLS" and six minutes, thirty seconds after launch, the Space Shuttle rotates from a 30° off-vertical climb to moving completely in the horizontal (90° off-vertical) orientation. In this "heads down" position, the orbiter appears to hang upside down from the external tank. The astronauts in the orbiter sit, strapped tightly in their seats, with their heads pointing toward the ground.

Six minutes after launch, the shuttle moves in the horizontal direction at a velocity of close to 20,000 kilometers (12,400) miles an hour. This speed, as fast as it may seem, still falls short of the velocity necessary to sustain an orbit. To gain additional velocity, the main engines on the orbiter remain thrusting as the shuttle begins an extremely shallow two-minute descent back toward the Earth. During this descent, the shuttle loses between 4 and 16 kilometers of altitude, but increases its velocity to about 28,160 kilometers (17,500 miles) an hour. NASA engineers refer to this strategy as *lofting*. The reason for this seemingly counterintuitive maneuver is the minimization of propellant usage. Given that the shuttle must climb above the dense parts of the lower atmosphere before accelerating to orbital velocity, it is more efficient to overshoot the target altitude and then descend back than to gradually ascend toward the target altitude. Keep in mind that the main engines on the orbiter, not the 4 to 16 kilometer descent, provide the majority of the velocity increase during this phase of the flight.

During the shallow descent, the external tank weighs much less than at launch time due to the consumption of most of the propellant. As a result, the main engines gradually throttle down to keep the acceleration at a tolerable level. Remember that Newton's laws state that decreasing the mass (fuel in the external tank) while keeping the force (main engine thrust) constant results in an increased acceleration. Decreasing the amount of thrust provided by the main engines allows the astronauts to experience a maximum force of 3g's, an amount that makes them feel three times heavier than normal. On missions during the 1960s, astronauts routinely experienced 7 to 8g's near the end of the ascent.

The orbiter's computers issue a command to extinguish the main engines eight minutes and thirty seconds after launch. This event, called *main engine cutoff* (MECO), inserts the Space Shuttle into a shallow elliptical orbit with a 65-kilometer perigee and 296-kilometer apogee. Insertion takes place at a point in the orbit slightly after perigee. Some astronauts describe the sensation of entering a weightless state of free fall after experiencing 3g's of thrust for eight and a half minutes as comparable to driving over a bump on the road at freeway speed and never coming back down.

Figure 47: *How the Shuttle Gets into Orbit (Part 2)* ◄

Which One? Today, Space Shuttle ascents use the OMS-2 Only Scenario exclusively. In the early to mid 1980s, the OMS-1 then OMS-2 Scenario was used. The OMS-1 then OMS-2 can also be used if the velocity after the ascent is lower than the expected value.

Solid arrow indicates position of the burn and points in the direction of the velocity change

Hollow arrow indicates position of the burn and points in the direction that the rocket engines fire to perform the burn

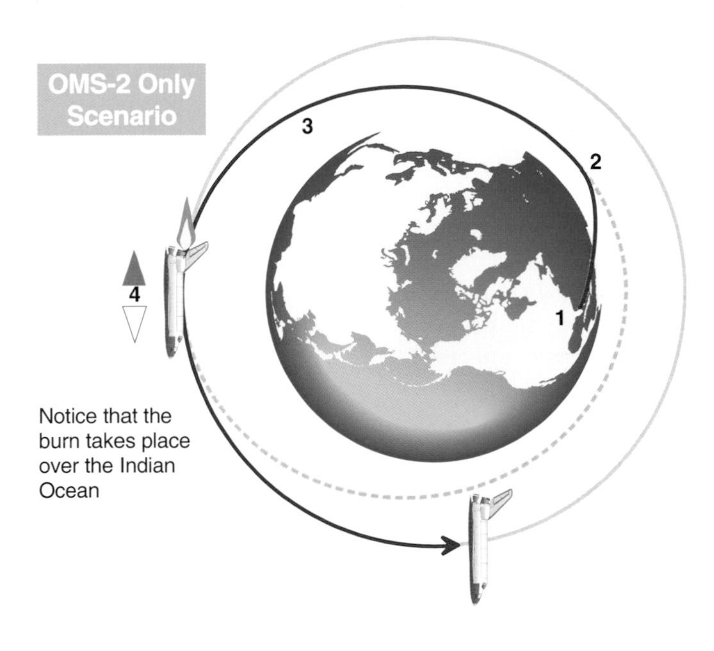

1 Launch and ascent (see "How the Shuttle Gets Into Orbit - Part 1).

2 Main engine cut-off (MECO). At this time, the orbiter is inserted into a shallow elliptical orbit with a low perigee and an apogee at the desired final altitude. Insertion takes place just after the perigee point of the shallow orbit.

3 Orbiter coasts upward to the apogee of the initial shallow orbit.

4 At apogee (45 minutes after the launch), OMS engines fire in the posigrade direction to increase the velocity and circularize the orbit. Flight controllers refer to this burn as the OMS-2 burn. Without it, the orbiter will continue on the shallow orbit.

OMS-1 then OMS-2 Scenario

4

5

3

2

1

Relative Size of the Earth, Orbits, and Shuttle Not Drawn to Scale

1 Launch and ascent (see "How the Shuttle Gets Into Orbit - Part 1).

2 Main engine cut-off (MECO). At this time, the orbiter is inserted into a very shallow elliptical orbit (inside orbit shown on the left) just before perigee.

3 At perigee (10 minutes after lift-off), the OMS engines fire in the posigrade direction to increase the apogee of the orbit to the desired final altitude. This is called the OMS-1 burn.

4 Orbiter coasts upward toward the apogee of the middle orbit.

5 At the apogee of the middle orbit, the OMS engines fire in the posigrade direction for the second time to circularize the orbit at the apogee altitude. This is called the OMS-2 burn.

Historical Fact

Why did NASA choose an altitude of 296 kilometers for their standard Space Shuttle orbit? The reason involves units of measurement other than the metric system. When the first mission plans were sketched out in the late 1970s, NASA used the nautical mile (1.15 normal or statute miles) to measure distance. 296 kilometers expressed in nautical miles turns out to be an even 160.

OMS stands for *orbital maneuvering system.* This system provides two engines that allow the orbiter to change its orbit in space. An OMS burn 45 minutes after launch marks the last event of the ascent to orbit.

RULE OF THUMB

Although the shuttle takes about eight minutes to reach orbit, television cameras that track the shuttle for news reports lose sight of the vehicle between four and five minutes after launch. On a rare, clear day, television coverage for six minutes is possible.

Small explosives disintegrate the bolts that hold the external tank and orbiter together about 20 seconds after MECO. Moments later, small thrusters on the tank cause it to tumble downward through the upper atmosphere. The tank eventually burns up due to frictional heating somewhere high above the Pacific Ocean. Meanwhile, the orbiter coasts up toward the apogee of its orbit and reaches the high point (296 kilometers) over the Indian Ocean sometime between 40 and 50 minutes after launch. At this time, two maneuvering thrusters on the back of the orbiter, called the *orbital maneuvering system* (OMS), fire to increase the spacecraft's velocity by about 240 kilometers (150 miles) an hour. This rocket burn transfers the orbiter to a circular orbit at an altitude of 296 kilometers. NASA engineers call this maneuver the "OMS-2 burn." Without the burn, the orbiter will coast down the other half of the ellipse and return to the point where main engine cutoff occurred. Such an orbit is too low to maintain because of atmospheric drag and frictional heating. After the burn, the ascent is finished. Now, the astronauts can unstrap themselves from their seats and begin their work in space.

Incidentally, the number two in "OMS-2" does not mean that two maneuvering thrusters are used during the burn. It means *second* OMS burn. What happened to the first burn? Between the early to mid-1980s, main engine cutoff occurred with the orbiter traveling at a velocity fractionally slower than that required to maintain an orbit that reached an altitude of 296 kilometers. An OMS-1 burn, occurring about two minutes after external tank separation, compensated for this shortfall. Then, the OMS-2 burn occurred 45 minutes after launch as described previously. Sometime in the late 1980s, NASA engineers figured out how to use the same amount of propellant in a more efficient manner. This discovery allowed main engine cutoff to occur with the orbiter traveling at a slightly faster velocity than previously possible. As a result, the OMS-1 burn is no longer needed. Figure 47 on page 187 illustrates the differences between OMS-2 and OMS-1 then OMS-2 scenario.

Many people often wonder why not use the main engines to directly insert the orbiter into its final orbit? The reason is that the propellant in the external tank runs out and causes MECO to occur at an altitude of about 113 kilometers (70 miles). In theory, a larger external tank that holds more propellant would allow MECO to occur at the final orbital altitude of 296 kilometers. However, a larger and heavier tank would take more energy to accelerate off the ground than the current design. Also, it creates inefficiency from a staging perspective. Remember that rocket efficiency increases when the rocket jettisons as much dead weight (empty fuel tank and structure) as early as possible. Finally, sending the external tank back into the atmosphere to burn up over unpopulated areas presents a difficult challenge from 296 kilometers above the surface of the Earth.

The times, altitudes, and velocities for all the events in the ascent described in this section represent the parameters for a standard mission where the goal is to reach a circular orbit at an altitude

of 296 kilometers and a 28.5° inclination. Most missions utilize this type of orbit because it maximizes the amount of payload mass deliverable to space from KSC. Each mission requires a slightly different MECO velocity, insertion orbit, and different final orbit. Despite this fact, the parameters for a different type of target orbit are similar.

5.4 What Happens During a Launch Emergency

Abort profiles deal with unexpected events during ascent that may require the termination of a flight. Such events include the failure of one or more main engines, or a sudden air leak in the crew cabin. Intact modes and contingency modes represent the two basic types of aborts for Space Shuttle ascents. An intact abort provides a preplanned method for a safe return of the orbiter to a designated landing site. On the other hand, a contingency abort occurs after a severe failure where an intact abort is no longer possible. In this case, the astronauts attempt to land the orbiter wherever possible and hope for the best. Both abort modes assume that the orbiter remains in reasonable functioning condition. Catastrophic failures such as the *Challenger* accident fall into the "not survivable" category. As such, no contingency plans exist for these types of occurrences.

The earliest intact abort option available takes the form of a return to the launch site (RTLS). If a single main engine on the orbiter fails during the first four minutes of the ascent, the orbiter's chances of reaching orbit vanish. The plan in this case involves returning back to Kennedy Space Center and landing at a runway several kilometers from the launch pad (see Figure 46 on page 182). An RTLS abort initiation must wait until after solid rocket booster separation if it has not yet occurred. Once the boosters separate, the two functional main engines continue to power the Space Shuttle upward in order to burn off a significant amount of propellant in the external tank. Then, the shuttle spins around so that it points back toward KSC with the orbiter on top of the tank. This maneuver is called the *powered pitch around* (PPA).

After the PPA, the two main engines must continue to burn off propellant until less than 5% of the original propellant load remains in the external tank. Jettisoning a non-empty tank will cause the propellant to slosh. In turn, the sloshing may cause the tank to collide with the orbiter. After the propellant runs out, main engine cutoff occurs, followed by external tank separation. The orbiter, now powerless without propellant to feed the main engines, glides to the Kennedy Space Center's runway. An RTLS abort can only occur within the first four and one-half minutes of flight. After that, the orbiter's velocity and downrange distance from KSC makes an RTLS impossible. NASA engineers call this point in time "negative RTLS."

If a main engine fails between the time of negative RTLS and about six minutes after launch, the Space Shuttle still will not be able to reach orbit. However, after four minutes of flight, the shuttle has

Shuttle Abort Modes

Return to Launch Site (RTLS)
Mandatory during the first few minutes of flight if one main engine fails. Shuttle jettisons the SRBs, burns out the propellant in the ET, then turns around and lands at KSC.

Trans-Oceanic Abort Landing (TAL)
After a few minutes of flight, the orbiter will have gained enough velocity to cross the Atlantic Ocean and land in either Spain or Africa in the event of a main engine failure. This mode is safer than an RTLS.

Abort Once Around (AOA)
Results from a main engine failure late in ascent but before the shuttle has gained enough velocity to achieve a low orbit. In this mode, the shuttle uses its maneuvering thrusters to enter an extremely shallow, unstable orbit and lands in California or Florida after less than one revolution around the world.

Abort to Orbit (ATO)
Preferred option for a main engine failure late in ascent. In this mode, the orbiter's velocity is fast enough to achieve a safe, but lower than planned, orbit.

Contingency Abort
Extreme emergency situation that does not fall into one of the previous four categories and where no plans exist for an intact landing. In this case, pilots make best attempt to land and hope for the best. All failures involving the solid rocket boosters fall into this category.

RULE OF THUMB

If two out of three main engines fail during the first four minutes of flight, a return to launch site abort may not be possible. The reason is that the powered pitch around maneuver that turns the shuttle back toward the launch site requires at least two functional main engines.

The Dark Side of Space

It was an unseasonably bitter cold winter night at Launch Complex 39B. The temperatures had fallen below freezing, and icicles had formed on the metal launch tower next to Space Shuttle *Challenger*. However, the countdown was proceeding smoothly, and the weather forecast for a morning launch looked favorable.

As the sun rose over the Kennedy Space Center on 28 January 1986, seven astronauts were busy preparing to board *Challenger* for America's 25th Space Shuttle mission, code-named 51-L, according to the official NASA manifest. The crew consisted of (top row, left to right) mission specialist Ellison Onizuka, payload specialist Sharon Christa McAuliffe, payload specialist Gregory Jarvis, mission specialist Judith Resnik, (bottom row, left to right) pilot Michael Smith, commander Dick Scobee, and mission specialist Ronald McNair. This crew represented American diversity at its best. Onboard were five men and two women. Among them were a Japanese American, an African American, and an elementary school teacher.

Unknown to NASA engineers, the bitter cold weather had critically damaged one of the solid rocket booster's O-rings. These rings functioned like giant rubber washers, and were designed to keep the hot exhaust gas from leaking out of the solid rockets at the joints where the booster segments stack on top of each other to form the complete booster. About sixty seconds into the flight, the right solid rocket booster began to leak, and a small flame emerged from the leaking joint (see small bright spot near bottom of the booster, top photograph on opposite page).

The flame gradually cut through and weakened the structure of the external tank for the next twelve seconds. Then, the lower attachment that held the booster to the tank failed and caused the booster to rotate into and collide with the tank. Within milliseconds, the external tank's structure completely failed, and a cataclysmic fireball of exploding hydrogen and oxygen propellant instantly enveloped *Challenger* and tore it to bits (see middle photograph, opposite page). Immediately before the explosion, pilot Michael Smith uttered a surprised, "uh-oh." They were the last audible words transmitted from *Challenger*. There were no survivors.

Several seconds after the explosion, only an expanding cloud of smoke and gas remained (see bottom photograph, opposite page). Many looked toward the sky and hoped that *Challenger* would miraculously emerge from the fireball. It never did. For hours, fire and debris rained down from the sky into the Atlantic Ocean. One of

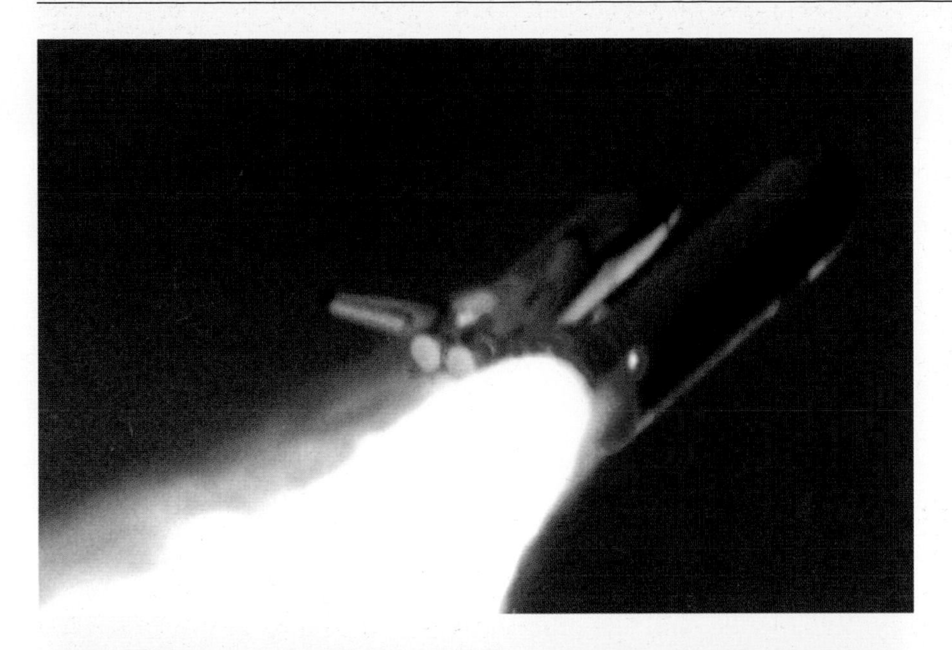

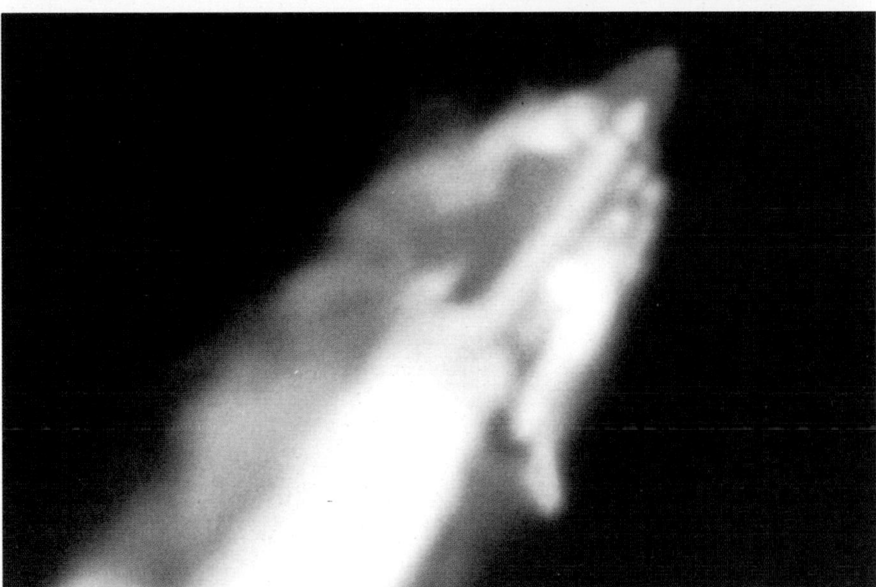

those pieces was *Challenger's* crew cabin with the seven astronauts still strapped in their seats. During the next several weeks, Navy divers managed to recover about 35% of the orbiter, 50% of the solid rockets, and 50% of the external tank from the seabed of the Atlantic.

Experts initially believed that the astronauts died during the explosion. However, upon examination of the recovered crew cabin, no signs of explosion, heat, or fire damage were found. Most of the damage to the cabin occurred when it struck the surface of the Atlantic. All of this evidence indicates that some of the astronauts may have been alive until the moment the cabin hit the water, but they almost certainly lost consciousness when the orbiter broke apart due to a rapid loss of air.

Former Mercury astronaut John Glenn, who made the first American orbital flight in 1962, said about *Challenger*, "We always knew there would be days like this." The amazing fact was that more than fifty American space missions had flown before 51-L without loss of life. This amazing record had caused much of America to become oblivious to the hazards faced by astronauts. Glenn himself almost burned up during reentry when the heat shield on his Mercury capsule came loose in 1962; Neil Armstrong and David Scott faced a near-death situation in 1966 when their Gemini capsule spun out of control in Earth orbit; and Jim Lovell, Jack Swigert, and Fred Haise almost suffocated in 1970 when an oxygen tank in their Apollo spacecraft exploded on the way to the Moon. Spaceflight will always remain extremely dangerous, despite the best of planning. With *Challenger*, NASA's luck simply ran out.

The lesson to be learned from the *Challenger* accident can be summarized by a quote from astronaut Gus Grissom, who perished during a training accident in 1967. He said, "If we die, we want people to accept it. We're in a risky business and we hope if anything happens to us, it will not delay the program."

Most of America felt the same way after the initial shock wore off. President Ronald Reagan immediately appointed former Secretary of State William Rogers to head a panel of distinguished scientists, astronauts, and engineers to investigate the accident. Less than two years after the flight of 51-L, Space Shuttles were flying again, complete with redesigned solid rocket boosters. The *Challenger* crew's dream was to explore space. Continuing with space exploration efforts keeps their dreams alive. ❑

RULE OF THUMB

In an emergency, it is not possible to land the orbiter until after jettisoning the solid rocket boosters and external tank. In addition, many engineers feel that it is not safe to jettison the solid rocket boosters while they are still thrusting.

Historical Fact

Although shuttle abort modes deal with main engine failures during ascent, a main engine has never failed in more than 70 launches. However, during the launch of STS-51F on 29 July 1985, one of the main engines on *Challenger* shut down shortly before reaching orbit. Eventually, the failure was attributed to the flight computer shutting the engine down due to a "warning light" from a faulty sensor, not a malfunction in the engine itself.

enough velocity to reach the other side of the Atlantic Ocean and land in Africa or Spain. This type of abort, called a *transoceanic abort landing* (TAL), can occur at one of several runways. For a due east launch from KSC to reach a 28.5° inclination orbit, TAL runways exist in the African cities of Ben Guerir in Morocco, Dakar in Senegal, and Banjul in Gambia. Launches that depart KSC traveling northeast to reach high inclination orbits (40° to 57°) utilize the United States Navy's airstrip in Rota, Spain as the TAL site.

After initiation of a TAL abort, the orbiter continues to ascend using the propellant from the external tank. However, the final altitude and velocity at main engine cutoff will fall extremely short of the target values for a normal ascent. Before jettisoning the external tank, small thrusters on the orbiter roll the entire Space Shuttle into a position where the orbiter rides on top of the tank. After jettisoning the tank, the orbiter glides to a landing at the designated location. NASA considers the TAL a safer option than an RTLS. If an abort situation develops after four minutes but before the time of negative RTLS, the flight controllers at Mission Control in Houston will probably instruct the astronauts to perform a TAL. Five minutes and thirty seconds after launch, a TAL becomes possible even if two main engines fail.

Six minutes after launch, an option called an *abort once around* (AOA) becomes feasible if one main engine fails. Like the TAL scenario, the final altitude and velocity when the external tank runs out of propellant will fall short of the target values for a normal ascent. However, by the time an AOA becomes feasible, the Space Shuttle's velocity allows it to achieve a very low orbit through the upper atmosphere. Normally, these types of orbits rapidly decay due to atmospheric drag and air friction. To compensate, the orbiter uses one or more burns from its two small orbital maneuvering thrusters (OMS) to sustain the temporary orbit before gliding to a landing at Edwards Air Force Base in Southern California, White Sands Test Facility in New Mexico, or the shuttle runway at the Kennedy Space Center. Seven minutes after launch, an AOA becomes feasible even if two main engines fail.

The final option, called an *abort to orbit* (ATO) becomes feasible six minutes and thirty seconds after launch if only one main engine fails. Like the AOA, the Space Shuttle's velocity when the external tank runs out of propellant will fall short of the desired value. In this case, the velocity differential will be small enough for the astronauts to use the two orbital maneuvering thrusters to make up some or all of the difference.

Seven minutes after launch, an ATO becomes feasible if two main engines fail. During the ascent, mission control informs the astronauts that ATO is possible at seven minutes into the flight by saying, "Single engine press 109." At seven and one-half minutes into the flight, they say, "Single engine press 104." These two phrases mean that the orbiter can still achieve orbit even if two main engines

fail by throttling the remaining engine at 109% and 104% of the rated thrust, respectively.

Of the four intact abort options, ATO has occurred once during a real mission. The others have only occurred during computer simulations. Extreme emergencies such as the failure of three main engines, problems with the solid rockets, ruptures in the external tank, or structural failure require a contingency abort and may not be survivable. Keep in mind that a safe landing, either intact or contingency, requires jettisoning the solid rockets and external tank. Both actions must wait until all the propellant burns off. Looking back at the *Challenger* accident when a solid rocket booster leaked and caused the external tank to explode, many people often ask if it would have been possible to jettison the still thrusting boosters if somebody found out in time. Such an action is theoretically possible, but only with a slim chance for a safe result.

5.5 Who Flies in the Space Shuttle?

At least two, and no more than seven, astronauts crew each Space Shuttle mission. Shuttle astronauts fall into three different categories: pilots, mission specialists, and payload specialists. The total number of astronauts and the number of each type of astronaut assigned to a flight varies depending on the mission objectives. The table in this section lists the number and types of astronauts typically assigned to fly on various types of shuttle missions.

Two pilots fly on every Space Shuttle mission. One of the pilots, called the *commander*, leads the entire crew, takes responsibility for overall mission execution, and makes decisions deemed necessary to preserve crew safety and vehicle integrity. The other pilot astronaut, called the *pilot*, ranks second in command of the mission, and assists the commander in the conduct of the mission. Although the orbiter's computers control the flight during critical phases of the mission, the two pilot astronauts take responsibility for executing the flight plan on the computer and monitoring the flight systems and instruments. During some mission phases such as the last phase of the landing, the commander manually flies the orbiter and the pilot serves as the commander's copilot.

Up to five mission specialists, typically trained scientists with doctorate degrees, conduct the scientific and engineering aspects of the mission once the orbiter achieves Earth orbit. Some of their typical duties include conducting scientific experiments, monitoring payload systems, and deploying payloads carried into space by the orbiter. Also, mission specialists put on space suits and venture outside the orbiter to repair satellites or to test future space station construction techniques when necessary. Both pilots and mission specialists work for NASA as a career and usually receive the opportunity to serve on several missions.

Typical Shuttle Crew Sizes

Flight Test Mission
1 Commander
1 Pilot
Total of 2

Satellite Deployment Mission
1 Commander
1 Pilot
2 or 3 Mission Specialists
Total of 4 or 5

Satellite Rescue, Repair, or Retrieval
1 Commander
1 Pilot
3 to 5 Mission Specialists
Total of 5 to 7

Dedicated Scientific Research
1 Commander
1 Pilot
3 Mission Specialists
2 Payload Specialists
Total of 7

Multipurpose (Science & Satellite Deploy)
1 Commander
1 Pilot
3 Mission Specialists
2 Payload Specialists
Total of 7

Space Station Construction
1 Commander
1 Pilot
3 to 5 Mission Specialists
Total of 5 to 7

Historical Fact

Typically, seven astronauts represent the maximum number that NASA will assign to a single mission. However, on STS-71 in July 1995, *Atlantis* launched with seven astronauts toward the Russian *Mir* space station. The orbiter delivered two Russians to the station and returned with two other Russians and one American from *Mir*. Therefore, although seven went up, eight came down.

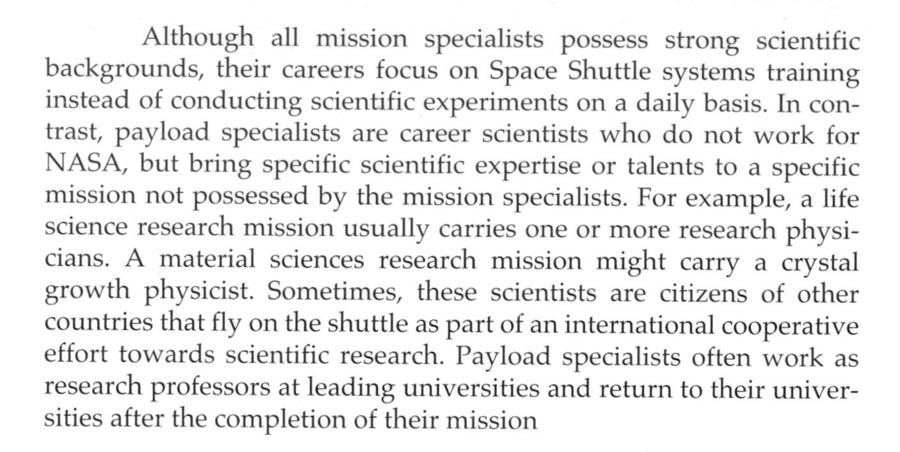

Although all mission specialists possess strong scientific backgrounds, their careers focus on Space Shuttle systems training instead of conducting scientific experiments on a daily basis. In contrast, payload specialists are career scientists who do not work for NASA, but bring specific scientific expertise or talents to a specific mission not possessed by the mission specialists. For example, a life science research mission usually carries one or more research physicians. A material sciences research mission might carry a crystal growth physicist. Sometimes, these scientists are citizens of other countries that fly on the shuttle as part of an international cooperative effort towards scientific research. Payload specialists often work as research professors at leading universities and return to their universities after the completion of their mission

American men and women of all races serve as Space Shuttle astronauts. However, to this date, only one woman has flown aboard the orbiter as a commander. This exclusion is not indicative of a NASA policy to exclude women from command rank. NASA requires extensive high-performance jet pilot expertise of all orbiter pilots. As a result, almost all shuttle pilots come from the military, where men fly almost all of the high-performance jets. Although prior military service is not a requirement for employment as a shuttle pilot, very few employers in the world offer training in high-performance jets. The situation in the military is changing, and in the future, both men and women will pilot orbiters.

5.6 Living and Working in Space

One of the first tasks on the astronauts' "to do upon entering orbit" list involves opening the payload bay doors. This task requires no more effort than opening a garage door with an automatic door opener. Figure 43 on page 153 shows what the orbiter looks like when the doors are open. Notice that keeping the bay doors open exposes the entire bay and all of its contents to the vacuum of space. The inside lining of the payload bay doors serves as a device that radiates the heat generated by the orbiter's electrical and life support systems away from the orbiter. A shuttle mission requires an abort eight hours after reaching orbit due to internal overheating if the payload bay doors remain closed.

Next, the astronauts go about transforming the inside of the crew cabin from ascent mode to orbital operations mode. During the launch and landing phases of a mission, the commander, pilot, and two mission specialists sit on the top deck of the cabin, while the other mission specialist and the payload specialists sit on the lower deck. After reaching orbit, the astronauts fold and place all the seats in storage except for the commander's and pilot's seats. Removing most of the seats in the orbiter gives the astronauts more room in which to work. Chairs serve no useful purpose in the weightless environment caused by free falling in orbit.

RULE OF THUMB

Do You Have the "Right Stuff?"

NASA is always on the lookout for qualified individuals to serve as shuttle astronauts. All applicants must have at least a bachelor's degree from an accredited university in engineering, biological science, physical science, or mathematics followed by three years of professional experience. Advanced degrees such as master's and doctoral degrees are desirable and often necessary due to the intense competition from other potential applicants.

Pilot candidates must have at least 1,000 hours pilot-in-command time in jet aircraft, 20/50 or better uncorrected vision correctable to 20/20 in each eye, and be no taller than 193 centimeters (6 feet, 4 inches).

Mission specialist candidates must have an uncorrected vision of 20/100 or better, correctable to 20/20 in each eye.

Astronauts are expected to be team players with just the right amount of individuality and self-reliance to be effective crew members. Thousands of people apply, but few make it. NASA accepts applications on a continuous basis and selects new candidates about every two years. Those who are selected must commit to work for NASA for at least several years even if they drop out of the astronaut corps.

Interested? Write to:

Astronaut Selection Office
Mail Code AHX
NASA Johnson Space Center
Houston, TX 77058

Opposite Page Top:
"Fish-eye" view of what the flight deck of an orbiter looks like as seen from the front end. The commander is on the right and the pilot is on the left.

Opposite Page Bottom:
View of what the mid-deck of an orbiter looks like. The "hole" near the bottom right is the hatch to the airlock.

NASA

▶ Figure 48: *How to Control the Orbiter's Attitude*

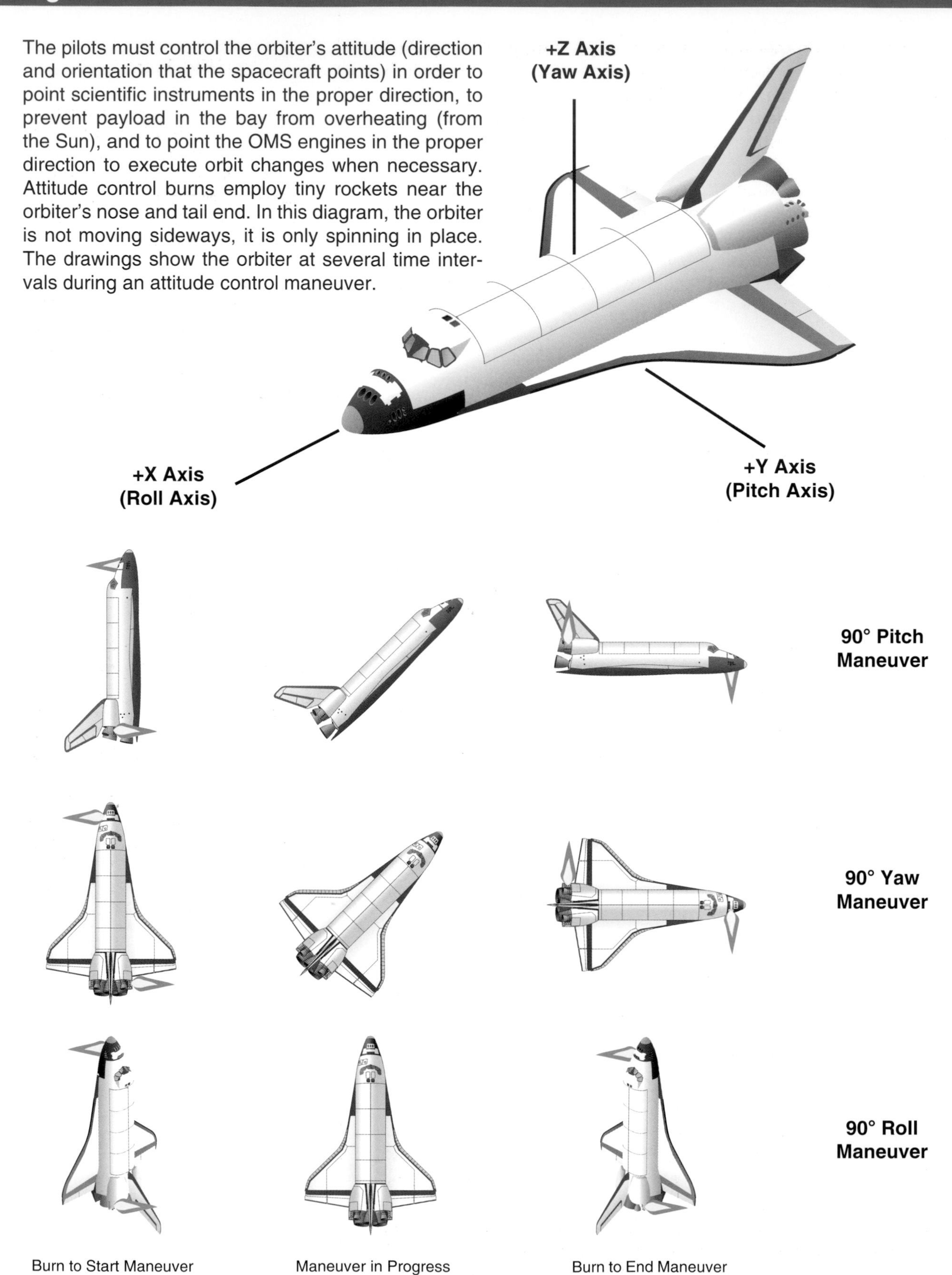

Space inside the two-level crew cabin is tight. The top deck, called the *flight deck,* contains the controls necessary to *pilot* the orbiter during ascent, landing, and in orbit. An airplane-like cockpit occupies the front of the flight deck. The cockpit seats allow the commander and pilot unrestricted access to all of the instruments necessary to control the orbiter in space and to fly the orbiter during the landing phase. During flight, the cockpit windows provide the pilots with a view of the external surroundings. Behind the cockpit, on the back wall of the flight deck, controls provide the astronauts with the means to operate systems in the payload bay. Two windows on this wall allow astronauts to see out into the payload bay and into space when the bay doors are open.

The bottom deck, called the mid-deck, contains storage areas for provisions such as food, crew sleeping stations, a personal hygiene station, and the airlock that gives the astronauts access to the unpressurized payload bay. On the launchpad and on the runway after a landing, the crew access hatch allows the astronauts to move between the outside world and the mid-deck. This hatch stays tightly shut during all phases of the flight. An inter-deck access hatch allows easy access between the mid- and flight decks. In order to understand how crowded the orbiter can be during a mission, imagine spending a week with seven of your friends in a Winnebago-type camper vehicle.

The entire crew cabin remains pressurized during every phase of a Space Shuttle mission. In fact, the chemical composition of the air and the pressure level of the air inside the crew cabin resemble that on Earth at sea level. Astronauts in orbit wear normal clothing instead of pressure suits and do not require oxygen masks to breathe. However, they do wear pressure suits during ascent and descent in the unlikely case of a sudden cabin decompression.

Attitude Control

Remember that upon entering space, the orbiter enters a state of continuous free fall that eliminates the need to fly nose first like an airplane in the atmosphere. Free fall motion in a vacuum always works regardless of which way the orbiter's nose points. However, proper orientation of the orbiter's nose in space plays an important role in activities such as pointing scientific instruments in the payload bay in the proper direction, shielding the contents of the payload bay from intense sunlight, taking pictures of the Earth through the orbiter's windows, or preparing for orbit changes.

Attitude control describes the process of controlling the direction in which the orbiter points. The pilots change the orbiter's attitude by performing a roll, pitch, yaw, or a combination of these three maneuvers (see Figure 48 on page 196). A roll occurs when the orbiter spins around its longitudinal axis (the long axis that runs from front to back); a pitch occurs during a spin around the transverse axis (the axis from left to right assuming that the nose is in front); and a

Scientific Fact!

About half of the Space Shuttle astronauts experience space adaptation syndrome upon entering orbit. The symptoms of this affliction include a feeling of flu-like malaise, nausea, and possible vomiting. In general, space adaptation syndrome feels like motion sickness or being seasick on a ship rocking in the waves. Scientists believe that the cause of this strange affliction comes from a sensory conflict between the inner ear and the eyes. The inner ears tell the brain that body is falling, but the eyes do not because the orbiter falls at the same rate as everything inside it. Typically, astronauts adapt and begin to feel better after a day in space.

RULE OF THUMB

RCS stands for *reaction control system.* Orbiters possess two types of RCS thrusters. One type, called a vernier thruster, allows the pilots to gently roll the spacecraft into a new orientation. They use a hand controller similar to an airplane's control stick. The computers detect motion in the hand controller and instantly figure out which vernier thrusters to fire. Another type of RCS thruster, called a primary thruster, is used for rapid rotations or for moving the orbiter sideways in its orbit. From the inside of the orbiter, this type of RCS burn sounds like a cannon being fired.

Above:

While in darkness, Space Shuttle *Discovery* fires its reaction control system (RCS) rockets to rotate itself into a new orientation. The exhaust from these tiny rockets can clearly be seen as bright streams emanating from the tail of the orbiter. This photograph was taken in May 1991.

yaw occurs during a spin around the vertical axis (the axis from top to bottom assuming that the payload bay is on top). In a weightless free fall condition, a spinning object must spin around its center of mass.

Tiny rockets, called the *reaction control system* (RCS), burn a hypergolic combination of hydrazine and nitrogen tetroxide to control the orbiter's attitude (see Figure 43 on page 153). One front RCS pod and two rear RCS pods each contain different rocket nozzles that direct thrust up, down, left, and right. Figure 48 on page 196 illustrates in which directions the RCS thrusters must fire to effect a roll, pitch, and yaw. In space, nothing acts to stop the orbiter from spinning once the RCS thrusters shut off. To stop the orbiter from spinning, the pilot must command the thrusters to fire in the opposite direction that sets the orbiter spinning in the first place. Notice (in the figure) that each spin maneuver requires an equal and opposite force on both sides of the orbiter's center of mass. A force acting on only one side of the center of mass causes the orbiter to both rotate and move (translate) in the direction of the applied force.

Earlier sections in this book mentioned that firing thrusters or rockets in space changes the orbit. When an orbiter spins around its center of mass, no net changes in forward (horizontal) velocity or radial (up and down) velocity occur. Both orbital velocity components remain the same. The only difference is that the orbiter either faces a different direction and moves with the same velocity, or continues to spin, but moves with the same velocity. As a result, no net change in the orbit occurs. Maneuvers that cause the orbiter to both spin and translate tend to change both the horizontal and radial velocity components. The new orbital elements depend on how much the RCS affects each velocity component.

RULE OF THUMB

In space, the orbiter need not fly nose first like a plane in the atmosphere. Whether the orbiter flies backward, upside down, or sideways, the laws of orbital motion will work the same because no air exists in space. Consequently, there is no need for a streamlined, aerodynamic shape. The orbiter only looks like an airplane because it must eventually glide through the atmosphere to a landing at a runway.

Orbit Changes

Orbit changes represent an integral part of every Space Shuttle mission. Every ascent from the launchpad ends with an orbit change from the orbit achieved at main engine cutoff to the final (usually circular) orbit. Final in this case means the final orbit achieved during the ascent phase. During the orbital phase of the mission, astronauts frequently change the orbit to accomplish different tasks, such as deploying or rescuing satellites. Every mission ends with an orbit change to allow the orbiter to descend into the atmosphere and toward the runway.

After external tank jettison, the main engines lose their propellant supply and serve no useful purpose during the rest of the mission. Instead of the main engines, the two orbital maneuvering system engines (OMS) produce the necessary thrust to effect orbit changes (see Figure 43 on page 153). Each engine develops a thrust of 27,000 Newtons (6,000 pounds) using a hypergolic combination of monomethyl hydrazine and nitrogen tetroxide. The orbiter carries a very limited amount of OMS propellant. All of the propellant avail-

able to both engines contains the potential to produce a total of 305 meters per second (1,000 feet per second or 682 miles an hour) worth of velocity change. Mission planners carefully distribute this capacity among the OMS-2 burn after main engine cutoff, all the orbit changes during the mission, and the deorbit burn to come home.

Both of the OMS rockets point in about the same direction as the orbiter's nose. When they thrust, they propel the orbiter roughly in the same direction that the nose points. Before an OMS burn, the pilots use the RCS to roll, pitch, and yaw the orbiter so that the nose points in the direction of the intended OMS thrust. The orbiter must point forward for a posigrade OMS burn to raise an orbit, and it must point backward for a retrograde OMS burn to lower an orbit. In reality, the OMS engines point 15° below the longitudinal axis. These engines also gimbal to provide some steering. Consequently, the orbiter's nose does not necessarily point in the exact OMS thrust direction during a burn.

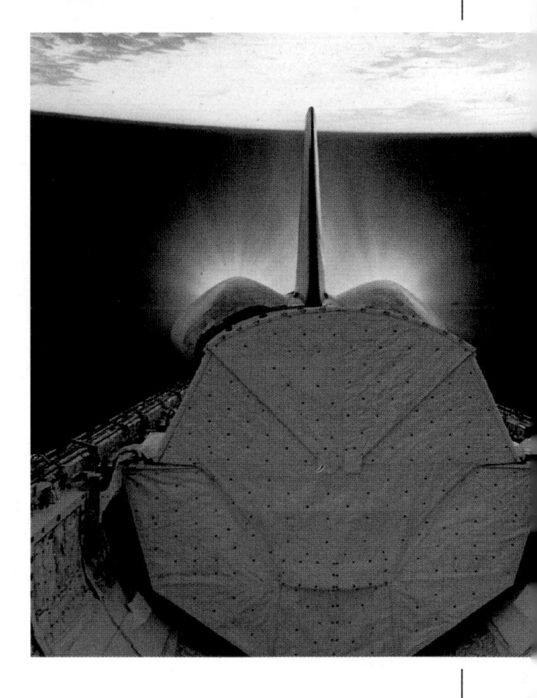

Above:

November 1982 – Commander Vance Brand fires the orbital maneuvering system engines (OMS) to change *Columbia's* orbit on mission STS-5. The bright glow emanating from the rear of the orbiter is exhaust from the OMS burn. This photograph was taken from inside *Columbia*, looking back toward the rear.

Satellite Deployment

The Space Shuttle frequently carries satellites in the payload bay for deployment in space. Some of these satellites require placement in low Earth orbit. In these cases, the pilots fly the orbiter into the satellite's destination orbit by using the OMS, then the mission specialists release the satellite from the payload bay by remote control at the proper time. Most of the other satellites that hitch a ride in the payload bay require placement at a geosynchronous altitude of 35,786 kilometers (22,236 miles) above the ground. Unfortunately, the amount of fuel in the OMS tanks limits the orbiter to low Earth orbit. The solution to this problem involves attaching a small rocket, generically called an upper stage, to the geosynchronous satellite while both are still on the ground. The orbiter then delivers the coupled pair to low Earth orbit. There, the upper stage provides the thrust to insert the satellite into the low Earth to geosynchronous transfer orbit.

RULE OF THUMB

The Space Shuttle is limited to low Earth orbit. In order to send a satellite to geosynchronous orbit using the shuttle, NASA must attach an "upper-stage" rocket to the satellite and then launch the pair from the orbiter.

In the 1980s, NASA frequently used a Payload Assist Module (PAM-D) upper stage on Space Shuttle satellite deployment missions. The "D" stands for "Delta" because this upper stage also flies on the Delta 2 rocket as its third stage. Most upper stages like the PAM consist of nothing more than a spherical solid-propellant fuel tank connected to a rocket nozzle. The top of the PAM's fuel tank attached to a "mating ring" type device that connected the rocket to the bottom of the satellite.

PAM stands for *Payload Assist Module*. In the 1980s, NASA used these rockets to send small to medium-size communications satellites to geosynchronous orbit from the shuttle.

The PAM and satellite formed a vertical stack with the PAM's nozzle at the bottom of the stack and the top of the satellite at the top. This stack sat upright inside a protective white shroud located near the back of the orbiter's payload bay. The shroud acted as a protective pair of jaws. In the closed position, the jaws completely encapsulated the stack to protect the satellite from overheating due to solar exposure while awaiting deployment in low Earth

Above:

Two-photograph sequence shows Space Shuttle *Discovery* deploying a communications satellite from low Earth orbit in late August 1985. The satellite is the dark cylindrical object, while the PAM upper stage is the spherical bulb with rocket engine nozzle below the satellite. Before deployment, both the satellite and PAM sat in the protective white cradle (bottom middle of left photo). In both photographs, the PAM has not yet begun to thrust. Typically, an upper stage will wait for about 45 minutes before igniting its engine to send its satellite to geosynchronous orbit. This delay gives the shuttle plenty of time to move far enough away for safety.

orbit. During satellite deployment missions, Space Shuttle orbiters had the capability to carry as many as three PAMs and three satellites in three separate shrouds.

Before the deployment, a phonograph turntable–like device within the protective shroud spun the stack around its vertical axis. This action, called *spin stabilization,* helped the PAM point in the proper direction in space without the aid of attitude control thrusters. Half an orbit before the scheduled ignition of the PAM, a mission specialist deployed the satellite from a control console on the flight deck, located behind the orbiter's cockpit. Deployment required nothing more than flipping a switch. The electrical circuit attached to the switch controlled the uncoiling of several springs beneath the PAM. The force of the uncoiling springs gently ejected the spinning stack from the shroud.

During the next half-orbit (about 45 minutes), the pilots maneuvered the orbiter away from the PAM. The astronauts needed to move far enough away for safety, but also needed to stay close enough to perform a rescue and repair in case of ignition failure. In the 45 minutes, the orbiter drifted into a position roughly 10 kilometers (6.2 miles) behind and above the PAM. Astronauts typically

chose a rear position because a front position invited a catastrophic collision if the PAM malfunctioned and veered off course. Remaining several kilometers behind minimized contact between the orbiter and the aluminum oxide particles in the PAM's exhaust.

A PAM-D deployed from a Space Shuttle orbiter delivered roughly 1,250 kilograms (2,756 pounds) to geosynchronous orbit. Heavier satellites use the more powerful Inertial Upper Stage (IUS). In reality, the IUS is a two-stage rocket capable of boosting satellites that weigh more than 3,000 kilograms (6,614 pounds) from the shuttle's orbit to geosynchronous orbit. In the late 1980s, NASA also used the IUS to boost several scientific spacecraft into trajectories bound for other planets in the solar system. Unlike the PAM, the IUS fits in the payload bay horizontally because its length plus a typical large satellite's length exceed the depth of the bay.

Above:

2 August 1991 - The Tracking and Data Relay Satellite (TDRS-E) is gently ejected from the payload bay of Space Shuttle *Atlantis* a mere six hours after launch. In this photograph, the Inertial Upper Stage (IUS) booster is visible near the top. The TDRS is connected to the top of the IUS but is not visible in the picture. Later, the IUS will fire its engines to boost the TDRS to geosynchronous orbit, 35,768 kilometers above the ground.

Robot Arm

Many people who know little about the Space Shuttle itself know about the existence of the famous robot arm. Officially, the arm is called the *remote manipulator system* (RMS) and was manufactured in Canada as part of an international cooperative effort to get other countries involved with the shuttle program. From inside the crew cabin on the flight deck, astronauts can control the arm to grab and move large objects in and out of the payload bay. Each orbiter, with the exception of the *Enterprise,* has a robot arm installed in the payload bay.

The robot arm design closely models a human arm and includes an upper arm, a lower arm, and an end effector. A shoulder joint attaches to the upper inside wall on the port (left) side of the payload bay and allows the arm to move up, down, left, and right, similar to a human shoulder. An elbow joint connects the upper and lower parts of the arm. Like a human elbow, the RMS elbow joint acts as a hinge and restricts the lower arm's movement between bending the lower arm towards the upper arm and bending the lower arm away from the upper arm so that the arm extends fully. A fully extended RMS arm measures 15.24 meters (50 feet). However, a major difference exists. On a fully extended human arm with the palm of the hand facing upward, the elbow bends upwards. In contrast, the elbow bends downward on a fully extended RMS.

A wrist joint connects the lower arm to an end effector device that serves as the RMS's hand. This joint allows the end effector to rotate, pitch, and yaw just like a human wrist. Rolling the end effector is analogous to rolling a human wrist between the palm up and the palm down position; pitching the end effector is analogous to positions between a stiff wrist and a limp wrist, while yawing the end effector is analogous to moving a hand left and right while keeping the wrist stationary.

Historical Fact

TDRS stands for *Tracking and Data Relay Satellite*. In the early days of spaceflight, it was only possible to communicate with the spacecraft during the short time periods when it was in range of a ground station. Today, the use of two TDRS satellites in geosynchronous orbit allows video and audio contact with a Space Shuttle orbiter for nearly 85% of the time it is in orbit.

Above:

Astronauts George "Pinky" Nelson and James van Hoften work on repairing the *Solar Max* satellite (floating above the back of the payload bay) in April 1984. One of the two is standing at the end of the robot arm. The arm's shoulder is not visible in the picture. However, the "elbow" is visible at the top right-hand corner. The bright glow at the top comes from the Earth. This photograph was taken from the front of the *Challenger* looking toward the rear.

The similarities between the human arm and the RMS end with the end effector. This device looks nothing like a human hand and lacks finger-like digits with which to grab objects. Instead, it looks like a hollow cylinder about the size of a large coffee can. When the astronauts want to grab an object, they move the arm and place part of the object inside the end effector. Three wires connect to the outside shell of the end effector. The other ends of the wires connect to a ring that fits snugly along the inside like a rubber washer in a pipe. When the inside ring rotates, each wire stretches across the center of the end effector's opening and forms a tight-fitting triangular snare over the object.

Keep in mind that the RMS can only move large objects designed with the end effector in mind. A sturdy item not likely to bend or break off must protrude from the large object so the end effector can fit over it. Because the end effector lacks fingers, it can not grab onto edges of objects, solar panels, or long poles and rods. Many people often wonder why NASA uses this funny scheme instead of providing robot fingers to the RMS. Keep in mind that the robot arm design dates back to the early 1970s. During that time period, the RMS design represented a breakthrough in robotic technology. Today, crude fingers exist on a few experimental arms in engineering laboratories. However, engineers still do not know how to construct dexterous, lightweight, and strong fingers. Each one of these three qualities is essential for efficient operation in space.

Astronauts control the RMS from the flight deck at a console located behind the cockpit. Two arcade-game type joysticks provide manual control of the arm. In addition to the joystick, the computer can move the arm into a specific orientation and location in space. However, the computer does not accept commands like, "Get the satellite and move it into the payload bay." Typically, mission specialists operate the arm and view the operations through the rear windows on the flight deck. In addition, television cameras on the wrist and elbow provide additional visual coverage of RMS activities.

Above:

This photograph shows a close-up view of the robot arm's wrist and end effector against the Earth in the background. Notice that the end effector does not have fingers. Instead, it looks like a canister. Grabbing an object involves putting a part of it inside the opening at the end of the end effector. Then, a device inside the end effector snares the object with a set of three wires. This photograph was taken from *Discovery* in April of 1985.

Extra Vehicular Activity and Satellite Rescue

Astronauts sometimes venture out into the vacuum of space to perform tasks such as repairing broken satellites or testing space station construction techniques. Officially, NASA uses the term *extra vehicular activity* (EVA) to describe these tasks. However, everybody uses the colloquial term *space walk*. In reality, this term is misleading because astronauts float instead of walk during an EVA. Try walking underwater in a swimming pool to understand the difficulty of physically walking in space.

The crew-side hatch on the mid-deck provides a means to enter and exit the orbiter on the ground or at the launchpad. However, if an astronaut left the orbiter for an EVA by way of the side hatch, all the air inside the pressurized cabin would rush out into the vacuum of space. Such a scheme would force all the crew members to put on pressurized space suits even if only one crew member performed an EVA. Fortunately, a simpler solution exists.

Scientific Fact!

RMS stands for *remote manipulator system.* It is the technical term for the "robot arm." The delicate design and construction of the RMS would cause it to collapse under its own weight if somebody tried to use it at the surface of the Earth. However, during the weightless condition experienced in orbit, the RMS can move objects with relative ease.

Most EVAs occur in the payload bay, which remains exposed to the vacuum of space throughout the entire orbital phase of a Space Shuttle mission. A small cylindrical room, called the airlock, provides access between the crew cabin's mid-deck and the payload bay. This room contains one door that leads into the crew cabin and another door that leads out into the vacuum of space. EVA-bound astronauts exit the orbiter by first entering the airlock and closing the door to the crew cabin. In the airlock, they put on their pressurized space suits,

Above:

16 September 1994 - On the STS-64 mission, astronaut Mark Lee floats freely as he tests the new Simplified Aid for EVA Rescue (SAFER) unit. This system is a jet-backpack similar to, but smaller and simpler than, the Manned Maneuvering Unit (MMU) used by shuttle astronauts in the 1980s. NASA is currently developing the SAFER system for use by astronauts on the new *International Space Station*, scheduled for construction in the late 1990s.

Scientific Fact!

EVA stands for *extra vehicular activity*. It is the technical term for "space walk." Before every EVA, astronauts must breathe pure oxygen for about three hours to purge all the nitrogen gas from their bloodstream. WIthout this purge period, a painful and possibly fatal condition called dysbarism may result. This problem manifests itself when nitrogen bubbles collect in the body's joints where the pressure around the body is reduced. Deep-sea divers refer to this problem as the "bends."

start a motor that slowly pumps out the air, and then open the door leading out into the payload bay. The space suits provide enough power and oxygen for an astronaut to work for about eight hours without returning to the airlock.

The weightless environment of microgravity prevents astronauts from walking along the floor of the payload bay. To compensate, hand rails conveniently located along the full length of the bay allow space walkers to float from one end to another by climbing hand over hand. A tether attaches to the space suit and clips to a wire that runs alongside the hand rails. These wires prevent astronauts from floating away from the orbiter, but allow freedom of movement in the bay at the same time. When the orbiter flies over the night side of the Earth, three floodlights on each side of the payload bay illuminate the bay.

Sometimes, an astronaut may need to venture away from the confines of the payload bay. In this case, he or she straps on a large backpack-like device called a manned maneuvering unit (MMU) while in the payload bay. This backpack converts the astronaut into a self-propelled satellite. The MMU's 24 tiny nitrogen gas thrusters allow an astronaut to fly at least 200 meters away from the orbiter and return safely. In addition, these thrusters provide roll, pitch, and yaw control. A normal flight speed rarely exceeds 1.6 kilometers (1 mile) per hour relative to the orbiter.

Mission specialists perform all EVAs. NASA likes to have two astronauts in the payload bay at all times for safety reasons. If one astronaut becomes incapacitated, the other one can lend a hand. The STS-49 mission in May 1992 set a world's record as three mission specialists on the *Endeavour* spacewalked at one time to perform a satellite capture. Although the air lock was designed for two astronauts, the crew used their ingenuity to fit in three at one time.

The most exciting use of the robot arm, astronauts performing EVA, and possibly the MMU, occurs during a satellite rescue and repair mission. Satellites in low Earth orbit sometimes break down because of faulty electronics. Some satellites destined for geosynchronous orbit never leave low Earth orbit because their upper stage malfunctions or misfires. In these cases, astronauts can rendezvous with the crippled satellite and repair it, store it in the payload bay for a return to Earth, or attach a good upper stage depending on the nature of the breakdown. Keep in mind that the crippled satellite must orbit the Earth in a low orbit with an inclination between 28.5° and 57°. The orbiter cannot reach satellites in polar or geosynchronous orbit.

The first step in the rescue mission involves the rendezvous. The pilots, with the aid of mission control, spend one or two days catching up with the satellite. Because the orbiter lacks the fuel to make drastic changes to its orbit in space, it is launched into an orbit of roughly the same inclination and altitude as the crippled satellite. Thus, most of the rendezvous occurs during ascent. Once the orbiter

catches up with the satellite, mission specialists bring it into the payload bay for a repair.

Several major problems exist that make satellite capture extremely difficult. The first deals with spin. Many satellites spin around their vertical axis for stability while they orbit the Earth. Astronauts need to find some way to stop the satellite from spinning before they bring it into the payload bay. The reason is that a spinning satellite in the bay creates a dangerous situation. One seemingly obvious solution involves using the RMS to grab onto the satellite and stop it from spinning. Although satellites in space spin slowly, they do not spin slow enough for an astronaut to maneuver the RMS end effector over a grapple or some other sturdy object not likely to break off the satellite.

Some people have suggested that astronauts just grab onto satellites to stop them from spinning. This is not a practical idea. Many satellites lack fixtures to hold onto. An astronaut could grab onto a solar panel, or an edge of the satellite. However, these sharp edges have the potential to puncture the gloves on a space suit. Even if an astronaut succeeds in grabbing onto the satellite, how does he or she stop the spin? Consider the case of a playground car-

Above Left:

7 February 1984 - Astronaut Bruce McCandless makes history by performing the first test flight of the Manned Maneuvering Unit (MMU). This rocket-powered backpack uses nitrogen thrusters to allow an astronaut to become a free-flying satellite. Before, space walkers always remained connected to their spacecraft by tethers to keep them from floating away. This photograph was taken from Space Shuttle *Challenger*.

Above Right:

9 February 1984 - Astronaut Robert Stewart uses his MMU to glide to within a few meters of Space Shuttle *Challenger* during the second test flight of the "flying backpack" during the mission.

Opposite Page Top:

November 1984 - Astronaut Dale Gardner uses the Manned Maneuvering Unit to fly from Space Shuttle *Discovery* to the *Westar 4* satellite. He docked with the satellite using a circular-shaped device that attached to the satellite's rocket nozzle. Then, he used the MMU's thrusters to stop the satellite from spinning.

Opposite Page Bottom:

November 1984 - Astronauts Dale Gardner (left), and Joe Allen (right) attempt to move the *Westar 4* satellite into the payload bay of *Discovery*. Allen is standing on the end of the "robot arm."

ousel. A person stops the spin by grabbing onto the carousel and then planting both feet firmly on the ground. In space, there is no ground. An astronaut who grabs a spinning satellite will begin to spin with the satellite.

One solution to de-spin satellites involves an astronaut in an MMU flying to the satellite, latching onto it with a device that mechanically attaches the satellite to the MMU, and then using the MMU's thrusters to stop the spinning. This scheme has been tried several times with varying levels of success. The *Solar Max* repair in April 1984 (mission 41-C) marked the first satellite rescue attempt in the history of spaceflight. During the initial capture attempts, the physical act of mechanically docking the MMU to the satellite caused it to wobble and tumble out of control. Eventually, *Challenger* astronauts succeeded in capturing, repairing, and returning *Solar Max* to orbit.

On another satellite rescue in November 1984 (mission 51-A), an astronaut flew to the satellite using an MMU. Then, he inserted a pole-like device, called a *stinger*, into the rocket engine nozzle at the bottom of the satellite. After insertion, the astronaut tripped a switch allowing prong-like claws to fly open and attach to the inside of the engine. One end of the stinger grappled the satellite and the other end connected to the MMU. This connection allowed the MMU's thrusters to stop the satellite's spin. Astronauts on *Discovery* successfully captured two satellites, called *Westar* and *Palapa*, and returned them to Earth on mission 51-A. A failure of each satellite's PAM upper stage had stranded both of these satellites in low Earth orbit several months earlier.

If all else fails, ground controllers can attempt to command the satellite's attitude control thrusters to fire and stop the spin. This solution can fail because a broken satellite typically responds improperly to commands radioed from the ground. Also, a broken satellite may tumble and wobble out of control in addition to spinning. Stopping a wobble, tumble, and spin at the same time requires enough propellant to shorten the satellite's useful lifetime. Remember, a satellite without attitude-control fuel is useless because it can no longer point its antennas or instruments in the proper direction. However, a working satellite low on attitude-control propellant is better than a nonfunctional satellite with a full load of propellant.

After astronauts stop the satellite's spin, they use the RMS to bring the broken spacecraft into the orbiter's payload bay. Unfortunately, most satellites lack fixtures for the RMS to grab. The stinger solves this problem. One end attaches to the satellite and the other end provides a fixture for the RMS end effector. NASA also uses a variation of the stinger called the grapple bar. This bar allows the RMS to lift the satellite like a soda can on a spatula. Unfortunately, attaching these devices to a satellite always presents difficulties. The astronauts on mission STS-49 can attest to this fact.

Satellite Rescue Missions

STS-41C *(Challenger, launched 6-Apr-84)*
Captured *Solar Maximum* scientific satellite and replaced faulty electronics

STS-51A *(Discovery, launched 24-Jan-85)*
Captured *Westar* and *Palapa* communications satellites stranded in low Earth orbit and returned the two to Earth

STS-51I *(Discovery, launched 27-Aug-85)*
Repaired *Syncom* communications satellite stuck in low Earth orbit during STS-51D

STS-32 *(Columbia, launched 9-Jan-90)*
Recovered *LDEF* scientific satellite left in Earth orbit on STS-41C

STS-49 *(Endeavour, launched 2-May-92)*
Attached new rocket motor to *Intelsat 6* communications satellite to allow it to reach geosynchronous orbit

STS-61 *(Endeavour, launched 2-Dec-93)*
Routine service call on *Hubble Space Telescope*, also repaired faulty optics

Waiting for the Doctor

Ever since NASA launched the less than perfect *Hubble Space Telescope* in April 1990, the collective voice of the American media had blasted the beleaguered space agency for gross incompetence. Never mind the fact that the flaw amounted to a sizing error in the telescope's main 2.4-meter- (94.5-inch-) wide reflecting mirror of less than 2% the width of an average human hair. Never mind the fact that despite the flaw, astronomers still rated *Hubble's* visual acuity at a level greater than five times that of the largest ground-based telescopes.

Most of the media's criticism centered on NASA's supposed "ineptness" in the management of large-scale, complicated projects. Only a few chose to mention the numerous scientific discoveries made possible by the space telescope during its first three years of operation. Even fewer chose to mention 45,000 astronomical images taken for more than 1,000 astronomers during the same time span.

NASA's solution? Send seven astronauts and Space Shuttle *Endeavour* on a mission to rendezvous with and repair *Hubble.* However, the mission took on much greater significance than restoring the space telescope's full visual capability. Many considered the telescope repair mission as a mission to repair and restore NASA's

plummeting image in the eyes of the public. Eight years after the death of seven astronauts during the explosion of Space Shuttle *Challenger* two minutes after launch, NASA projects still remained plagued with technical problems that tended to overshadow brilliant accomplishments. Shuttle launches delayed by equipment glitches, a space probe with a crippled communications antenna en route to Jupiter, and another satellite that mysteriously disappeared three days before reaching Mars all contributed to a growing misconception that American space ventures never amount to anything except failure.

Unfortunately, the seven astronauts whom NASA charged with fixing its image, and also the images coming out of the slightly nearsighted space telescope, faced the most difficult task assigned to a space crew since Apollo astronauts landed on the Moon in the late 1960s and early 1970s. The mission plan called for the "Dr. Goodwrenches," a nickname given to the seven by mission control in Houston, to move large blocks of equipment in and out of cramped quarters, to make sure no loose screws, nuts, and bolts floated off and damaged the telescope's mirrors, and to replace and install small, delicate electronic circuits in the guts of the telescope. Sound easy? If so, consider the mandatory space walking dress code of puffy full-body pressure suits and thick, cumbersome space gloves. Finally, imagine performing all of those tasks hundreds of kilometers above the ground, in a freezing vacuum, and while in free fall around the Earth at over twenty-five times the speed of sound.

Who on Earth would possibly volunteer for this high-profile "Mission Impossible"? Probably every able-bodied American with an interest in space would step forward and sign up, given the chance. Instead, the space agency's management assigned the most experienced crew ever to venture into space on a single flight. Mission Commander Richard Covey from the Air Force and Pilot Kenneth Bowersox from the Navy drew the assignment of maneuvering the 100-ton Space Shuttle *Endeavour* from the ground in Florida to a pinpoint rendezvous with the *Hubble Space Telescope*. While in orbit, the job of venturing out into the vacuum of space to fix the telescope fell on Mission Specialists Story Musgrave, Jeffrey Hoffman, Kathryn Thornton, and Tom Akers. Mission Specialist Claude Nicollier rounded out the crew. He received the unglamorous but vital assignment of staying inside the shuttle and using *Endeavour's* robot arm to assist the four space walkers.

Every astronaut on the repair mission had flown in space at least once prior. Together, their previous combined experience totaled 16 shuttle missions. Kathryn Thornton, who holds a doctorate in physics, and Tom Akers, a lieutenant colonel in the Air Force, brought prior satellite rescue experience to the team. Together, they had previously served on the STS-49 mission that rescued the crippled *INTELSAT 6* communications satellite in May 1992. However, the honorary title of "Renaissance Astronaut" belonged to Story Musgrave. He joined the astronaut corps in 1967 with the hope of walking on the Moon during the Apollo program. Unfortunately, that dream never materialized. In the meantime, Musgrave kept himself busy by flying on four shuttle missions, and earning a degree in literature to complement his other degrees in mathematics, statistics, operations analysis, chemistry, medicine, physiology, biophysics, and computer programming.

Despite their raw talent and vast space-faring experience, each member of the crew endured more than 1,000 grueling hours training for the mission. The four astronauts assigned to make the space walks spent much of their training time underwater in a large swimming pool. There, they used the buoyant effect of water to simulate weightlessness as they practiced the repair routines time and time again on a full-scale waterproof model of the *Hubble*. NASA management also asked Claude Nicollier to train underwater. His assignment on the mission involved no space walking, but NASA wanted him to be ready to fill in for one of the other four.

The appointed moment finally arrived at 4:27 A.M. EST on Thursday, 2 December 1993. As most of America rested soundly asleep, *Endeavour* lit up the dark sky along the Florida coast in a picture-perfect but rare night liftoff. A mere eight minutes after launch, *Endeavour* arrived in space to begin its orbital catch-up chase with the *Hubble*. Shortly after midnight that Saturday morning, and two days after the launch, mission controllers in Houston shrugged off sleepy eyes to watch their monitors with ner-

CONTINUED ON NEXT PAGE

Opposite Page:

4 December 1993 - The *Hubble Space Telescope*, backdropped over Madagascar, is firmly anchored in Space Shuttle *Endeavour's* payload bay after a two-day orbital chase to catch up with the telescope. The two rectangular objects on the side of *Hubble* are solar panels that provide the telescope with electricity.

Left:

7 December 1993 - Astronauts Story Musgrave and Jeff Hoffman prepare to install a new camera into the *Hubble*.

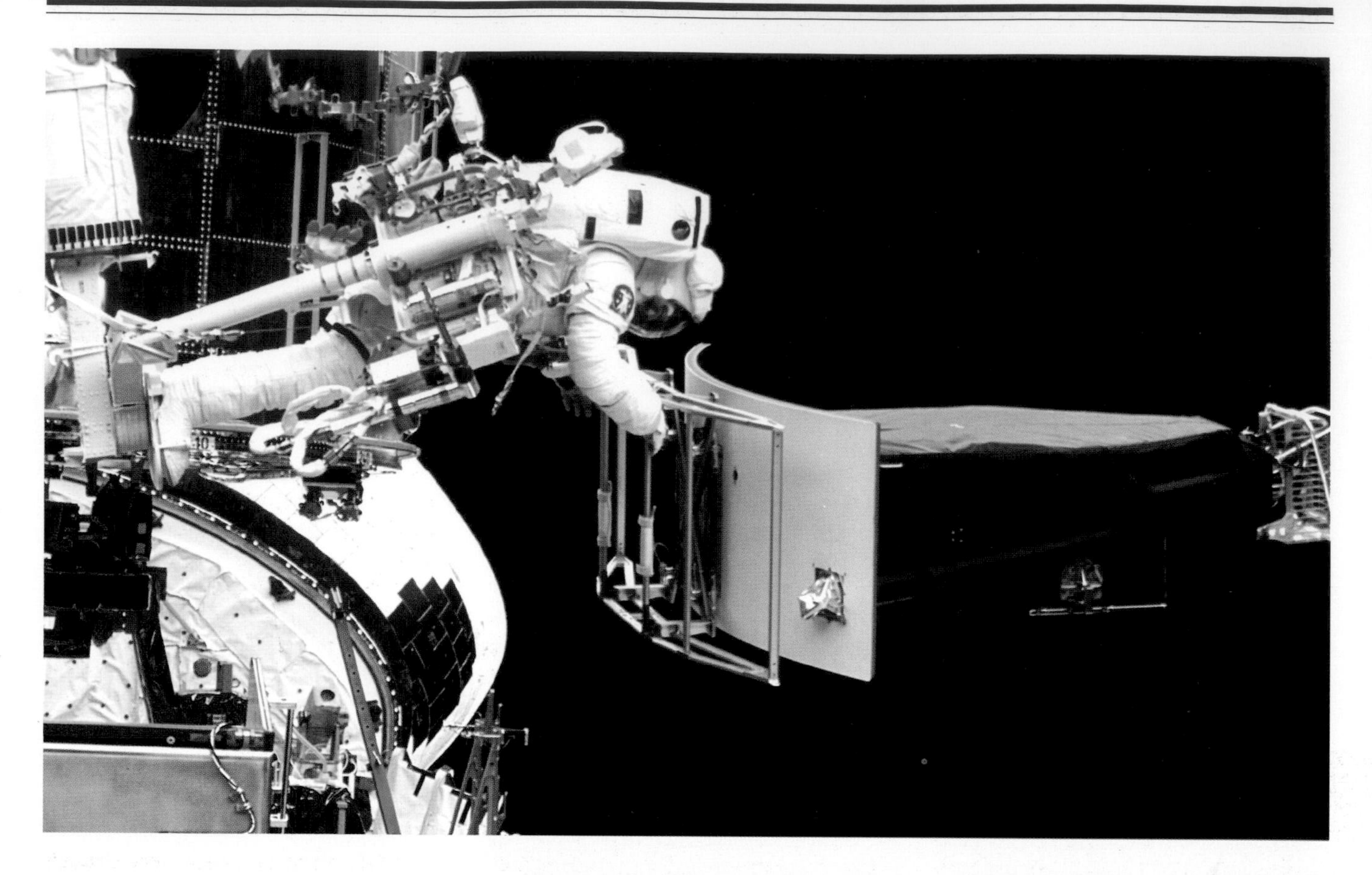

Above:

7 December 1993 - Astronaut Jeff Hoffman carefully holds on to a device called the Wide Field and Planetary Camera. Hoffman and Musgrave took this instrument out of the *Hubble* and replaced it with a new and improved version.

CONTINUED FROM PREVIOUS PAGE

vous optimism as *Endeavour* caught up with *Hubble* east of Australia, over the South Pacific. "Houston, *Endeavour* has a firm handshake with Mr. Hubble's Telescope," announced Commander Dick Covey to mission control after Claude Nicollier unceremoniously snagged the telescope with the shuttle's robot arm. Shortly afterward, *Hubble* was safely and firmly anchored in the shuttle's payload bay.

Now came the hard part. Theoretically, NASA planners thought that the astronauts could complete the orbital "house call" over a five-day period with a schedule of one two-person, six-hour space walk per day. The teams of Musgrave and Hoffman, along with Akers and Thornton, would work on the telescope on alternating days, using the "in-between" days to rest. Realistically, would the plan work? Musgrave and Hoffman began to answer that question late in the night on Saturday, 4 December 1993, as they donned the puffy pressure suits, ventured out into space an hour earlier than originally scheduled, and approached the space telescope in the shuttle's payload bay.

One of the major tasks on the first night's space walk agenda involved replacing a set of faulty gyroscopes used by *Hubble* to maintain its sense of balance and position. Although the complexity of the gyroscope replacement ranked relatively low, the astronauts' heavy breathing and intense concentration attested to the fact that while in weightlessness, most tasks become chores. This observation became painfully evident to personnel in mission control as they heard the astronauts tersely count off each turn of a bolt or screw. Then, every time the two astronauts changed posture, they left a floating stream of custom made cordless power tools, spare gloves, and other parts in their wake.

Most of the activities required to replace the gyroscopes flowed smoothly. At one time, Hoffman anchored himself to the end of *Endeavour's* robot arm, and dangled Musgrave into the telescope upside down. This unusual position gave Musgrave easy access to change the failed gyroscopes. The only problem arose toward the end when the access doors leading to the gyroscope mounts failed to latch shut.

"Uh-oh, we're in trouble," Hoffman muttered after discovering the problem. Mission control urged the two to move on to other tasks first, but Musgrave insisted on improvising a creative solution. After some careful thought, he simply used a spare strap from his toolbox as a foothold from which to use his body to shoulder the door shut. This unorthodox answer to an unforeseen contingency earned the space-walking duo a loud round of applause from all the Houston-based ground controllers monitoring the flight. However, the night was far from over, and the two worked outside the shuttle until 5:30 in the morning.

Late Sunday night, while the first two space walkers rested, Akers and Thornton ventured out into *Endeavour's* payload bay to replace *Hubble's* twin set of shaky solar panels. NASA described the problem as a case of thermal jitters. Every time the telescope sped in and out of daylight along its 96-minute orbit, the alternating heating and cooling cycles caused the panels to vibrate the telescope, slightly smearing any photographs taken at the time. The new panels came equipped with a set of shock-absorbing springs to dampen the vibrations.

Before the second space walk, ground controllers sent a command to the telescope asking it to roll up the old solar panels like a window shade. One panel rolled up. The other one, slightly kinked in the middle from sources unknown, refused to retract. Originally, NASA

Left:
8 December 1993 - Astronaut Tom Akers maneuvers inside a compartment within the *Hubble* that was slated to house the telescope's corrective optics. This photograph was taken prior to the installation of the optics.

wanted Akers and Thornton to store the old rolled-up panels in the shuttle's payload bay for the trip home. Unfortunately, once removed, the fully extended panel did not fit in the bay. No problem, said NASA. Akers simply disconnected the panel, then gently handed it to Thornton who rode the shuttle's robot arm to the edge of the payload bay and dumped the panel overboard.

Next, *Endeavour's* commander, Dick Covey, fired tiny rocket thrusters on the shuttle to slowly nudge it away from the jettisoned solar panel. As the rectangular piece of mangled space junk began to flap in the artificial breeze caused by the thruster's plume, Thornton excitedly exclaimed, "It looks like a bird!" However, the two astronauts had little time to sight-see as the mission plan called for them to spend the next several hours installing the new solar panels. They finished at 4:00 in the morning.

On Monday night, 6 December 1993, NASA turned the show over to Musgrave and Hoffman again. Their assignment involved removing a device called the Wide Field and Planetary Camera (WFPC) and replacing it with a new and vastly improved version. The camera, built by NASA's Jet Propulsion Laboratory in Pasadena, California, is about the size of a baby grand piano, and is designed to take detailed pictures in the visible, infrared, and ultraviolet wavelengths. When installed, the new WFPC will allow *Hubble* to peer deeper into the universe in the ultraviolet range, a region normally hidden to astronomers on the Earth by the atmosphere.

One of the night's many tense moments came right before the installation of the new camera. At that time, Musgrave removed a bright red protective cover from the mirror designed to channel light from the telescope's main mirror into the WFPC. Any accidental contact with the camera's mirror might have ruined the sensitive reflecting surface, or thrown the optics out of alignment. Engineers at the Jet Propulsion Laboratory had worried about preserving the pristine state of the camera's mirror from the beginning. During assembly, they even asked the laboratory's director to reduce air contamination by stopping road paving and building re-roofing projects on the laboratory's grounds.

Both astronauts worked more efficiently during the WFPC replacement than the ground planners' wildest dreams. They simply unbolted the original camera, gently slid it out along the guide rails, and then eased the new one into *Hubble* almost one hour ahead of schedule. Hoffman commented after finishing the installation, "Look at that baby . . . we'll be able to see some nice pictures with that."

So far, almost everything on the mission, code-named STS-61 on the official NASA manifest, had amazingly gone according to the script. However, the most crucial task came on Tuesday night as the team of Akers and Thornton took over for the tired Musgrave and Hoffman. The second team of space walkers, who had replaced *Hubble's* solar panels two nights prior, now faced the daunting task of precisely installing the "prescription lenses" designed to correct the tiny flaw in the telescope's main mirror.

"Fingertip it," Akers cautioned Thornton as he stood inside the telescope and assisted his partner in easing the 290-kilogram (640-pound) corrective optics box in place. The telephone booth sized box, called COSTAR by NASA engineers, contained ten mirrors. Each one was individually attached to a mechanical arm designed to deploy like an unfolding blade on a pocket knife. Tinsley Labs, hidden behind a shopping mall across the bay from San Francisco, California, built COSTAR for NASA, and designed the

CONTINUED FROM PREVIOUS PAGE

mirrors to refocus light from *Hubble's* main 2.4-meter-wide mirror into the telescope's instruments and cameras. Big corporations like Kodak, Hughes, and United Technologies declined to take on the COSTAR project because they could not meet NASA's strict fabrication standards. Tinsley, however, delivered the mirrors under budget, ahead of schedule, and with no flaws.

After Akers and Thornton completed their seven-hour, Tuesday-night-to-Wednesday-morning space walk highlighted by the successful installation of COSTAR, Musgrave and Hoffman ventured outside the shuttle one last time on Wednesday night, 8 December 1993. During the fifth space walk of the mission, Musgrave and Hoffman performed miscellaneous "wrap-up" tasks such as replacing electronic parts and placing protective covers over instruments at the top of the telescope. However, when the newly installed solar panels failed to unroll as planned, Musgrave once again demonstrated the value of using human creativity and adaptation to overcome unforeseen problems in space. He simply pushed the protective cradles holding the rolled-up panels, and they unfurled.

Shortly after midnight on Friday, 10 December 1993, Claude Nicollier used *Endeavour's* robot arm to pick up the rejuvenated 13-meter-(43-foot-) long telescope and place it back in orbit. Later that morning, President Clinton and Vice President Gore called the astronauts in orbit to congratulate the seven-member repair team. "I want to thank each and every one of you for what you did. You made it look easy," said Clinton. "It was an immense boost to the space program in general and to America's continuing venture in space," added the delighted president. The White House was counting on a successful repair to build support for NASA in Congress.

After completing 162 revolutions of the world, *Endeavour* landed in Florida shortly before midnight on Sunday, 12 December 1993. Although NASA hailed the repair mission as a complete success, skeptics of space exploration remained dutifully unimpressed. They were quick to claim that the mission and all of the dramatics would not have been necessary if the space telescope had been designed correctly from the outset.

Not so, said NASA. They countered with the argument that nobody ever drives a car for 15 years without bringing it in for periodic maintenance. NASA engineers point out that *Hubble* was designed to be serviced in space every three years, and the December 1993 mission would have occurred even if the optics had been designed perfectly the first time. The next "house call" on *Hubble* is set for 2002. *Hubble's* 15-year-design lifetime ends in 2005.

Although the seven astronauts performed the repair tasks flawlessly, it took NASA four weeks to determine if the repair scheme actually fixed *Hubble's* vision. Fortunately for eager scientists, the answer was yes. On 13 January 1994, NASA called a press conference to show off new, clear pictures taken by the rejuvenated *Hubble Space Telescope*. "We sent a message to the young people of this country that risks are definitely worth taking," said NASA administrator Daniel Goldin. ❑

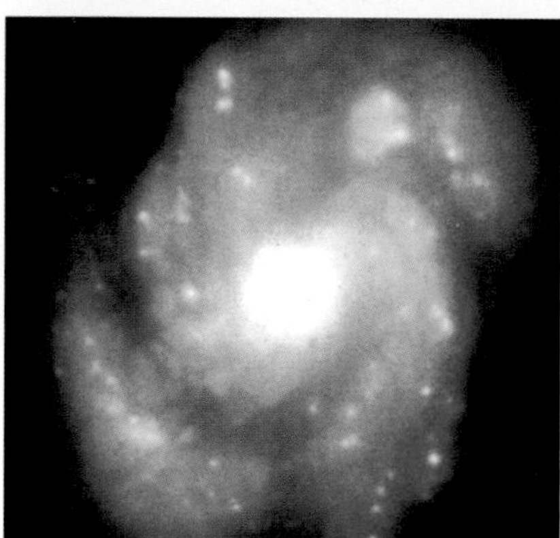

Above Top:

8 December 1993 - Astronaut Kathryn Thornton, known as K.T. to her closest friends, selects a power tool to use during the installation of *Hubble's* new corrective optics.

Above Middle:

Images taken before (left) and after (right) the telescope servicing mission.

Opposite Page:

9 December 1993 - Astronaut Story Musgrave, anchored to the end of *Endeavour's* robot arm, gets a ride to the top of the telescope.

JPL

NASA
HUGHES

On the maiden voyage of *Endeavour* (STS-49) in May 1992, the astronauts attempted to capture an *INTELSAT 6* communications satellite that was stranded in orbit because its upper stage failed. Before the capture attempt, Intelsat ground controllers managed to stop the satellite's spin without consuming an excessive amount of attitude-control propellant. A mission specialist then tried to attach a grapple bar to the satellite. To accomplish this feat, he rode on the RMS end effector to a position near the satellite and attempted to attach the grapple bar to the bottom of the satellite. Unfortunately, the satellite started to wobble almost every time he touched the satellite with the bar. On attempts where the *INTELSAT* did not wobble, the bar failed to latch properly. Eventually, in a daring move, the commander sent two other mission specialists outside to hold the *INTELSAT* while the third attached the grapple bar. A fourth mission specialist stayed inside to operate the RMS.

After the capture of a satellite, astronauts use the RMS to bring it into the payload bay to store for the trip home, or to perform a repair before deploying it in orbit again. The repair always proceeds smoothly compared to the capture because analysts on the ground spend months determining the cause of malfunction. Astronauts also spend months practicing and simulating the capture. However, no training facility in the world can faithfully simulate the weightless condition experienced in orbit. Although practicing the EVA underwater in a pool helps, the pool environment is much more forgiving than the space environment when it comes to how objects wobble and tumble after being bumped or nudged. In short, manipulating objects in the absence of gravity is an extremely difficult task.

Spacelab Research

Conducting scientific research represents the least dramatic but perhaps the most important of all the different types of Space Shuttle missions. Unfortunately, space inside the crew cabin is very tight and allows little room to store equipment for scientific research. Spacelab solves this problem by providing additional laboratory space in the orbiter's payload bay. This laboratory was designed and built by the European Space Agency for NASA in an effort to promote international cooperation for scientific research in space. Many distinguished international scientists have flown aboard the Space Shuttle during Spacelab missions as payload specialists.

The cylindrical-shaped laboratory measures about 5.5 meters (18 feet) long and 4 meters (13 feet) in diameter (see Figure 49 on page 216). A pressurized tunnel eliminates the need to perform an EVA to reach the laboratory by connecting the front end of the Spacelab to the orbiter's mid-deck. This tunnel connects to the crew cabin where the outer door of the airlock normally sits. Before Spacelab missions, ground engineers remove the airlock from inside the crew cabin and place it on top of the tunnel. Keep in mind that the Spacelab only appears in the orbiter during missions where scientific research dominates the mission itinerary.

Opposite Page:

13 May 1992 – Three astronauts from Space Shuttle *Endeavour* attempt to attach a grapple bar to the bottom of the *INTELSAT 6* satellite. The bar allows the shuttle's robot arm to grab onto it for the purpose of moving the satellite. Left to right are Richard Hieb, Thomas Akers, and Pierre Thuot. This photograph was taken from the front of the orbiter looking toward the rear.

Historical Fact

The first Spacelab mission flew on *Columbia* in December 1983. Who was in command? John Young, of course. He had previously commanded the first Space Shuttle flight, flew to the Moon twice and spent over 72 hours on the lunar surface on one of the trips, and flew on two Gemini missions, including the first Gemini flight.

RULE OF THUMB

On most Spacelab missions, the crew splits into two teams so that scientific research can proceed 24 hours a day.

► Figure 49: *A Look at the Spacelab*

What: Spacelab is a laboratory system that appears on the orbiter during missions where scientific research dominates the itinerary

Parts: One laboratory module for the astronauts, and up to three pallets that store equipment requiring exposure to the vacuum of space

Builder: European Space Agency (for NASA)

Length: The laboratory measures about 5.4 meters (17 feet) long, each pallet measures 2.9 meters (9.5 feet) long

Diameter: Roughly 4.1 meters (13.3 feet)

Missions: Study of gravity's role in the areas of life sciences, material processing, and physical sciences. Study of global climate change, ozone depletion, and astronomical phenomena

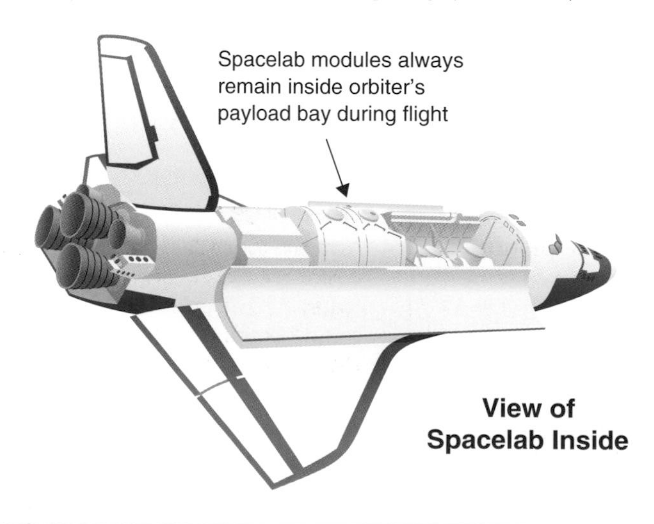

View of Spacelab Inside

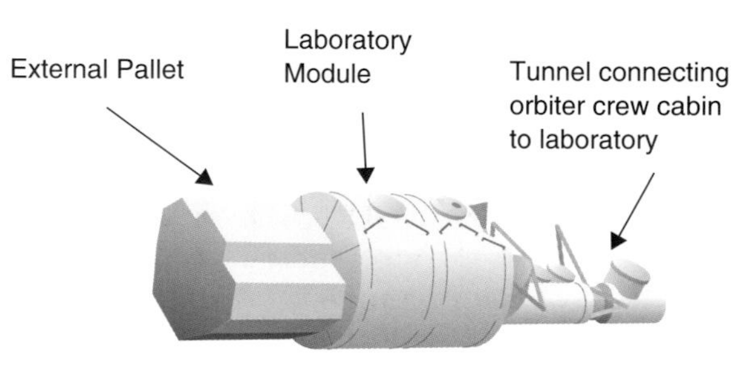

View of Spacelab Without Orbiter

Spacelab's interior resembles the inside of a metal trailer and provides spacious accommodations compared to the cramped crew cabin. On television, the laboratory looks like a rectangular hallway filled with electronic gadgets that line the walls from floor to ceiling. Six racks on each side of the laboratory contain compartments that store experiments and equipment with which to conduct the experiments. NASA engineers custom-build each rack to support the specific research goals of a mission. After use, the engineers disassemble the racks and use the parts to construct the racks for future Spacelab missions.

The Spacelab system also consists of devices called *pallets* that mount on the floor of the payload bay behind the laboratory. A Spacelab pallet provides a flat platform to mount experiments that must experience exposure to the vacuum of space. For example, these devices include instruments to study astronomical phenomena, instruments pointed at the Earth to take measurements of processes occurring in the atmosphere, and various types of high-resolution cameras. Some missions include both laboratory and pallet, some include just the laboratory, while others may consist of several pallets but no laboratory.

Many people often wonder why NASA spends millions of dollars to fly a laboratory into space when many laboratories exist here on the surface of the Earth. The answer involves the weightless condition that objects experience while in orbit around the Earth. Complete understanding of a physical, chemical, or biological process requires a complete understanding of how each aspect of the environment affects the process. Scientists need to know what happens to a process when different stimuli are added or removed. On the ground, scientists can selectively control every type of stimulus except for gravity. In space, scientists attempt to ascertain gravity's largely unknown role in affecting the outcome of many physical, chemical, engineering, and biological processes. Spacelab missions occur in different series, each with a different scientific emphasis related to microgravity.

The Spacelab Life Sciences (SLS) series examines the effect of weightlessness on physiological processes and methods to counteract the adverse effects. Some of the areas of research include space medicine, medical and surgical techniques in microgravity, calcium loss in human bones during spaceflight, effects of microgravity on white and red blood cell production, and vestibular system studies related to space motion sickness. SLS-1, the first mission in this series, flew on STS-40 in June 1991. A total of four more SLS missions are planned through the year 2000 to provide data that will help humans survive safely in microgravity for long periods of time.

The United States Microgravity Laboratory (USML) series primarily examines chemical and physical processes. USML-1, the first in the series, flew on STS-50 in June 1992. Some of the experiments on that mission included studies on the mechanics of fluid

Shuttle Spacelab Missions

STS-9 *(Columbia)*, Spacelab 1
28-Nov-83, Carried 71 experiments in astronomy, life sciences, atmosphere physics

STS-51B *(Challenger)*, Spacelab 3
29-Apr-85, Biological and materials processing experiments, carried 24 rats in cages

STS-51F *(Challenger)*, Spacelab 2
29-Jul-85, Mostly astronomy

STS-61A *(Challenger)*, Spacelab D-1
30-Oct-85, Joint materials processing experiments with West Germany

STS-35 *(Columbia)*, Astro-1
2-Dec-90, Astronomical telescopes on shuttle

STS-40 *(Columbia)*, SLS-1
5-Jun-91, Microgravity effects on humans, carried 30 rats and 2,478 jellyfish

STS-41 *(Discovery)*, IML-1
22-Jan-92, Materials and life sciences

STS-45 *(Atlantis)*, ATLAS-1
24-Mar-92, Study of Earth's atmosphere

STS-50 *(Columbia)*, USML-1
25-Jun-92, Materials processing and physics

STS-47 *(Endeavour)*, SLJ-1
12-Sep-92, Joint life sciences with Japan

STS-56 *(Discovery)*, ATLAS-2
8-Apr-93, Study of Earth's atmosphere

STS-55 *(Columbia)*, Spacelab D-2
26-Apr-93, Second Spacelab with Germany

STS-58 *(Columbia)*, SLS-2
18-Oct-93, Microgravity effects on humans

STS-65 *(Columbia)*, IML-2
8-Jul-94, Materials and life sciences

STS-66 *(Discovery)*, ATLAS-3
3-Nov-94, Study of Earth's atmosphere

Spacelab Acronyms:
IML– International Microgravity Laboratory
SLS – Spacelab Life Sciences
USML – United States Microgravity Laboratory
ATLAS – Atmospheric Laboratory for Applications and Science

Opposite Page:
Astronauts aboard Space Shuttle *Endeavour* conduct scientific experiments inside the Spacelab. This mission (STS-47) was code-named Spacelab-J and represented a joint effort by NASA and Japan to conduct science in space.

RULE OF THUMB

Most astronauts agree that the orbiter seems much less crowded on missions that include the Spacelab laboratory module. The reason is that the Spacelab contains enough interior volume to double the amount of work area normally available on the orbiter.

Scientific Fact!

Common procedures on Earth become extremely difficult in the weightless environment of microgravity. For example, without gravity, it is impossible to pour liquids from one container to another.

Historical Fact

NASA had a hard time keeping the thermal protection tiles on the orbiters during the early days of the Space Shuttle program. When *Columbia* was transported from its assembly plant in California to the Kennedy Space Center, more than 2,000 tiles fell off. During the first mission, about 30 tiles fell off *Columbia* during launch. Fortunately, no tiles in critical areas fell off.

flows, fluid drops, fluid surface tension, and the mechanics of flames and combustion in microgravity. Astronauts also took advantage of microgravity to grow superpure crystals. The results from the crystal growth experiment may someday enhance semiconductor and microchip technology. NASA plans to fly at least one more USML mission.

The ASTRO series focuses on studying the stars and the physical universe from the Space Shuttle. Instruments that sense and record astronomical phenomena from orbit provide better data than ground-based instruments because particles of dust and gas in the atmosphere degrade the quality of data recorded by ground-based instruments. ASTRO-1, the first in the series, flew on STS-35 in December 1990. Some of the major instruments on that mission included ultraviolet telescopes and the broadband X-ray telescope (BBXRT). Studying the energy output from distant objects in space may yield clues to the nature and formation of the universe. ASTRO flights do not require the laboratory module as the instruments mount onto a Spacelab pallet in the payload bay. NASA plans at least one more ASTRO flight.

The Atmospheric Laboratory for Applications and Science (ATLAS) series studies one of the most pressing concerns of our time. During the first ATLAS mission on STS-45 in March 1992, instruments mounted on a Spacelab pallet gathered solar and atmospheric data contributing to ongoing studies regarding the problem of global climate change and global warming. Atmospheric studies from ATLAS-1 included temperature, pressure, chemical composition, and the correlation between these atmospheric parameters and solar radiation. Many of these experiments focused on the shrinking ozone layer that protects the surface of the Earth from ultraviolet radiation. Like the ASTRO series, ATLAS only utilizes Spacelab pallets. NASA plans nine more ATLAS flights between 1993 and 2001.

5.7 Returning to Earth

Astronauts must perform many preparatory activities before returning to Earth. Some of these activities include cleaning up and storing loose objects in the crew cabin, reinstalling the mission and payload specialist seats, powering down scientific experiments and the Spacelab (if the orbiter is carrying one), and closing the payload bay doors. If the doors fail to close, a mission specialist must perform a contingency EVA to solve the problem because the orbiter can not fly through the Earth's upper atmosphere with the doors open. After completing these activities, the astronauts put on the pressure suits that they wore during the ascent. These suits are less bulky than the ones the mission specialists use during EVAs.

An OMS burn starts the de-orbit process. Approximately halfway around the world from the landing site, the pilots use the attitude-control thrusters to flip the orbiter into an orientation where the OMS rocket engines face forward. A rocket burn in this position

slows the orbiter and drops it into a lower orbit with an apogee at the burn point and perigee halfway around the world from the burn point but at a much lower altitude. Typically, the orbiter transfers from a 296-kilometer altitude circular orbit to an elliptical orbit with a 296-kilometer apogee altitude and a 10-kilometer perigee altitude.

After the de-orbit burn from the OMS engines, the orbiter enters the upper atmosphere as it descends towards the perigee of the new orbit. Under normal circumstances, a spacecraft will speed up as it drops toward perigee on its way down from apogee. However, because perigee lies deep within the atmosphere in this case, a gradually increasing number of air molecules begin to impede the orbiter's forward progress. Constant collisions with these air molecules slow the orbiter below orbital velocity and cause the spacecraft to plunge into the atmosphere at a much faster rate than the normal descent rate toward perigee.

About 30 minutes before landing and 30 minutes after the de-orbit burn, the orbiter passes below the imaginary space-atmosphere interface boundary (121,920 meters, 400,000 feet, or 76 miles). Below the interface altitude, the energy from all the microscopic collisions

Above:

Commander Vance Brand (helmet visible on the bottom left) guides Space Shuttle *Columbia* during its fiery descent through the atmosphere, above the Pacific Ocean. At this point in time, the shuttle is traveling at more than 15 times the speed of sound on its way to a landing in the California desert at Edwards Air Force Base. This photograph was taken by mission specialist Joe Allen, who sat behind the two pilots. The bright glow in the cockpit windows is caused by the shuttle flying into the molecules of the Earth's atmosphere at extremely high speeds.

Above Top:

As shown in this close-up view of an orbiter's nose area, most of the shuttle is covered by small rectangular tiles for protection from the heat of reentry during landing. The tiles fit on the orbiter's surface like pieces of a jigsaw puzzle.

Above Bottom:

This photograph shows a close-up view of tiles near the rear of an orbiter and illustrates the size of an average tile relative to a human hand.

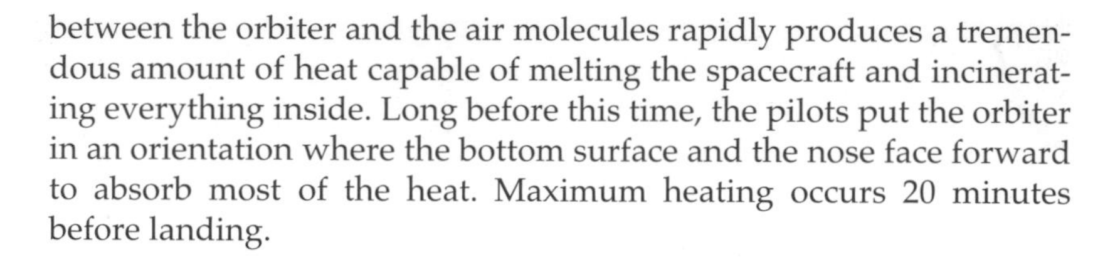

between the orbiter and the air molecules rapidly produces a tremendous amount of heat capable of melting the spacecraft and incinerating everything inside. Long before this time, the pilots put the orbiter in an orientation where the bottom surface and the nose face forward to absorb most of the heat. Maximum heating occurs 20 minutes before landing.

A safe landing critically depends on the orbiter's ability to survive the intense heat generated during the descent through the atmosphere. Space capsules in the past employed an ablative heat shield that melted during reentry into the atmosphere and gradually charred off to dissipate heat from the capsule. Unfortunately, resurfacing the entire orbiter with a new heat shield for every flight would cost a lot of money and would consume too much time. Instead, the orbiter employs a reusable non-ablative heat shield system known colloquially as the "tile system." These tiles cover the orbiter's aluminum structure and keep the structure's temperature to under 177° Celsius (350° Fahrenheit) during reentry.

Approximately 32,000 small opaque silica-glass tiles cover 70% of the orbiter's exterior surface area. Each unique tile requires custom fabrication and hand placement onto the orbiter. Most measure between 15 centimeters by 15 centimeters square (6 inches by 6 inches) to 20 centimeters by 20 centimeters (8 inches by 8 inches) with a thickness of between 1 centimeter (0.4 inches) and 9 centimeters (3.5 inches). A tile weighs very little for its size, looks like smooth Styrofoam, feels like a cross between ceramic and ordinary chalk, and breaks easily in the grasp of human hands.

Roughly 20,000 tiles, called high-temperature reusable surface insulation (HRSI), cover areas likely to encounter intense heat, such as the bottom of the orbiter and most of the nose. These black-colored tiles resist temperatures up to 704° C (1,300° F) by radiating 90% of the absorbed heat back into the atmosphere. The remainder of the tiles, called low-temperature reusable surface insulation (LRSI), cover areas that encounter intense heat but to a lesser degree than the bottom side. These white-colored tiles resist temperatures up to 649° C (1,200° F) and primarily cover the portions of the upper side of the wing, and the sides of the orbiter closest to the nose. Together, the HRSI and LRSI tiles give the orbiter its black-and-white appearance.

Two other types of orbiter thermal protection exist. A material called reinforced carbon-carbon (RCC) covers the very tip of the nose and the front edges of the wing. These blunt surfaces, called *leading edges*, encounter temperatures greater than 1,260° C (2,300° F). The remainder of the orbiter, including the top of the payload bay doors and most of the top of the wings, only encounter mild heating effects as compared to the rest of the orbiter. A thin layer of white insulation, called flexible reusable surface insulation (FRSI), covers these areas. FRSI protects up to 371° C (700° F).

The heat of reentry causes another major effect besides warming the outer surface of the orbiter. All the energy that causes the heating also strips electrons away from the air molecules that strike the orbiter. Normally, air particles are electrically neutral. Removing electrons (ionizing the particles) causes the air to acquire a positive charge. This sheath of ionized air envelops the bottom and sides of the orbiter and prevents direct radio contact with the ground. The period of communications blackout begins about 25 minutes before landing at an altitude of 80.5 kilometers (50 miles) and ends 12 minutes prior to landing.

Sixteen minutes before landing, the orbiter steers through the first of four S-turn maneuvers. During the last 4,000 kilometers (2,500 miles) to the runway, the orbiter follows a long winding path similar to the path a person takes when skiing down a snow-covered slope.

Above:

8 May 1989 - Space Shuttle *Atlantis* glides toward a landing in the Southern California desert after successfully sending the space probe *Magellan* on its way to a mission at Venus. NASA prefers to land shuttles at the Kennedy Space Center in Florida. However, bad weather often forces the shuttle to its alternate landing site at Edwards Air Force Base in the Mojave Desert.

Stations in Space

At this point, NASA scientists might not be surprised to discover a large, black, rectangular monolith shallowly buried in the crater Tycho on the Moon's surface. While this idea of alien artifacts may seem quite a bit far fetched, the other two of Arthur C. Clarke's visions of the future, geosynchronous communications satellites and a large space station, are reality. Nearly 35 years after director Stanley Kubrick introduced America to the futuristic concept of orbital outposts in Clarke's *2001: A Space Odyssey*, NASA is poised to undertake the world's most ambitious construction project by building a space station that will be nearly the size of a football field when fully assembled.

Ideas proposed in works of science fiction seldom turn into reality by the dates envisioned by their authors, and almost never in the manner that they speculate. Clarke's *Odyssey*, however, seems to be the uncanny exception with respect to the idea of space stations. Like its cinematographic counterpart, the NASA station will be in use by 2001, though not fully assembled. In addition, the Kubrick-directed film featured a single station with an "American Zone" and a "Russian Zone." Once again, reality reflects science fiction as Russia will be working with NASA in a joint effort to construct this gigantic orbiting outpost. Fortunately, this cooperation was fostered in the spirit of international peace, rather than in the film's Cold-War view of the world's two superpowers forever dividing all territory from Europe to outer space.

Despite the fact that the construction of a space station is difficult even with today's technology, this is not the first time that NASA has attempted such a project. *Skylab*, launched on 14 May 1973, was the first major American attempt at providing a home in space for astronauts to conduct scientific research on a semi-constant basis. Over the course of its lifetime, *Skylab* was home to three separate three-astronaut crews. Each crew made the journey from Florida to the space station aboard Apollo-style capsules similar to the ones that carried astronauts to the Moon and back.

In total, the three crews spent more than 3,000 hours conducting experiments, logged more than 42 hours walking in space, collected more than 200,000 photographs for Earth observation experiments, and closely observed the comet Kohoutek. *Skylab* demonstrated that humans could work effectively in space for prolonged periods of time. Long after the last crew departed, the abandoned laboratory fell victim to atmospheric drag and reentered the Earth's atmosphere on 11 July 1979. Although most of

Skylab burned up in the upper atmosphere, a few sizable chunks fell in the outback of Australia.

Like NASA, Russia also experimented with the construction of space stations in the post-Apollo era. Their attempts began in 1971 with *Salyut-1*. Early missions to this station were plagued with failure. For example, the crew of *Soyuz 10*, the first spacecraft sent to *Salyut-1*, was unable to enter the station because of a problem with the docking port on the *Salyut*. Later stations, such as *Salyut-3*, *Salyut-4*, and *Salyut-5*, were improved and supported a total of five crews. The experiments and observations from the cosmonauts on these crews led to improved designs for second-generation Russian space stations. These later attempts, beginning with *Salyut-6* in 1977, resulted in longer missions and additional improvements important toward the design of the world's most famous and successful orbital outpost to date, the *Mir*.

This laboratory, launched in February 1986, was both the first space station inhabited on a constant basis, and the first with a modular design that supported expansion. Over its lifetime, *Mir* facilitated thousands of experiments in space and has housed many crews, some of which consisted of international teams of cosmonauts. Russia finally abandoned *Mir* in August 1999, after facing enormous pressure to redirect its financial resources toward the joint space station effort with NASA.

The drive toward this *International Space Station* was ignited by a mandate from President Ronald Regan in 1984. An all-American design was initially crafted, but revised significantly in the spring of 1985 when Japan, Canada, and the European Space Agency (ESA) each signed a bilateral agreement with the United States for participation in the project. NASA's plan for construction involved the use of over 50 Space Shuttle launches to ferry pre-fabricated modules into low-Earth orbit. The launch of the first element of this station, named *Freedom* by Regan, was significantly delayed in the wake of the *Challenger* accident that grounded the shuttle fleet.

In the years following *Challenger*, significant budget cuts led to many delays and changes to the design of *Freedom*. Under direction from newly elected President Bill Clinton early in 1993, NASA and its contractors were ordered to formulate a scaled-back design to reduce costs and to include more international involvement, or face outright cancellation of the grand project. Major changes involved an agreement with the Russians to purchase their Soyuz spacecraft to serve as a "lifeboat" for use in emergencies, and for the former Cold-War adversary to supply major additional hardware elements central to the core of the station. In part to avoid Congressional association with the more expensive *Freedom* design, the new station became know as the *International Space Station*.

The first phase of the new space station campaign began in 1995 with the Shuttle-Mir program. This joint effort involved more than two years of continuous stays by American astro-

CONTINUTED ON NEXT PAGE

Opposite Page:

This artist's conception shows NASA's *International Space Station* passing above the Straits of Gibraltar and the Mediterranean Sea after all assembly has been completed. NASA is targeting the completion date for 2004.

Bottom:

December 1988 - The first American-built module of the *International Space Station* is hoisted from the cargo bay of shuttle *Endeavour*. The entire assembly will take nearly 40 shuttle flights.

nauts aboard the Russian *Mir* station and nine docking missions between the shuttle and *Mir*. Although no construction took place during this phase of the project, valuable precursor knowledge was gained by the two countries in the areas of international space operations and joint scientific research.

After more than a decade of planning and replanning, construction of the *International Space Station* began in November 1998 when a Russian Proton rocket launched the *Zarya* control module into low-Earth orbit. This Russian-built module provides the station with electrical power and rudimentary rocket propulsion capability during the early assembly stages. Space Shuttle *Endeavour* followed *Zarya* into orbit one month later with the American-built *Unity* module in the cargo bay. *Unity's* main function is to serve as a connection node for the addition of larger modules that will be added later in the station's construction process.

On shuttle *Endeavour's* 12-day mission, code named STS-88 on the official NASA flight manifest, the crew rendezvoused with and captured *Zarya*. Following the capture, astronauts Jerry Ross and James Newman completed the connection of the first two modules by performing a three-hour space walk. Then, in a historical moment on 10 December 1998, shuttle commander Robert Cabana and Russian cosmonaut Sergei Krikalev opened the hatch from *Endeavour* and became the first occupants to float into the new, but incomplete, space station.

Over the course of the assembly, NASA and Russia will combine their efforts to conduct 46 launches to carry modules and supplies into low-Earth orbit. The United States, with 37 shuttle launches in the current plan, holds the major responsibility for developing and ultimately operating the major elements and systems aboard the space station. Some of these include thermal control, life support, guidance, navigation, data handling, power systems, communications and tracking, and mission control facilities. In addition, NASA will supply a scientific laboratory module, four large solar panels, and a habitation module.

International partners are also contributing key additional elements. Canada is providing a 16.8-meter (55-foot) robotic arm that will be used for assembly and maintenance tasks. Both Japan and the European Space Agency are building scientific laboratory modules for the station. Last, but certainly not least, Russia is providing two research laboratory modules with independent life-support systems, a power platform that will provide 20 kilowatts of solar-array power for science experiments, and missions to resupply the station.

Above:

December 1988 - During shuttle mission STS-88, Astronaut James Newman makes the final connections between the first two modules of the *International Space Station*, the American-built *Unity* and the Russian-built *Zarya*.

Opposite Page Top:

December 1998 - After being connected together by the crew of shuttle *Endeavour*, the first two modules of the *International Space Station* are released back into orbit.

Opposite Page Bottom:

This artist's conception shows the fully assembled *International Space Station*. The station will orbit at an average altitude of about 400 kilometers (250 miles).

NASA plans call for assembly operations to be completed by 2004. When finished, the *International Space Station* will weigh more than 450,000 kilograms (1 million pounds), and will use over one acre of solar panels to provide power to six state-of-the-art scientific laboratories. Scientists around the world will benefit from studies that will include biology, medicine, materials fabrication, Earth observation, and technologies for living in space. ❑

UNITY

NASDA

Each S-turn maneuver helps to slow the orbiter. The last of the four S-turns occurs five and one-half minutes prior to landing with the orbiter still traveling at a velocity over twice the speed of sound.

The orbiter produces sonic booms when it travels through the air faster than the speed of sound. Two loud booms, one right after the other, announce the orbiter's arrival at the runway about three minutes prior to landing. Each boom sounds like a giant explosion in the air and lasts for less than half a second. Often, the booms wake people from bed more than 50 kilometers away. At the runway, the orbiter becomes visible to the eye approximately one minute before landing and looks like a white brick falling out of the sky. The landing gear wheels touch down on the runway at a velocity of 364 kilometers (215 miles) an hour. This velocity is more than twice as fast as a standard commercial jet landing.

Consider the descent of a standard passenger airplane to gain insight into the shuttle's descent rate from orbit. Five minutes prior to the landing of a passenger jet, the captain probably turns on the "fasten seat belt sign," the flight attendants collect empty drink cups, and the aircraft is probably no more than 3,050 meters (10,000 feet) off the ground and traveling at 200 kilometers (124 miles) an hour. In contrast, during the last 16 minutes of a Space Shuttle flight, the orbiter travels a horizontal distance almost equal to the distance from San Francisco to New York, and drops a vertical distance of almost 64 kilometers (40 miles). At five minutes prior to landing, the orbiter's altitude of 25,000 meters (83,000 feet) exceeds the normal cruising altitude of a passenger jet more than twice over.

Today, most landings occur at the Shuttle Landing Facility (SLF) located eight kilometers from Complex 39 at the Kennedy Space Center. This 4.6-kilometer (15,000-foot) concrete runway is surrounded by swamps infested with hungry alligators. In contrast, the length of a runway at a large commercial airport rarely exceeds three kilometers (10,000 feet). If the weather at KSC violates the mission rules for a safe condition at the landing site (rain, mostly cloudy skies, or high winds), mission control will instruct the astronauts to land at Edwards Air Force Base in Southern California. In contrast to KSC, the runway at Edwards is a dry lake bed.

Keep in mind that the orbiter lacks jet engines and must glide to a landing like an engineless brick with wings. As a result, the commander must navigate the shuttle to the runway on the first try. There is no second chance if the shuttle misses. Essentially, the astronauts ride the world's heaviest glider from orbit to the runway. ❏

Opposite Page Top:

On 29 June 1995, astronauts on Space Shuttle *Atlantis* took this photograph as they approached the Russian space station *Mir.* This mission marked the first time in history that an American shuttle docked with a Russian space station. During the several days that the station and shuttle were docked, the American astronauts and Russian cosmonauts carried out joint scientific experiments to demonstrate the practicality of international cooperation in space exploration. The shuttle carried five Americans and two Russians to the station. There, they dropped off the two Russians and picked up two other Russians and an American for the trip back to Earth. That American, Dr. Norm Thagard, had been conducting science experiments on *Mir* since the beginning of the year. He was the first American to fly on a Russian rocket. NASA carried out several more shuttle-*Mir* dockings before the turn of the century.

Opposite Page Bottom:

Starting in 1998, NASA was joined by Russia, Japan, and the European Space Agency to build the world's largest space station in low Earth orbit. When fully assembled, the station will span close to 110 meters (361 feet) in length, will weigh over 400,000 kilograms (882,000 pounds), and will support a full-time crew of six astronauts to conduct scientific research around the clock. NASA figures show that construction will require 27 shuttle launches and 44 Russian rocket launches. This station was first proposed by President Ronald Reagan in 1984. The photographs are computer simulations of what the station will look like.

Final Frontiers

Chapter 6: How to Reach Other Planets

Opposite Page:
Photographic montage generated at NASA's Jet Propulsion Laboratory shows all of the planets in the Solar System (not to scale) except for Pluto. Clockwise from top to bottom are Mercury, Venus, Earth, the Moon, Mars, Jupiter, Saturn, Uranus, and Neptune.

Since the beginning of the space age in 1957, more than 100 humans have rocketed into space. Of those, only twenty-four journeyed as far as the Moon. Nobody yet has ventured any farther. On a celestial scale, the Apollo lunar voyages represented trips from home scarcely farther than a small step out the front door. Currently, humans delegate the responsibility of exploring Earth's local neighborhood in the universe to the electronic eyes and instruments of automated satellites. Their job involves probing the hundreds of exotic worlds that orbit the Sun. These celestial bodies range in size from irregular-shaped chunks of rocks scarcely larger than Manhattan to gigantic spheres of hydrogen gas hundreds of times more massive than the Earth. Many of them hide scientific secrets that will certainly yield clues into the nature and formation of the Earth and the universe, and may yield clues to the formation of life on Earth and the existence of life beyond the Earth. This chapter describes some of our neighbors in space and how spacecraft can break out of Earth orbit to visit other worlds.

6.1 What Is the Solar System?

Every destination in space reachable by modern spacecraft exists in a tiny section of the universe called the Solar System. This section consists of nine planets including the Earth, millions of comets, and a countless number of asteroids and other chunks of rocks that orbit the Sun. Also, a total of about seventy moons orbit the nine planets. These moons are small worlds that orbit the planets at the same time that the planets orbit the Sun. Several of these small moons in orbit around the larger planets are actually larger than the two smallest planets in the Solar System.

Planets are spherical worlds that do not shine by themselves. Instead, they depend on reflected sunlight to make them visible from telescopes on the Earth and cameras on spacecraft. Eight of the nine planets fall into two general groups: four relatively small rocky worlds that orbit close to the Sun, and four large gaseous spheres that orbit further out. The names of the planets in order of increasing distance from the Sun are Mercury, Venus, Earth, Mars, Jupiter, Saturn, Uranus, Neptune, and Pluto.

Mercury

Mercury orbits much closer to the Sun than any other planet in the Solar System. In fact, Mercury orbits so close to the Sun that the bright solar glare always obscures the planet. For years, Mercury's

Above:

Mercury, the closest planet to the Sun, looks like the Moon. This photograph was taken by *Mariner 10* in 1974.

Opposite Page, Top Left:

A planet-wide thick layer of clouds perpetually covers Venus and prevents spacecraft cameras from directly seeing the surface. This photograph was taken by *Mariner 10* in 1974.

Opposite Page, Top Right:

4 May 1989 - Astronauts aboard Space Shuttle *Atlantis* launch the *Magellan* spacecraft (communications and radar antenna at the top, and rocket engine for the interplanetary transfer at the bottom). Shortly after, *Magellan's* rocket was fired to propel the spacecraft out of Earth orbit and toward Venus. It arrived there 15 months later on 10 August 1990.

Opposite Page, Bottom:

Magellan imaged Venus by transmitting radar signals through the clouds, and listening to the radar echoes that bounced off the surface. This photograph was generated by a computer using *Magellan* radar data. It shows Ishtar Terra, part of a 3.5-kilometer (2-mile)- high plateau region the size of Australia in Venus' northern hemisphere. The mountain in the upper center of the image rises over 1.5 kilometers (1 mile) above the plateau.

close proximity to the Sun kept the best telescopes on Earth from revealing the planet as anything but a small object lacking any surface detail. Consequently, astronomers knew very little about the tiny planet until 29 March 1974, when NASA spacecraft *Mariner 10* flew past it on a scientific reconnaissance mission.

Mariner 10 revealed Mercury as an ancient, heavily cratered planet, not much larger than the Moon and extremely similar in appearance. Today, astronomers know that Mercury ranks as the second-smallest planet in the Solar System. The *Mariner 10* photographs also showed gigantic cliffs crisscrossing the surface. Some measure as high as 3 kilometers (1.9 miles) high and 300 kilometers long. Scientists speculate that these cliffs formed long ago in Mercurian history, possibly as long ago as four billion years. Then, Mercury's inner core consisted of molten iron. As the core cooled over a period of a billion years, its size shrank by up to four kilometers. The shrinking buckled the planet's crust, creating the cliffs. In essence, Mercury's outer crust or "skin" is literally too big for the planet's current size.

Mercury's surface bakes and freezes at the same time. At closest approach to the Sun, the surface temperature on the sunlit side reaches 470° C (878° F) and dips to -180° C (-292° F) on the night side. The planet's close distance to the Sun and long day length of 59 Earth days contribute to the scorching day-side temperatures. At night, the lack of a sizable atmosphere keeps Mercury from retaining heat gained during the day. In fact, the atmosphere is almost nonexistent, about a trillionth the density of Earth's atmosphere.

Venus

Some astronomers refer to Venus as Earth's twin sister because the two planets appear almost identical in size, physical composition, and density. At that point, the similarities end. Data returned from Russian and American spacecraft over the last thirty years have dispelled the "twin" myth. A traveler visiting Venus would suffer from acid burns and be subject to suffocation, a roasting, and a crushing. Many astronomers frequently refer to this planet as the closest thing to hell in the Solar System.

Venus orbits the Sun farther out than Mercury, but closer than the Earth. However, Venus' surface temperature exceeds that of Mercury partially because of the presence of a dense carbon dioxide atmosphere 95 times thicker than Earth's nitrogen-oxygen atmosphere. The carbon dioxide combined with a perpetual global layer of sulfuric acid clouds formed from past volcanic activity acts to trap solar heat reaching the surface. Essentially, conditions on Venus represent a runaway "greenhouse effect." Current surface temperatures exceed 480° C, hot enough to melt lead on contact.

Despite hellish conditions today, some astronomers speculate that Venus experienced a more benign past similar to Earth conditions today. Scientific data returned from space probes that flew by

USA

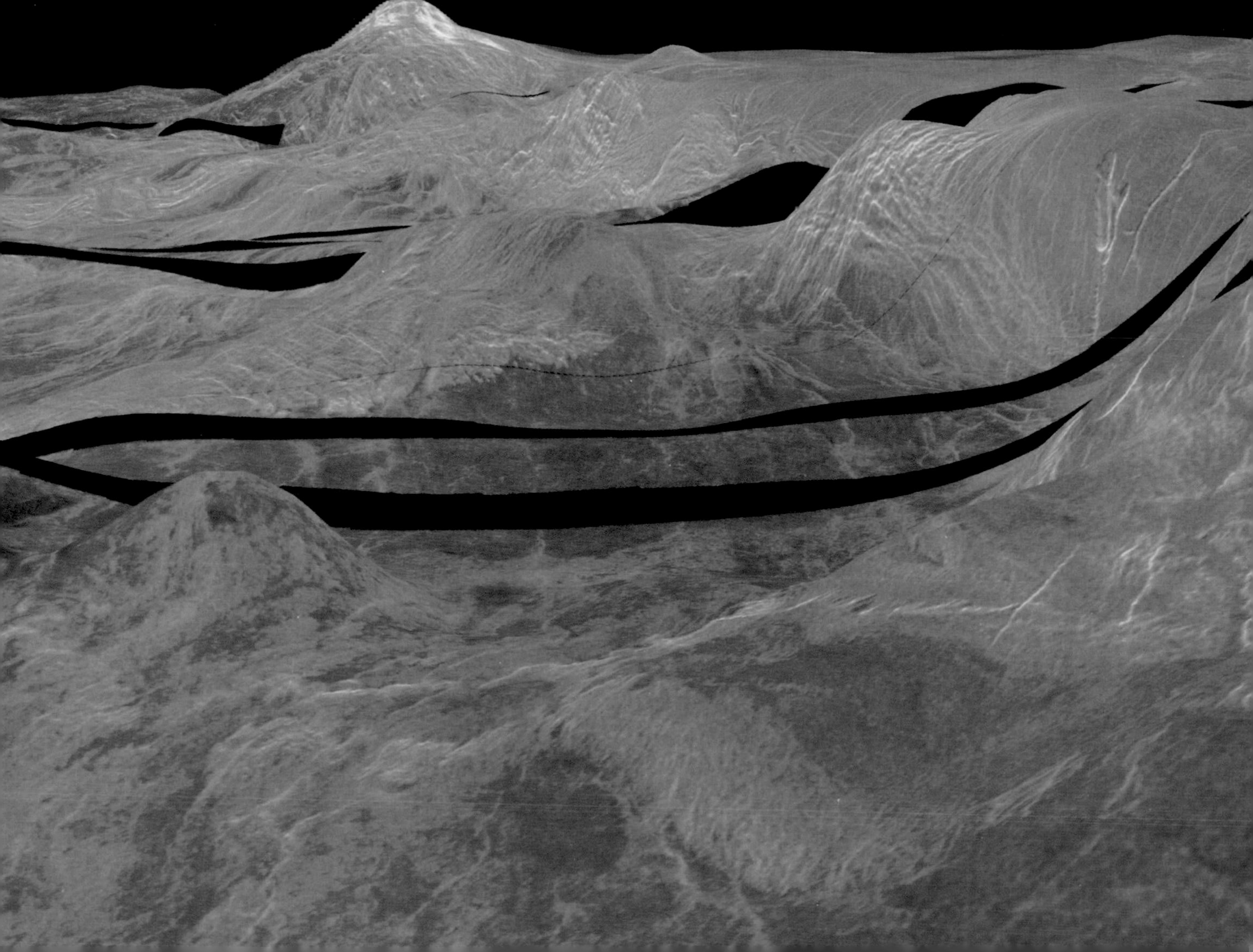

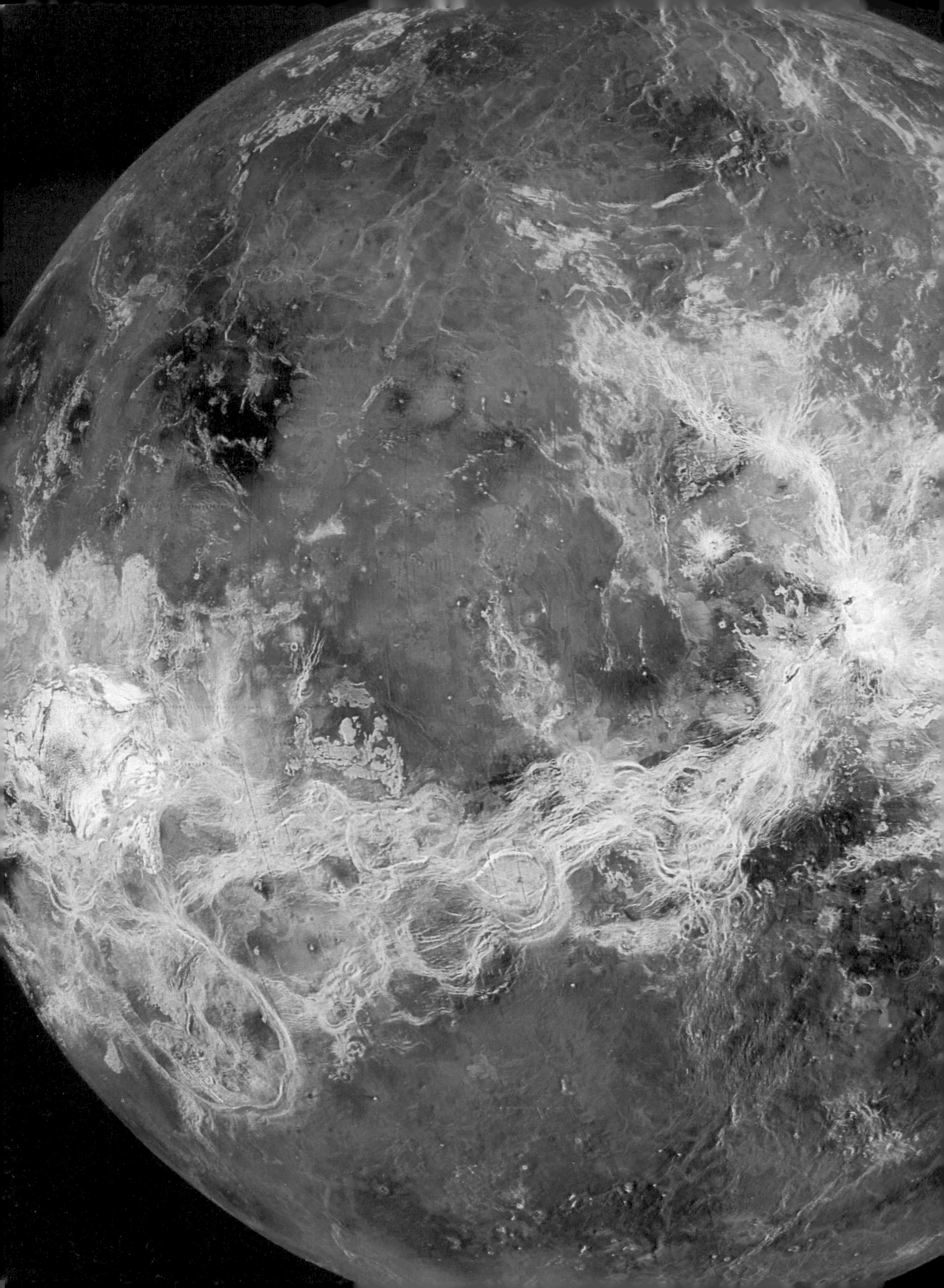

Venus shows evidence of concentrated deuterium atoms in the atmosphere. This element is a heavy form of hydrogen. Unlike normal hydrogen atoms with no neutrons, deuterium atoms contain a single neutron. Normally, a deterministic, but small percentage of all hydrogen takes the form of deuterium. On Venus, scientists detected a higher percentage than normal. This finding suggests that Venus once supported lush, warm oceans like the Earth. As global warming accelerated the greenhouse effect, the water evaporated. Eventually, a greater proportion of the lighter hydrogen atoms in the water vapor escaped from the atmosphere into space, leaving a higher than normal concentration of deuterium behind. An intriguing question is that if conditions on Venus once resembled conditions on Earth, could Earth's climate someday head in Venus' direction?

In 1989, NASA launched the *Magellan* spacecraft to Venus with the hope of unlocking many of the planet's mysterious secrets. One of the problems with studying the planet is that a permanent layer of sulfuric acid clouds perpetually shields Venus' surface from telescopes on Earth or imaging cameras on spacecraft. *Magellan* overcame that problem by using a technique called synthetic aperture radar. The spacecraft's 3.7-meter (12-foot) radar dish emitted thousands of radio pulses every second that bounced off Venus' surface. By analyzing the reflections with computers back home, scientists constructed a picture map of the surface that revealed details with a resolution of 250 meters (820 feet). Over a period of several years, *Magellan* circled Venus every three hours and radar-mapped over 98% of the surface.

Mars

Mars orbits the Sun slightly farther out than the Earth. This small, red-colored, rocky planet ranks in size as larger than Mercury, but only half as large as Venus and Earth. Despite its small size, planetary scientists consider Mars as the prime location in the Solar System to search for extraterrestrial life. Sensors aboard various NASA spacecraft launched to Mars over the last thirty years have shown that advanced life forms almost certainly do not exist on Mars today. However, many people feel that the planet may hide bacterial forms of life or their fossil remains.

Scientific optimism regarding the discovery of extraterrestrial life results in part from the fact that surface temperatures on Mars resemble the Earth's climate more than any other planet in the Solar System. Daytime temperatures average well below freezing even during the Martian summer, and dip much lower at night. However, during extremely rare heat waves some locations near the equator may warm up to as high as 25° C (77° F) at noontime on a hot summer day. Although these conditions almost seem semi-inviting, the composition of the atmosphere leaves much to be desired from a human perspective. Most of the atmospheric gas consists of carbon dioxide, similar to Venus. The major difference between the atmospheres of Mars and Venus is the density. While the average baro-

Above:

This image, taken from the *Hubble Space Telescope* in low Earth orbit, shows a cold, cloudless day on Mars. The average temperatures on the surface of Mars are colder than Antarctica. Like on Earth, clouds are normally present in the Martian atmosphere, though not to the same extent as on Venus.

Opposite Page:

This photographic montage, generated using radar data from *Magellan*, shows what Venus would look like from space if all the clouds were removed. Venus has the unique distinction that almost all of its geological features are named after women.

Scientific Fact!

Both Venus and Mercury orbit between the Earth and the Sun. Consequently, they cannot be seen at night. On the other hand, they cannot be seen in the day because of the bright light from the Sun. The only time these two planets can be seen is right after sunset, or just before dawn. When Venus is visible, it shines brighter than any object except for the Moon. In addition, both Venus and Mercury are the only two planets in the Solar System that do not have at least one moon.

Above:

In 1976, two NASA spacecraft named *Viking 1* and *2* landed on Mars and sent back many photographs of the surface. Panoramic images of the Martian landscape returned by the *Viking 1* lander (top photo) reveal rolling plains, littered with small boulders and marked by rippled sand dunes. The large boulder on the left, named Big Joe, measures 1 by 3 meters (3.2 by 9.8 feet) and sits about 7.9 meters (26 feet) from the lander. The vertical white object near the center of the photograph is a boom that contains weather-sensing instruments. *Viking 2's* landing site (bottom photo) reveals a similar landscape, but without sand dunes. The objects near the bottom of the picture are part of the *Viking 2* lander.

metric pressure on Venus measures more than ninety times times greater than on Earth, the corresponding pressure at the surface of Mars is more than ninety times less than on Earth.

In order to determine whether life ever existed on Mars, scientists need to solve the mystery of missing water on the planet. Currently, almost all of the known water on the surface of Mars remains trapped in its southern polar ice cap. In addition, several pictures radioed back to Earth from the surface of Mars show patches of water-ice frost covering the rocks during the Martian winter, and photographs taken from orbit show clouds in the atmosphere. However, liquid water cannot exist on the surface because the atmospheric pressure is too low. Consequently, melting ice sublimes directly into water vapor.

Despite the fact that liquid water cannot exist on the Martian surface today, photographic evidence taken from NASA spacecraft that orbited Mars revealed a network of thousands of kilometers of flood channels and flood plains on the surface. These photographs

suggest that the Martian past supported more temperate conditions with a thicker atmosphere and deep oceans filled with water. Today, Mars lies trapped in a global ice age, similar to ice ages that blanketed the Earth in the past. Understanding the mystery of water on Mars may also yield clues into Earth's geological past and future.

Geologically, Mars also ranks as one of the most interesting planets in the Solar System. Despite the planet's small size, it contains a canyon deeper than the Grand Canyon and longer than the United States, and four monstrous, but extinct volcanoes that make Mount Everest appear tiny in comparison. In addition, every Martian summer, in the southern hemisphere, torrential winds that clock faster than 200 kilometers per hour howl across the planet, creating global dust storms that block the Sun for weeks at a time. Mars also holds two moons, Phobos and Deimos, within its gravitational grasp. Both of these look like potato-shaped rocks scarcely larger than Manhattan Island in New York.

NASA sent six spacecraft and two landers to explore Mars between 1965 and 1976, but none in the two decades following. This trend changed in 1996 with the launch of *Mars Pathfinder* and *Mars Global Surveyor*. *Pathfinder* landed on the surface and deployed a miniature rover to drive around and collect data, while *Surveyor* remained in orbit to conduct the most detailed scientific survey of Mars ever attempted. Read the next chapter to learn more about Mars and Mars exploration in more detail.

Jupiter

This planet orbits the Sun more than twice as far out as Mars and ranks as the largest planet in the Solar System. In terms of orbital distance from the Sun, Jupiter is the first of four planets that consist primarily of lightweight gasses densely compacted around a small solid core about the size of the Earth. This type of internal structure offers little in terms of a solid surface, in sharp contrast to the structure of the small, rocky inner planets. However, the internal pressure deep below the upper atmosphere of these gas giants compresses the gas into a liquid form. At locations near the core, the gas becomes so compressed that it looks and behaves like a metal.

Jupiter contains more mass than all of the other planets in the Solar System put together. In fact, over 1,300 Earth-size worlds could fit inside Jupiter if it were hollow. Almost all of this internal space contains hydrogen and helium in the same proportions as the Sun. Consequently, many scientists believe that Jupiter is essentially a ball of Sun-type material that never ignited to become self-luminous. Despite this fact, Jupiter is still a violent place. Consider the fact that this planet gives off twice as much heat as it receives from the Sun, and its red, orange, white, and brown clouds that streak across the upper atmosphere show signs of influence from storm fronts much larger than the diameter of the Earth.

Big Moons of the Solar System

#1 - Ganymede (orbits Jupiter)
5,262 kilometers in diameter
Largest moon in the Solar System, composition of up to 50% ice, heavily cratered

#2 - Titan (orbits Saturn)
5,150 kilometers in diameter
Only moon in Solar System with a sizable atmosphere, surface may contain the chemical building blocks of amino acids

#3 - Callisto (orbits Jupiter)
4,800 kilometers in diameter
One of the most heavily cratered objects in the Solar System, moon's size equal to Mercury

#4 - Io (orbits Jupiter)
3,648 kilometers in diameter
Surface contains at least nine active volcanoes that spew sulfur

#5 - Moon (orbits Earth)
3,476 kilometers in diameter
Largest of the big moons relative to the size of the parent planet it orbits

#6 - Europa (orbits Jupiter)
3,138 kilometers in diameter
A thick layer of surface ice may hide deep oceans beneath, almost no craters on surface

#7 - Triton (orbits Neptune)
2,700 kilometers in diameter
Contains active volcanoes that spew ice and black dust, only large moon that orbits with a retrograde orbit

#8 - Titania (orbits Uranus)
1,610 kilometers in diameter
Contains a 1,600-kilometer long canyon, lots of small craters, and many rough rocks

#9 - Oberon (orbits Uranus)
1,550 kilometers in diameter
A very ancient, icy, heavily cratered surface reveals little sign of geological activity

#10 - Rhea (orbits Saturn)
1,528 kilometers in diameter
A heavily cratered body with bright wispy markings, over 70% water-ice in composition

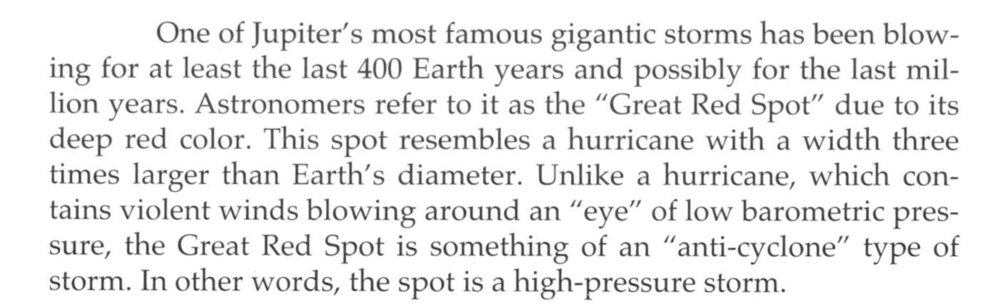

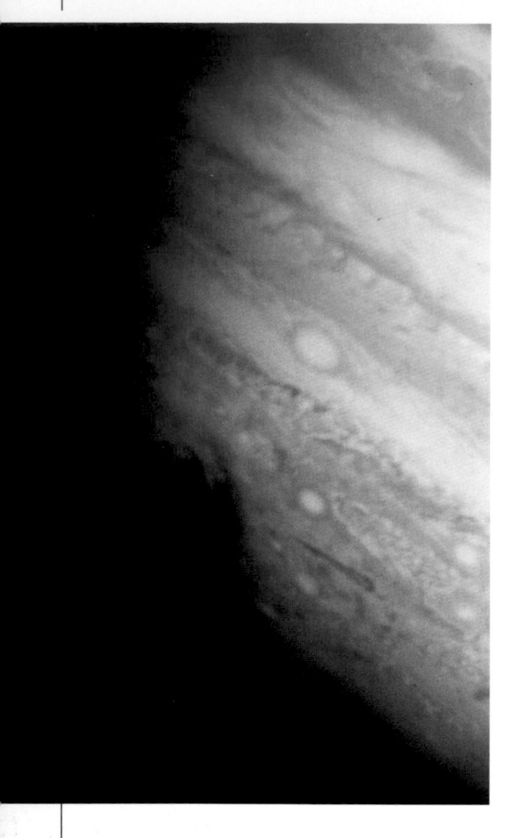

Above:

Early in the morning on 18 July 1994, Fragment G of comet Shoemaker-Levy 9 collides with Jupiter. The *Hubble Space Telescope* in low Earth orbit took this photograph of the aftermath of the collision. The comet hit at a speed in excess of 60 kilometers per second and caused an explosion more powerful than 1,000,000 nuclear bombs. The dark spot and ring (center of the photo) was caused by the collision and is about the size of the Earth.

Opposite Page:

This photographic montage shows the planet Jupiter (the large object a little up and to the right of the picture's center), along with its four largest moons. The moons and Jupiter are not shown to scale, but they are in their relative positions: Io (upper left), Europa (center), Ganymede (lower left), and Callisto (lower right). Notice the spot that looks like an eye in Jupiter's southern hemisphere. Astronomers call this feature "The Great Red Spot." It is a hurricane-type storm that is larger than the Earth and has been blowing for at least the last 400 years. A spacecraft named *Voyager 1* transmitted these images to Earth in early March 1979.

One of Jupiter's most famous gigantic storms has been blowing for at least the last 400 Earth years and possibly for the last million years. Astronomers refer to it as the "Great Red Spot" due to its deep red color. This spot resembles a hurricane with a width three times larger than Earth's diameter. Unlike a hurricane, which contains violent winds blowing around an "eye" of low barometric pressure, the Great Red Spot is something of an "anti-cyclone" type of storm. In other words, the spot is a high-pressure storm.

Many astronomers refer to Jupiter and its 16 moons as a "mini Solar System." Most of these moons look like odd-shaped pieces of rock scarcely larger than 200 kilometers across. However, four of the moons are among the largest and most interesting in the Solar System. For example, at least nine active volcanoes on the moon Io spew sulfur and lava into space on a regular basis, a surface completely glazed over with ice may hide deep oceans of water on Europa, and Callisto and Ganymede respectively equal and exceed the size of the planet Mercury. Both Io and Europa compare in size to the Earth's Moon.

Between 1972 and 1981, four NASA robotic spacecraft have conducted quick flybys of Jupiter to take photographs and gather scientific data. However, the most detailed scientific investigation of Jupiter began in December 1995. At that time, a large NASA spacecraft called *Galileo* arrived in orbit around Jupiter and spent more than two years studying the chemical composition and physical state of the planet's atmosphere, the structure and dynamics of the planet's gigantic magnetic field, and the composition and physical state of the many moons. One of the many highlights of the *Galileo* mission was the atmospheric probe. This mini-spacecraft was released from *Galileo* and descended into the Jupiter atmosphere with the aid of a parachute. During descent, the probe gathered data about the composition of the atmosphere for one hour before being crushed by the intense pressure.

Saturn

Saturn, the next planet out from Jupiter and the second largest overall, also contains densely compressed hydrogen and helium in its interior. But in contrast to the vivid red colors and wild turbulence found in the atmosphere of its gas-giant neighbor, Saturn's butterscotch-colored atmosphere looks almost placid in comparison. This calm appearance disguises some of the fastest winds in the Solar System. These violent jet streams race around equatorial regions at over 1,770 kilometers (1,100 miles) per hour.

Most of Saturn's fame comes from the countless number of frosted rocks and chunks of ice in orbit over the equator. These rocks form a spectacular set of thousands of rings that surround the entire planet. Although only tens of meters thick, the major ring set spans a distance of close to 300,000 kilometers from one edge to the opposite edge. Each ring is divided into literally thousands of minor ringlets,

Above:

NASA spacecraft *Voyager 1* recorded this image of Saturn on 18 October 1980 when it was about 34 million kilometers (21.1 million miles) from the planet. A set of thousands of rings that surrounds Saturn turns an otherwise bland-looking planet into a spectacular sight. At the distance this photograph was taken, the many rings blend together and give the false impression that only two or three exist. Despite the fact that Saturn's gigantic diameter measures more than nine times that of Earth's, Saturn would float on water if there existed an ocean large enough to hold it.

similar to grooves on a phonographic record. Nobody knows how the rings formed. Some scientists speculate that they might have resulted when a moon or a passing body ventured too close to the planet. If so, the unlucky object would have been ripped apart by Saturn's enormous gravity. Others theorize that the rings consist of raw material that never coalesced to form a complete moon due to the strong influence from Saturn's gravity. Another possibility is that the object was shattered by collisions with other objects in orbit around the planet. Although all the gas-giant planets possess rings, Saturn's outshine them all.

In addition to the ring particles, Saturn also holds a large family of moons captive within its gravitational grasp. In fact, Saturn holds the record for the planet with the most number of moons in the Solar System. At last count, astronomers have confirmed the existence of twenty-one known moons, with another six suspected but not

confirmed. Many of these moons measure less than 50 kilometers (31 miles) across, are icy in composition, and orbit near the rings. Astronomers call them *shepherd moons* because their gravity "shepherds" the ring particles into line like a flock of sheep.

The size of Saturn's icy shepherd moons pales in comparison to Titan, the planet's largest moon. Titan's size as compared to other moons in the Solar System is exceeded only by Ganymede in orbit around Jupiter. One of Titan's many interesting features, other than the fact that its size exceeds the planets Mercury and Pluto, is its thick atmosphere of nitrogen and methane gas. Titan is the only known moon in the Solar System to retain an atmosphere. Scientists suspect that the gasses in Titan's atmosphere may support the formation of simple organic molecules and other building blocks of amino acids.

Scientists must wait until June 2004 to study Titan further. At that time, a NASA spacecraft called *Cassini* will reach Saturn after having completed a six-year journey that begin on 15 October 1997 from Florida. *Cassini* will spend nearly four years in orbit around the ringed planet to radio data back to Earth regarding the planet and its many Moons. One of the major highlights of the mission will feature a probe that *Cassini* will drop into Titan's atmosphere. As it descends, the probe will gather data concerning the chemical composition of the atmosphere. If it survives the landing, the probe will also relay data back to Earth regarding surface conditions on Titan. Previously, three spacecraft have conducted quick flyby missions of Saturn. *Cassini* will become the first to orbit the planet.

Uranus

Uranus, the seventh planet from the Sun and third largest in the Solar System, orbits almost twice as far out as Saturn and takes eighty-four Earth years to complete one orbit. This planet is the farthest planet from Earth visible without the aid of binoculars or telescopes. Cloud formations in this gas giant's atmosphere take on a dull greenish blue color because methane gas complements the standard hydrogen and helium composition. Scientific data suggests that approximately 8,000 kilometers (4,971 miles) under Uranus' cloud tops, a scalding ocean of water and dissolved ammonia exists.

One of the most interesting features of this world is that the planet lies tipped on its side, with its polar regions facing the Sun. All of the other planets in the Solar System orbit with their equatorial regions facing the Sun. Astronomers speculate that an Earth-size object collided with Uranus long ago, knocking the green gas giant over onto its side. Today, this odd tilt exaggerates the length of the Uranian seasons. Each polar region receives forty-two years of sunlight followed by forty-two years of complete darkness. However, because not much sunlight reaches Uranus, the summer and winter temperatures fall within 2° C of each other. Average temperatures at the cloud tops dip to a chilling -210° C (-346° F).

Top Photograph:

This close-up view of Saturn's B-ring shows that the rings are actually composed of millions of tiny "ringlets." The black marks across the rings are called *spokes* and consist of dark particles of ice or rock. This ring spans about 100,000 km from edge to edge. *Voyager 2* recorded this image in 1981.

Bottom Photograph:

Mimas, one of Saturn's moons, contains a large crater 135 kilometers across. How Mimas survived the impact is unknown, considering the moon measures only 390 kilometers across. In this picture, Mimas resembles the Death Star from the movie *Star Wars*.

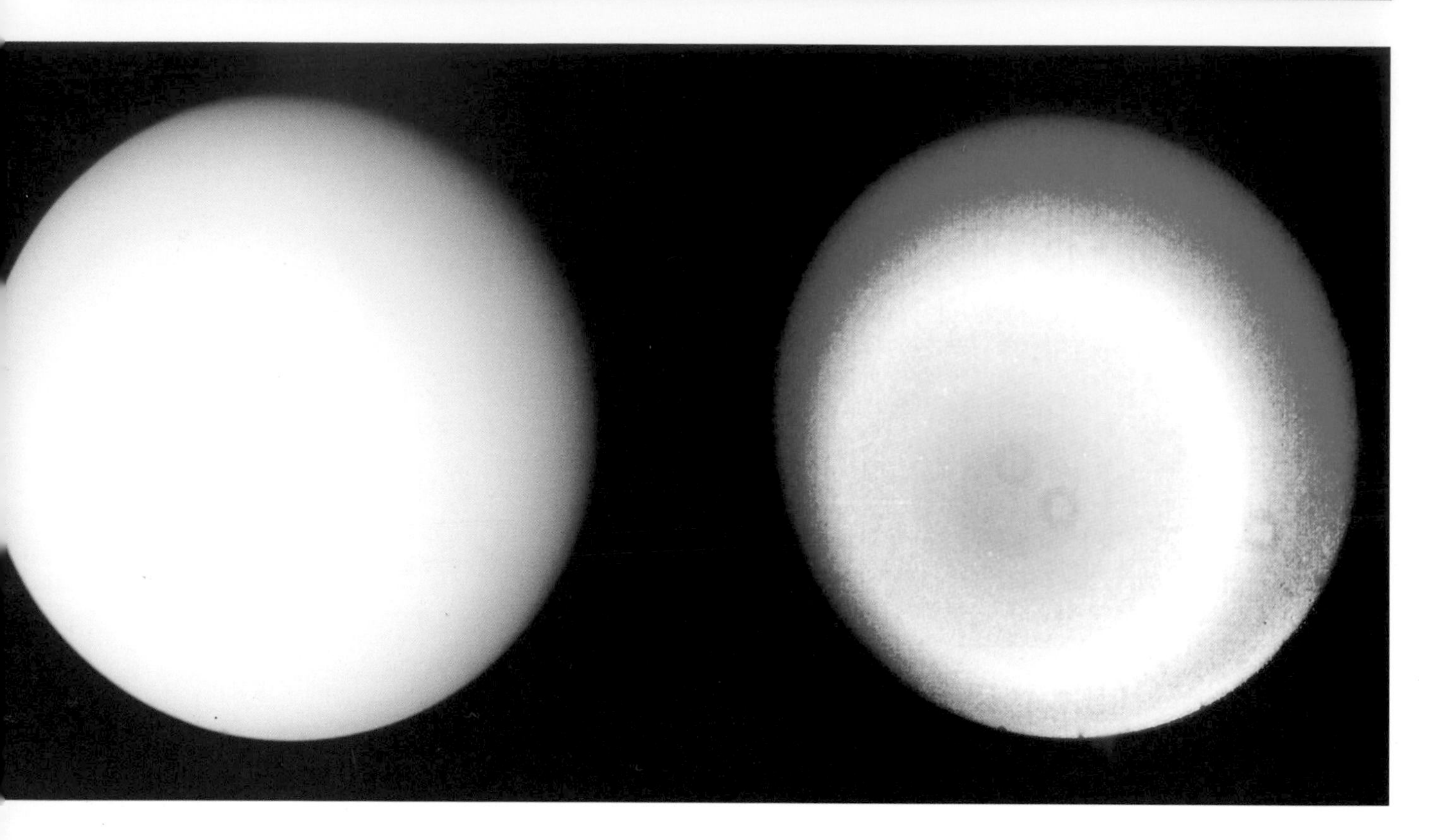

Above (Left):

Uranus' clouds appear virtually featureless as seen from spacecraft *Voyager 2's* cameras in January 1986. Acetylene and ethane in the atmosphere forms a smog-like haze that obscures more detailed observations. The photograph shows what observers would have seen if they had been on the spacecraft.

Above (Right):

After *Voyager 2* transmitted the Uranus images back to Earth, scientists used computers to enhance the images, revealing details not visible with the naked eye. This processed photograph shows the south pole of Uranus and different zonal bands in the atmosphere. Scientists speculate that these bands are probably composed of hydrocarbons generated by the electrons in Uranus' magnetic field. The small doughnut-shaped dots (from the center to the lower right-hand corner) are out-of-focus dust particles trapped on the camera's lens after seven years of traveling through space.

Like Jupiter and Saturn, Uranus also holds a large number of moons captive within its gravitational pull. All fifteen of the planet's moons received their names after characters in Shakespearean plays. The diameters of the largest four Uranian moons, Oberon, Titania, Umbriel, and Ariel, measure about one-third the diameter of Earth's Moon. The fifth largest, Miranda, contains a jumble of grooves, craters, and cliffs haphazardly thrown together. Although Miranda measures only 470 kilometers (292 miles) across, it contains canyons ten times deeper than the Grand Canyon in Arizona. Scientists think that Miranda was blown apart by an enormous collision with another moon long ago. The chaotic terrain formed when the pieces from the collision coalesced under influence from their own gravity. However, scientists hoping to study Uranus and Miranda further will have to wait a long time. Only one NASA spacecraft has flown by Uranus, and NASA has no plans to send another one in the near future.

Neptune

Neptune orbits farther from the Sun than any other gas giant and resembles Uranus almost exactly in size, chemical composition, and color. The winds in the upper atmosphere blow at more than 2,000 kilometers an hour, ranking as the strongest in the Solar System. Another interesting fact about this planet is that Neptune owes its discovery to the use of mathematics. After the discovery of Uranus in the 1800s, astronomers realized that an unknown gravitational force was pulling it slightly off its expected orbital path. Mathemati-

cians Couch Adams in England and Urbain Leverrier in France proceeded to calculate the location of this unknown object based on Uranus' deviation from its predicted orbit. In 1946, astronomer Johann Galle from Berlin turned his telescope to the predicted location and discovered Neptune. This gas giant takes 165 Earth years to complete an orbit and holds eight moons captive.

One of Neptune's eight moons, Triton, ranks as the seventh largest in the Solar System, with a 2,700-kilometer (1,678-mile) diameter, about three-fourths the size of Earth's Moon. Triton's most interesting features are ice-spewing volcanoes that resemble geysers on Earth. These volcanos spew nitrogen vapor mixed with black dust. Unfortunately, further study of Neptune and Triton probably will not occur for some time to come. Like Uranus, only one NASA spacecraft has flown by Neptune. NASA currently has no plans to send another in the near future.

Above:

NASA spacecraft *Voyager 2* recorded this image of the planet Neptune on 17 August 1989. This photograph shows two of the four cloud features that *Voyager* had been tracking for two months prior to mid-August. The large dark oval near the left side of the planet is an atmospheric disturbance, called the "Great Dark Spot." It circles the planet every 18.3 hours. The second dark spot near the lower right edge of Neptune circles the planet every 16.1 hours. Compared to the Earth, Neptune is about four times larger.

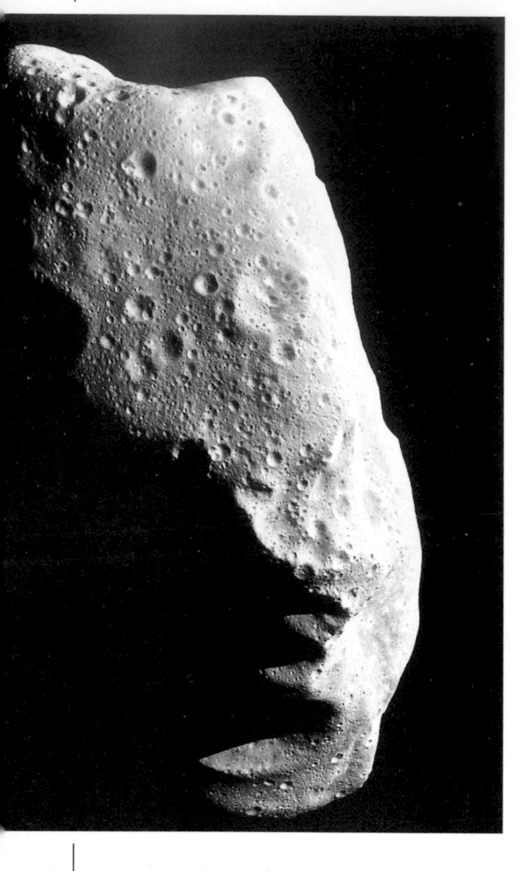

Above:

While on its way to Jupiter, spacecraft *Galileo* took this photograph of asteroid 243-Ida on 28 August 1993. Ida is a 52-kilometer (32-mile)-long potato-shape rock that orbits the Sun between Mars and Jupiter. Many astronomers speculate that the asteroids are material that never formed into a planet. Some feel that the strong gravity of Jupiter is to blame. This was only the second time in history that a spacecraft had taken a close-up picture of an asteroid.

Pluto

After astronomers discovered Neptune, they realized that its gravity alone did not account for enough force to yank Uranus away from its expected orbit. In addition, something else was pulling Neptune off of its predicted orbit. For seventy-five years, many astronomers searched for the missing ninth planet. Among the most dedicated of them was an American named Percival Lowell. Although he failed, his assistant, Clyde Tombaugh, continued the search and discovered Pluto in 1930, almost twelve years after Lowell's death. Pluto normally orbits beyond Neptune. However, its extremely elliptical orbit takes it inside the orbit of Neptune some of the time. In fact, Pluto's position in the late 1990s was closer to the Sun than Neptune and only began to move away after the year 1999.

Pluto does not resemble the other four gas-giant, outer planets at all. This planet's composition consists mainly of ice and rock. In addition, Pluto ranks as the smallest planet in the Solar System, about two-thirds the size of Earth's Moon. Because of its small size and funny orbit, some astronomers speculate that Pluto was a moon of Neptune that escaped long ago. In total, Pluto takes 248 years to complete one orbit around the Sun. Scientists know little about Pluto other than the fact that it has one moon about half its size called Charon. As of today, no spacecraft has yet photographed Pluto. However, NASA scientists hope to mount a robotic reconnaissance mission to Pluto in the near future.

Planet X?

Some time after the discovery of Pluto, astronomers realized that its small size could not possibly account for enough gravity to perturb Uranus and Neptune off their predicted orbital paths around the Sun. Consequently, some astronomers remain convinced that a tenth planet exists somewhere beyond Pluto's orbit. Theoretical mathematical models predict that "Planet X" is a frozen world about the same size as Uranus and Neptune, but orbits the Sun more than three times farther out than Neptune. Some astronomers have spent much of their time looking for this mystery planet, but without much success.

Asteroids and Comets

A countless number of asteroids orbit the Sun in between Mars and Jupiter in a zone called the *asteroid belt*. These orbiting rocks range in size and shape from irregular-shaped chunks less than 1 kilometer across to spherical globes close to 1,000 kilometers in diameter. Astronomers estimate that the mass of all the asteroids put together would equal less than half of the Moon's mass. Some asteroids orbit outside the main belt and fragments from these giant rocks occasionally graze the Earth's atmosphere. Technically, astronomers call them *meteors*. When these rocks strike the Earth's upper atmo-

sphere at high velocities, they burn up from frictional heating and create long streaks of light more commonly described as "shooting stars." Sometimes, part of the meteor survives this incineration and impacts the ground. In this case, astronomers call them *meteorites*.

Every day, many small meteors burn up in the upper atmosphere. Occasionally, larger meteors smash into the Earth. On average, the Earth collides with rocks in orbit around the Sun measuring 50 to 100 meters in diameter every few hundred years. Part of a meteorite this large will burn up in the upper atmosphere. However, a significant fraction will survive to impact the ground, resulting in an explosion more powerful than the yield of the largest nuclear weapon ever designed. Meteor Crater in the Arizona desert resulted from one of these Earth-meteor collisions. The impact crater measures over a kilometer across and hundreds of meters deep.

Mathematicians predict that about every 100 million years on the average, a meteor on the order of 10 kilometers (6.2 miles) in diameter will barrel into the Earth. Such a collision would release more destructive energy than 10,000 times the power of all the nuclear weapons on the Earth and cause a global catastrophe killing a significant fraction of all the life on Earth. The last such collision occurred roughly 65 million years ago near what today is the Yucatán Peninsula in Mexico. Scientists using sensitive instruments discovered this fact when they detected the remnants of an eroded crater the size of the Gulf of Mexico near the Yucatán. This catastrophe, called the Cretaceous-Tertiary impact, or K/T for short, is thought to have caused the extinction of the dinosaurs. Some mathematicians speculate that the probability of a K/T-type impact occurring in the life of a newborn child today may be as low as 1 in 2,000. Sound too ridiculous to believe? Take a look at any detailed photograph of the Moon and try to count the number of impact craters.

Millions of comets also orbit the Sun. Scientists speculate that comets resemble asteroids but contain more ice than rock. In general, comets are icy conglomerations of dust, rock, and frozen gas lumped into irregularly shaped nuclei about 10 to 30 kilometers across. Typically, they orbit the Sun in extremely elliptical paths that take them inside Mercury's orbit at closest approach and beyond Pluto at the other end. Heat from sunlight boils off part of the frozen material from the surface of the nucleus as a comet approaches the Sun. This melting creates a long glowing tail of charged particles that follows the nucleus through space. Comet tails sometimes exceed 100 million kilometers in length and can provide viewers on the Earth with a spectacular sight.

Nobody knows exactly where comets come from. Most orbit the Sun with periods of hundreds or thousands of years. Some take even longer. For example Delavan's Comet of 1914 will not return again for an estimated 24 million years. These long orbital periods suggest that comets originate from far beyond the orbit of Neptune. In 1950, Dutch astronomer Jan Oort postulated that comets reside in a

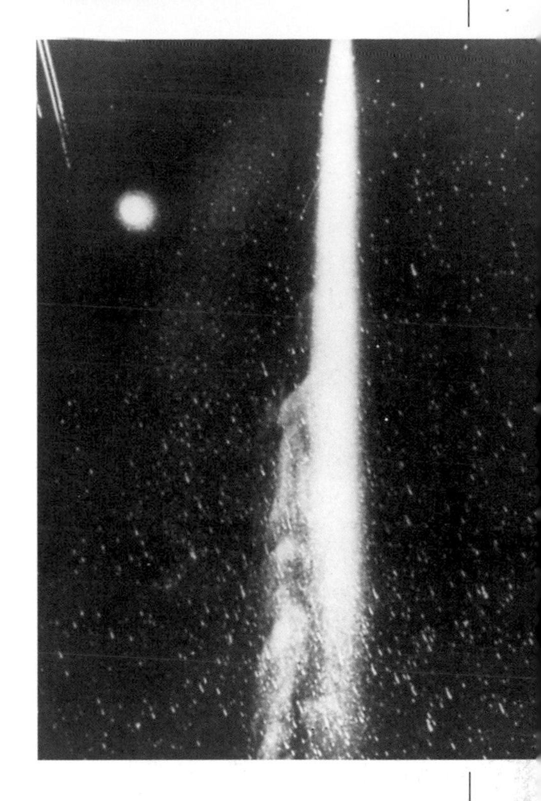

Historical Fact

Halley's Comet (shown above) probably ranks as the most famous of all comets. It appears in the vicinity of the Earth about every 76 years, with confirmed sightings dating back to 240 B.C. The last time was in 1986. The comet was named after Sir Edmund Halley, who plotted the comet's orbit in the 18th century. He was the first one to realize that historical records of comet sightings at 76-year intervals were of the same comet. Halley predicted the return of the comet in 1759 but did not live to see it. Astronomers who have studied past data believe that the Earth may have passed through Halley's tail during the comet's appearance in 1910.

	Mercury	Venus	Earth	Mars	Jupiter	Saturn	Uranus	Neptune	Pluto
Mean Dist. From Sun (AU)	0.39	0.72	1.00	1.52	5.20	9.53	19.18	30.08	39.34
Inclination of Orbit (deg)	7.00	3.40	0.00	1.86	1.31	2.49	0.66	1.77	17.14
Orbital Eccentricity	0.206	0.007	0.017	0.093	0.048	0.056	0.046	0.010	0.248
Orbital Speed (km/s)	47.89	35.00	29.79	24.13	13.06	9.64	6.81	5.43	4.74
Diameter (km)	4,878	12,103	12,755	6,790	142,796	120,660	51,118	49,528	2,284
Mass (Earth = 1)	0.06	0.81	1.00	0.11	318	95	14.5	17.14	0.002
Density (water = 1)	5.43	5.24	5.52	3.95	1.33	0.69	1.29	1.64	2.03
Length of Day	58.7 days	243 days	23 hr 56 min	24 hr 37 min	9 hr 56 min	10 hr 40 min	17 hr 14 min	16 hr 6 min	6 days 9 hr
Length of Year	87.97 days	224.7 days	365.26 days	686.98 days	11.86 years	29.46 years	84.07 years	164.8 years	248.6 years
Number of Moons	0	0	1	2	16 or more	21 or more	15 or more	8 or more	1
Temperature Range (°C)	-180 to 430	465	-80 to 60	-120 to 25	-150 (clouds)	-170 (clouds)	-210 (clouds)	-210 (clouds)	-220
Atmosphere	Almost None	CO_2	Nitrogen Oxygen	CO_2	Hydrogen Helium	Hydrogen Helium	Hydrogen Helium Methane	Hydrogen Helium Methane	None (?)

Above:

The table above provides a quick way to look up the key facts and statistics of the nine planets of the Solar System. Values given for the inclination and eccentricity are the values for the planet's orbit around the Sun. In order to compute the semi-major axis length of a planet's orbit, multiply the mean distance from the Sun value listed in the table by approximately 149,597,900 kilometers (93,000,000 miles). In order to compute the mass of a planet, multiply the value listed in the table by approximately 5.9742×10^{24} kilograms. The numbers given for length of a day and length of a year are expressed in "Earth" time.

gigantic cloud surrounding the Solar System that extends deep into space. He suggested that most comets orbit in this frozen abyss until gravitational perturbations from nearby passing stars bump them into new orbits that make them fall toward the Sun. Closer in, the gravity of larger planets such as Jupiter may perturb some of them further, trapping the comets in smaller orbits within the bounds of Neptune's orbit or closer. Both asteroids and comets consist of raw material left over from the formation of the Solar System. A study of both may yield clues into the origin of the Earth and the other planets.

The Sun

The Sun accounts for over 99.5% of the total mass in the Solar System, and its interior could easily contain the equivalent of 1,300,000 Earth-size objects or 1,300 Jupiters. Almost all of this vast internal space contains hydrogen gas. The pressure of the gas increases exponentially with increasing distance below the Sun's surface in a process similar to that which makes divers deep in the ocean feel more pressure than divers near the surface. At the center of the Sun, the gas is so hot and compressed that the hydrogen atoms fuse together to form helium. This thermonuclear fusion process releases a tremendous amount of energy and radiation that flows upward toward the surface of the Sun, and out into space to warm the Solar System. The heat and sunlight that leave the Sun travel at the speed of light (300,000 kilometers or 186,200 miles per second), take 8.3 minutes to reach the Earth, and about five hours to reach Pluto.

Scientific Fact!

The Sun is powered by the same process that gives a hydrogen bomb its deadly punch. However, it is so far away that If the Sun stopped shining this very instant, nobody on Earth would know for another eight minutes.

The temperature at the center of the Sun reaches about 14,000,000° C and the pressure exceeds 200 billion times the atmospheric pressure at the surface of the Earth. This combination allows the Sun to convert 4,000,000 tons of itself into energy every second. However, do not worry about the Sun burning out anytime soon. Astronomers estimate that the Sun has been burning for the last four

billion years and will continue to do so for at least the next four billion years.

Solar Orbit

All of the Earth orbit concepts presented previously in this book regarding orbital geometry, the dynamics of satellite motion on elliptical orbits, and space maneuvers also apply to solar orbits. For example, circular orbital velocity decreases with increasing distance from the Sun, satellites move faster when they are closer to the Sun than when they are farther away, and there are still 360° of true anomaly on every orbit. However, a few differences exist. One of them deals with terminology. Spaceflight engineers use the terms *perihelion* and *aphelion* to respectively describe a satellite's closest approach and farthest separation from the Sun. They also use the terms *periapsis* and *apoapsis* to describe the same concepts for orbits around planets other than the Earth. Remember that the corresponding terms for orbits around the Earth are *perigee* and *apogee.*

Another difference deals with the coordinate system. Spaceflight engineers use the heliocentric J2000 coordinate system instead of the geocentric J2000 coordinate system for orbits around the Sun. In this new system, the X axis still points toward the constellation Aries in deep space, but the XY plane contains the orbit of the Earth. Spaceflight engineers call this the *plane of the ecliptic*. One result of this definition is that the inclination of the Earth's orbit around the Sun measures 0° and the inclinations of all other solar orbits are measured relative to the ecliptic plane. This arbitrary definition of inclination simplifies the mathematics of computing orbital trajectories from the Earth to the other planets.

Describing the size of a solar orbit often requires the use of different units than the ones used to describe the size of orbits around the Earth. The reason is that meters and kilometers were not designed with the scale of the Solar System in mind. Instead, spaceflight engineers prefer to use a distance called the astronomical unit (AU) when describing the semi-major axis length of solar orbits. One AU equals the average distance between the Earth and the Sun, an amount equal to roughly 150,000,000 kilometers (93,000,000 miles). This unit allows people to visualize the size of a solar orbit by comparing it to the size of the Earth's orbit around the Sun. On this scale, the Earth orbits at a distance of one AU, while Pluto's perihelion and aphelion distances measure 30 and 50 AU, respectively. An interesting aspect of the AU is that light takes 8 minutes 20 seconds to travel 1 AU. At that rate, sunlight takes over five hours to reach Pluto.

All of the planets except for Pluto and Mercury travel around the Sun in orbits with extremely low eccentricities and inclinations. Consequently, spaceflight engineers often assume that the other seven planets orbit the Sun in circular coplanar (same inclination) orbits for the purposes of making rough calculations during the initial stages of planning an interplanetary mission. This type of calculation reveals

Scientific Fact!

One AU equals 149,597,900 kilometers (about 93,000,000 miles). This figure represents the average distance between the Earth and the Sun. To get an idea of the size of an AU, imagine the Sun shrunk to the size of a basketball. The Earth would be located about 27 meters (89 feet, approximately the length of a basketball court) away and would be the size of a tiny pebble scarcely larger than a pinhead.

Rule of Thumb

For the purpose of performing initial feasibility studies on a proposed interplanetary mission, it is acceptable to assume that all of the planets (except for Mercury and Pluto) orbit the Sun in circular orbits.

Historical Fact

On 27 August 1962, NASA launched the world's first spacecraft destined for another planet. This small satellite, called *Mariner 2*, flew by Venus about three months later on 14 December 1962. The onboard instruments collected about 42 minutes worth of scientific data.

RULE OF THUMB

Remember, ΔV stands for "delta V." It is a shorthand way of saying "a change in velocity." Most ΔVs result from rocket burns. The net effect is that the spacecraft moves to a different orbit. Also remember that the amount of rocket propellant required for a velocity change increases with increasing ΔV.

Historical Fact

The first spacecraft ever to escape the Earth's gravity was called *Pioneer 5*. On 11 March 1960, NASA launched this tiny probe into an orbit around the Sun. Over the next few years, *Pioneer* 5 returned data regarding the radiation and solar particles emitted by the Sun.

Scientific Fact!

Because the Moon is in orbit around the Earth, a spacecraft need not achieve escape velocity in order to reach the Moon.

whether or not a mission is feasible. In the later stages, they use computers to determine the exact trajectory based on the exact orbital elements of the planets.

6.2 How to Reach Other Planets

An interplanetary mission where a spacecraft flies on a straight-line trajectory from the surface of the Earth to another planet is impossible from an energy standpoint. Instead, booster rockets must first carry the spacecraft into low Earth orbit. Then, an upper stage fires at the proper time to insert the spacecraft onto a path that allows it to escape the grasp of the Earth's gravity and enter an elliptical orbit around the Sun that intersects the orbit of the destination planet. The rocket burn that allows the spacecraft to escape the Earth must be precisely timed so that both the spacecraft and destination planet arrive at the intersection point at the same time.

Escape Velocity

How does a spacecraft in a low orbit escape the Earth? Recall what happens when a spacecraft in a circular orbit increases its velocity by performing a posigrade burn. This extra velocity creates an elliptical orbit with a perigee at the burn point and an apogee higher than the original circular orbit altitude. The new apogee altitude depends on the amount of ΔV. A stronger burn allows the spacecraft to coast upward and away from the Earth for a longer time before gravity wins out and pulls the craft back down toward perigee. An escape trajectory results when the ΔV gives the spacecraft enough energy to coast away from the Earth forever.

A maneuver to create an escape trajectory out of a circular orbit requires a rocket burn that lasts for a few minutes on the average. After the rocket engines impart enough energy to raise the velocity past the escape threshold, they shut down and the spacecraft begins an unpowered coast away from the Earth. Gravity then starts to slow the spacecraft and continues to do so all throughout the escape journey. The Earth significantly slows the spacecraft's progress, but it is powerless to halt the escape once the escape velocity has been reached. Remember that nothing powers a spacecraft during the escape journey. Instead, it relies on the momentum supplied by the rocket burn that created the escape trajectory.

The exact magnitude of the escape velocity depends on the altitude of the orbit. Think about Earth escape as comparable to riding a bicycle from the bottom to the top of a valley in order to understand the escape velocity concept. People expend less energy riding to the top from halfway up the hill than from the bottom. Similarly, Earth escape from higher altitudes requires less energy than from lower altitudes. The required energy to escape amounts to increasing the forward velocity to at least 141% of the circular orbital

Figure 50: *How to Escape the Earth* ◄

Characteristics: A spacecraft on a hyperbolic or parabolic orbit has so much velocity at perigee that it will eventually escape the gravitational pull of the Earth and never come back. This figure shows how departing spacecraft can enter such "escape trajectories" from low-Earth orbit.

Solid arrow indicates position of the burn and points in the direction of the velocity change

Hollow arrow indicates position of the burn and points in the direction that the rocket engines fire to perform the burn

1 Spacecraft starts on a low-Earth circular orbit.

2 Posigrade burn (velocity increase) occurs here. Which orbit the spacecraft winds up on (3a, 3b, 3c, or 3d) depends on the amount of velocity increase.

3a Normally, the burn puts the spacecraft onto an elliptical orbit with a perigee at the burn point.

3b A stronger burn will put the spacecraft onto a larger elliptical orbit as long as the amount of velocity increase does not amount to more than 41% of the original circular velocity.

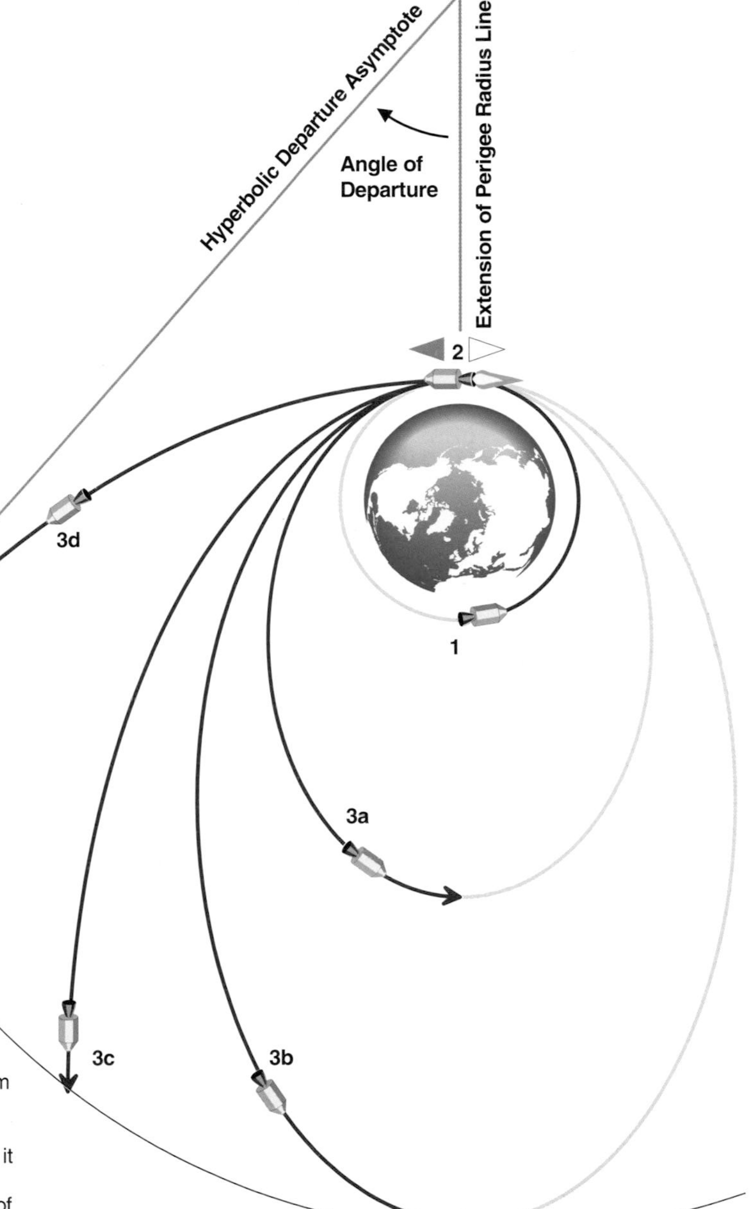

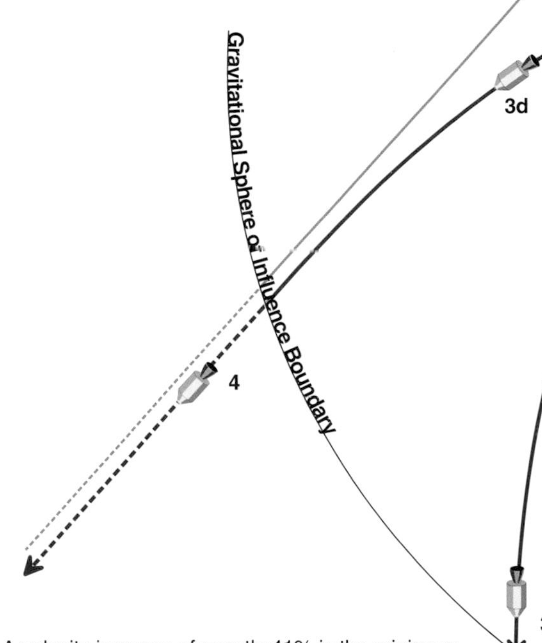

3c A velocity increase of exactly 41% is the minimum amount needed to create an escape trajectory. Such an increase puts the spacecraft onto a parabolic orbit. As it moves away from the Earth, it slows down. Eventually, it slows to zero speed (relative to the Earth) at the gravitational sphere of influence boundary.

3d A velocity increase greater than 41% puts the spacecraft onto a hyperbolic escape trajectory. As it moves away from the Earth, the spacecraft slows down. However, it has enough velocity at perigee that at the sphere of influence boundary, it keeps on going.

4 After the spacecraft passes the boundary, the Earth no longer slows it down. If the craft and the Earth were the only two objects in the universe, the craft would keep moving in the direction of the departure asymptote forever with a constant velocity equal to that at the boundary.

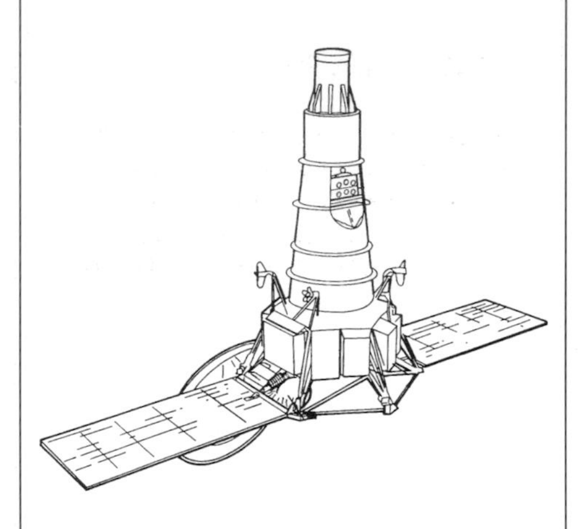

Ranger Missions to the Moon

Summary
The *Rangers* were a series of NASA spacecraft sent to photograph the Moon during the mid years of the 1960s. They were launched from Florida on a direct course to impact the lunar surface. During the several hours before impact, they used their radio transmitters to send back thousands of high-resolution television images of the lunar surface.

Dimensions	
Height	3.1 meters
Mass	366 kilograms
Launch Dates	
Ranger 7	28 July 1964
	Returned 4,306 pictures
Ranger 8	17 February 1965
	Returned 7,137 pictures
Ranger 9	21 March 1965
	Returned 5,814 pictures

Scientific Fact!

Who ever said, "what goes up must come down?" An object at the surface of the Earth propelled to a velocity of 11,200 meters per second will never come back. This "escape" velocity is more than 30 times faster than a speeding bullet.

velocity for the spacecraft's current altitude. A ΔV that increases the forward velocity by less than 41% creates an ellipse out of the original circular orbit.

Consider the ΔV differences between an escape from low Earth orbit (LEO) and geosynchronous orbit (GEO) to understand how the 41% rule works. A spacecraft on a low Earth circular orbit with a 300-kilometer (186-mile) altitude moves with a horizontal velocity of 7.7 kilometers (4.8 miles) per second. An Earth escape from this orbit requires a posigrade ΔV of 3.3 kilometers (2 miles) per second to increase the spacecraft's original circular velocity by 41% to a total of 11 kilometers (6.8 miles) per second. This extreme speed allows a spacecraft to cover a distance equal to that which separates New York and San Francisco in about six minutes.

An Earth escape from GEO takes a lot less energy than from LEO. The velocity of a spacecraft in a circular orbit at geosynchronous altitude amounts to 3.1 kilometers (1.9 miles) every second. Escape from this altitude requires a posigrade ΔV of 1.3 kilometers (0.8 miles per second) to increase the spacecraft's original circular velocity by 41% to a total of 4.4 kilometers (2.7 miles) per second. These numbers validate the fact that Earth escape requires less ΔV from higher altitudes than from lower altitudes. However, a spacecraft must first expend energy to get to a higher altitude before it can escape from a higher altitude. In fact, sending a spacecraft to GEO on a Hohmann transfer from LEO takes more energy than an Earth escape from LEO.

Departure Geometry

Figure 50 on page 247 shows the geometry of an escape trajectory as compared to a standard elliptical orbit. All escape orbits fall into either the parabolic or hyperbolic category. A parabolic escape orbit results when a spacecraft on a circular orbit increases its forward velocity by exactly 41%. The key characteristic of a spacecraft on this type of escape trajectory is that the Earth's gravity eventually succeeds in bringing the craft to a complete halt. However, the halt does not occur until the spacecraft reaches the "outer boundary" of the Earth's gravitational attraction. Notice that the parabolic path looks like the part of an ellipse between perigee and apogee except for the fact that apogee has been stretched to a distance infinitely far from the Earth.

A hyperbolic orbit results when the ΔV on the circular orbit increases the velocity by more than 41%. One of the main differences between a hyperbolic orbit and a parabolic orbit is that in a hyperbolic orbit, the Earth never succeeds in slowing the spacecraft to a complete halt. The initial additional velocity over the 41% escape minimum allows the spacecraft to have some velocity left over when it reaches the outer boundary. Spaceflight engineers refer to this "leftover velocity" as the spacecraft's hyperbolic excess velocity. Increasing the amount of ΔV at the time of the escape burn increases the final hyperbolic excess velocity.

Another major aspect of a hyperbolic departure involves the direction of escape. Orbiting objects tend to fly in curved paths around the Earth instead of straight lines because the world is round and gravity tends to pull objects directly toward the center of the Earth. This curving effect occurs on escape trajectories as well. Notice in the figure that a spacecraft flies in a curved path immediately after the burn that creates the hyperbolic orbit. However, gravity's ability to pull the spacecraft back by "curving the orbit" around the Earth slowly diminishes as the craft's distance from the Earth increases. As a result, the hyperbolic path begins to look less curved and more like a straight line as the spacecraft approaches and passes the outer boundary. This straight line marks the direction of the spacecraft's motion at the time of escape and is called the departure asymptote.

Departure asymptotes do not exist on parabolic orbits because a satellite on this type of escape trajectory has no excess velocity at the outer boundary, and therefore, no direction of motion. Think about the Earth escape analogy of riding a bicycle out of a valley in order to understand the excess velocity at the outer boundary concept. If the top of the valley represents the outer boundary, then a parabolic orbit works analogously to having just enough speed at the bottom to barely coast to the top without pedaling. On the other hand, a hyperbolic orbit is analogous to having enough speed to coast up to the top without pedaling, and then having the momentum to keep on coasting after the road levels off at the top.

By now, you might be wondering, "Where is this outer boundary?" Physicists will claim that no such boundary exists because every object in the universe gravitationally attracts every other object no matter the distance between the objects. As a result, spaceflight engineers set the boundary at a distance far enough away from the Earth that the Sun's gravity pulls on a spacecraft with much more force than the Earth. They assume that only the Earth's gravity pulls on a spacecraft before it reaches the boundary, and only the Sun's pull affects the craft outside the boundary. The assumption is slightly inaccurate because a region exists inside the outer boundary where the Sun and Earth influence a spacecraft equally. However, escaping spacecraft usually pass through this region rather quickly. This rapid passage allows engineers to make rough trajectory estimates without resorting to sophisticated computer programs that calculate orbits by taking both the Earth's and the Sun's gravity into account.

The area inside the outer boundary is called the *sphere of influence*. Larger planets tend to have larger spheres of influence than smaller planets. For example, the outer boundary of the Earth's gravitational pull occurs at a distance approximately 925,000 kilometers (575,000 miles) from the surface. On the other hand, the sphere of influence of Mars stretches for only 580,000 kilometers (360,000 miles) away from its surface. Notice from these two examples that a planet's gravitational outer boundary always lies at a distance extremely close

RULE OF THUMB

Spacecraft at higher altitudes can escape the Earth with a lower ΔV than spacecraft at lower altitudes. This difference is analogous to the fact that it takes less energy to climb to the top of a hill from the middle than from the bottom.

Scientific Fact!

The *hyperbolic excess velocity* measures how fast a spacecraft moves relative to the Earth at the time it escapes the Earth's gravity. A spacecraft on a parabolic escape trajectory has no excess velocity.

The departure asymptote points in the direction that the spacecraft moves at the time of escape.

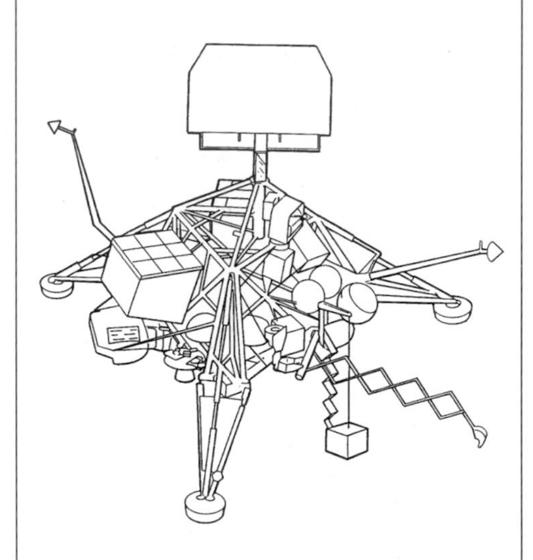

Surveyor Missions to the Moon

Summary

The *Surveyors* were a series of NASA spacecraft sent to soft-land on the Moon. One purpose was to determine if the Moon's surface was solid enough to support the heavy weight of astronauts and their landing vehicle.

Dimensions	
Height	3.0 meters
Mass	1,008 kilograms

Launch Dates	
Surveyor 1	30 May 1966
Surveyor 2	20 September 1966
Surveyor 3	17 April 1967
Surveyor 4	14 July 1967
Surveyor 5	8 September 1967
Surveyor 6	7 November 1967

RULE OF THUMB

A spacecraft is considered to have "escaped the Earth" when it is at a far enough distance away from the Earth that the Sun's gravity affects its motion much more than the Earth's gravity. At this point, engineers say that the spacecraft has "passed out of the Earth's sphere of influence."

to the planet compared to its average distance from the Sun. The relative small size of these spheres of influence results from the fact that the Sun's huge size allows its tremendous gravitational pull to dominate the entire Solar System.

Interplanetary Trajectories

A spacecraft automatically enters an orbit around the Sun after it passes the outer boundary. The key to calculating the size and shape of this new solar orbit involves understanding relative motion. Think about the following analogy. A train moving at 50 kilometers an hour passes by an old man sitting on the ground near the train tracks. Inside, a woman tourist walks from the rear to the front of the train at 5 kilometers an hour relative to the inside of the train. Passengers seated inside will see her move past them at 5 kilometers an hour. However, the old man sees her moving at 55 kilometers an hour because she moved forward inside a train moving forward. Now, suppose that the woman walks from the front to the rear. The old man will then see her moving at only 45 kilometers an hour because she moved to the rear of a train moving forward. To the old man, she still appeared to move forward because the train moves forward at a much faster rate than her movement to the rear.

How does the train scenario apply to an escaping spacecraft? Think of the speeding train as the Earth orbiting the Sun, the inside of the train as the Earth's sphere of influence, the woman as a spacecraft on an escape trajectory, and the old man as the Sun. Every object in the Earth's sphere of influence moves with the Earth, just as everything in the train moves with the train. Therefore, if a spacecraft escapes in exactly the same direction as the Earth's motion, the Sun will see the craft moving forward with the Earth, but at a slightly faster velocity. On the other hand, if a spacecraft escapes in the opposite direction, the Sun will still see the craft moving forward, but with a velocity slightly slower than the Earth's. This opposite direction of escape works exactly like the analogy where the woman walked to the back of a train moving forward.

Consider what a spacecraft's solar orbit looks like after it escapes the Earth in the same direction that the Earth moves in (see Figure 51 on page 251). At the outer boundary, the spacecraft's distance from the Sun equals the Earth's, but the craft's velocity exceeds that of the Earth. This faster velocity elevates the spacecraft into a solar orbit larger than the Earth's orbit. The escape point turns into the perihelion (closest point to the Sun). Aphelion (farthest point from the Sun) of the solar orbit occurs half an orbit later, 180° on the opposite side of the Sun from the Earth escape point.

Spacecraft use this type of maneuver to reach planets farther away from the Sun than the Earth. How far away from the Earth the spacecraft can travel depends on the aphelion distance of the solar orbit. In turn, this distance depends on the spacecraft's hyperbolic excess velocity, a quantity measuring how fast the craft moves rela-

Figure 51: *How to Reach the Outer Planets* ◄

Characteristics: A Hohmann transfer from the Earth to Mars or an outer planet (Jupiter, Saturn, Uranus, Neptune, Pluto) requires that the spacecraft move faster than the Earth around the Sun at the time of Earth escape.

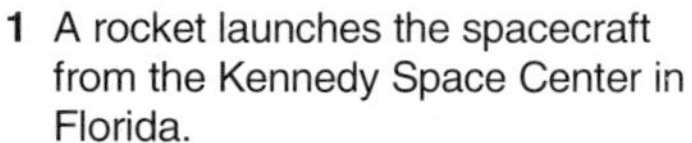

Solid arrow indicates position of the burn and points in the direction of the velocity change

Hollow arrow indicates position of the burn and points in the direction that the rocket engines fire to perform the burn

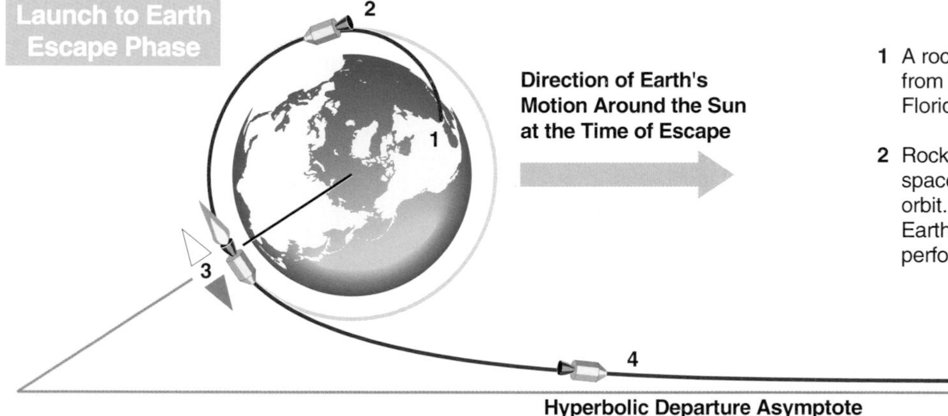

1 A rocket launches the spacecraft from the Kennedy Space Center in Florida.

2 Rocket engines shut down and the spacecraft enters a low-Earth orbit. The spacecraft circles the Earth until the time arrives for it to perform the escape burn.

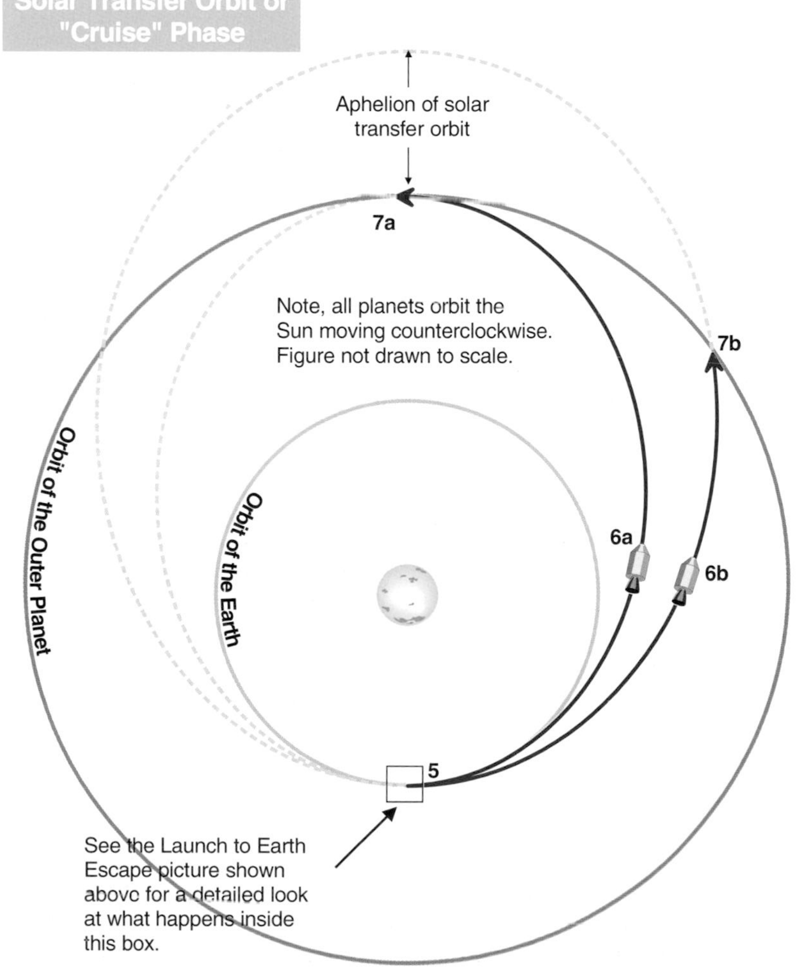

3 A posigrade burn (velocity speed-up) occurs here at the appointed time. This burn puts the spacecraft onto a hyperbolic escape trajectory.

4 Spacecraft moves away from Earth and to the departure asymptote.

5 Earth escape occurs here at the gravitational sphere of influence boundary. An escape in the same direction that the Earth orbits the Sun allows the spacecraft to travel around the Sun faster than the Earth.

6 This speed-up puts the spacecraft into an orbit around the Sun that has a perihelion at the escape point and aphelion at a greater distance from the Sun than the Earth. 6a is a Hohmann transfer, and 6b is a faster transfer. Note, reaching 6b instead of 6a requires the use of more propellant at the time of the initial escape burn (event #3).

7 Here, the spacecraft reaches the orbit of the destination planet. The initial burn to escape the Earth must be timed properly so that the spacecraft and destination planet arrive here at the same time. For a Hohmann transfer to Mars, there is only one opportunity to launch from Earth every 25 to 26 months.

► Figure 52: *How to Reach the Inner Planets*

Characteristics: A Hohmann transfer from the Earth to an inner planet (Mercury or Venus) requires a spacecraft to move slower than the Earth around the Sun at the time of Earth escape.

Solid arrow indicates position of the burn and points in the direction of the velocity change

Hollow arrow indicates position of the burn and points in the direction that the rocket engines fire to perform the burn

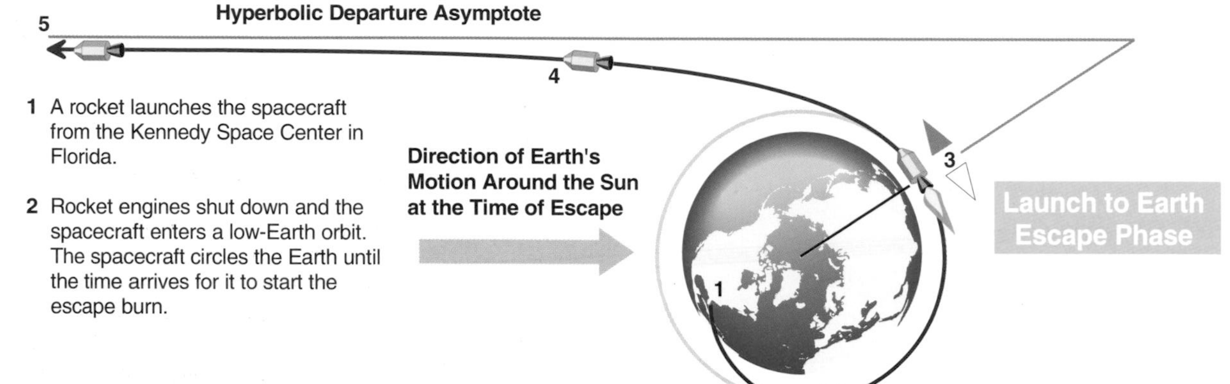

1 A rocket launches the spacecraft from the Kennedy Space Center in Florida.

2 Rocket engines shut down and the spacecraft enters a low-Earth orbit. The spacecraft circles the Earth until the time arrives for it to start the escape burn.

3 A posigrade burn (velocity speed-up) occurs here at the appointed time. This burn puts the spacecraft onto a hyperbolic escape trajectory.

4 Spacecraft slows down as it moves away from the Earth.

5 Earth escape occurs here at the gravitational sphere of influence boundary. As seen from the Earth, the escape is in the opposite direction that the Earth moves in. However, as seen from the Sun, the spacecraft is still moving forward with the Earth, but at a slower speed than the Earth.

6 The slow-down puts the spacecraft in an orbit around the Sun that has an aphelion at the escape point, and a perihelion at closer to the Sun than the Earth. 6a is a Hohmann transfer, and 6b is a faster transfer. Note, reaching 6b instead of 6a requires the use of more propellant at the time of the initial escape burn (event #3).

7 Here, the spacecraft reaches the orbit of the destination planet. The initial burn to escape the Earth must be timed properly so that the spacecraft and destination planet arrive here at the same time. For a Hohmann transfer to Venus, there is only one opportunity to launch every 19 months.

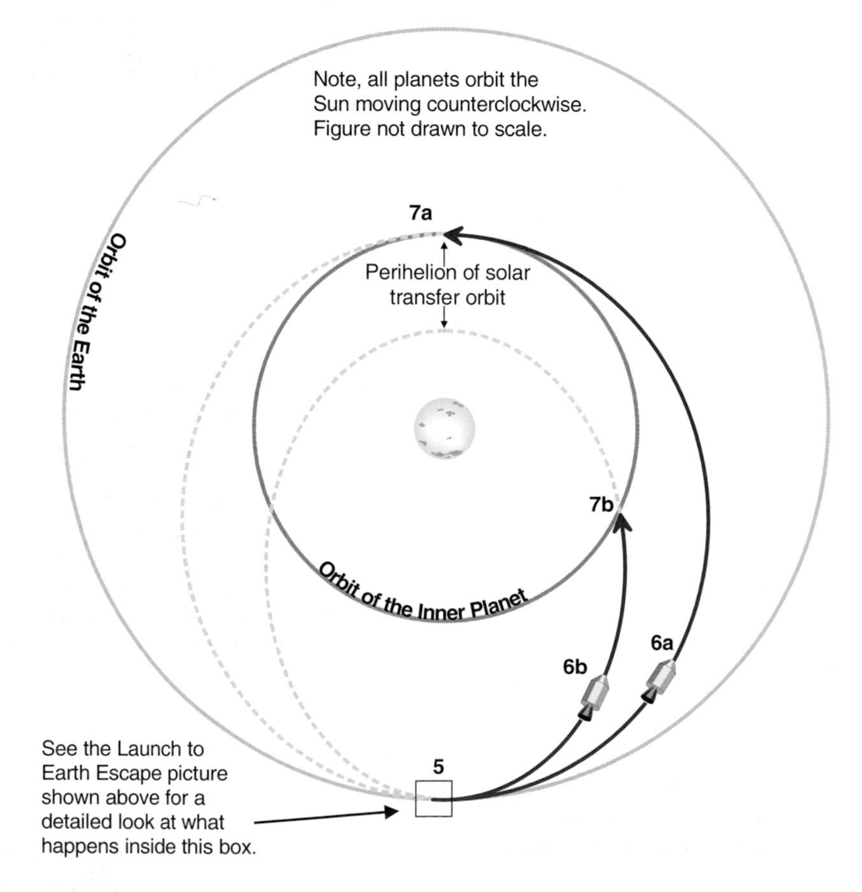

tive to (away from) the Earth at the outer boundary. Technically, spaceflight engineers use a number called C_3 (pronounced see-three) to measure the relative energy required to reach another planet. This parameter is the square of the hyperbolic excess velocity and represents the Earth escape energy. For example, a spacecraft on an Earth escape trajectory with a C_3 equal to 16.0 will theoretically move at 4.0 kilometers per second at the outer boundary.

As a rule of thumb, faster hyperbolic excess velocities (higher C_3 values) allow a spacecraft to enter an orbit around the Sun that takes it farther away from the Earth than slower excess velocities (lower C_3 values). The only way to increase C_3 involves increasing the strength of the burn that sends the spacecraft onto the Earth escape trajectory. This constraint dictates that for a spacecraft of constant mass, it takes much more propellant to reach faraway places like Neptune than closer locations like Mars. At a minimum, the rocket that performs the Earth escape burn must supply enough C_3 to guarantee that the aphelion distance of the solar orbit is as far out from the Sun as the destination planet's orbit.

Figure 52 on page 252 shows what a spacecraft's solar orbit looks like after an Earth escape in a direction opposite that of the Earth's motion. At the outer boundary, the spacecraft still moves forward, but at a slower velocity than the Earth. This slowdown drops the craft into an orbit shallower than the Earth's orbit. The point of escape becomes the aphelion (farthest point from the Sun). Perihelion (closest point to the Sun) occurs half an orbit later, 180° on the opposite side of the Sun from the Earth escape point. Spacecraft bound for planets closer to the Sun use this type of maneuver.

In this case, how far away from the Earth and close to the Sun the spacecraft can travel depends on the perihelion distance of the solar orbit. Again, faster hyperbolic excess velocities (higher C_3 values) allow a spacecraft to enter an orbit around the Sun that takes it closer to the Sun than slower excess velocities (lower C_3 values). This constraint dictates that for a spacecraft of constant mass, it takes much more propellant to reach Mercury than Venus. The reason is that Mercury is closer to the Sun and farther from the Earth than Venus. At a minimum, the rocket that performs the Earth escape burn must supply enough C_3 to guarantee that the perihelion distance of the solar orbit is as close to the Sun as the destination planet's orbit.

When the spacecraft's solar orbit barely touches the orbit of the destination planet, an interplanetary Hohmann transfer condition exists. In the business of Solar System exploration, using a Hohmann transfer orbit theoretically represents the method of sending a spacecraft to another planet that requires the least amount of rocket propellant during the Earth escape burn. The main reason is that an interplanetary Hohmann transfer solution results in the minimum-size orbit that touches both the Earth's orbit (departure point) and the destination planet's orbit (arrival point). Transfer orbits that extend

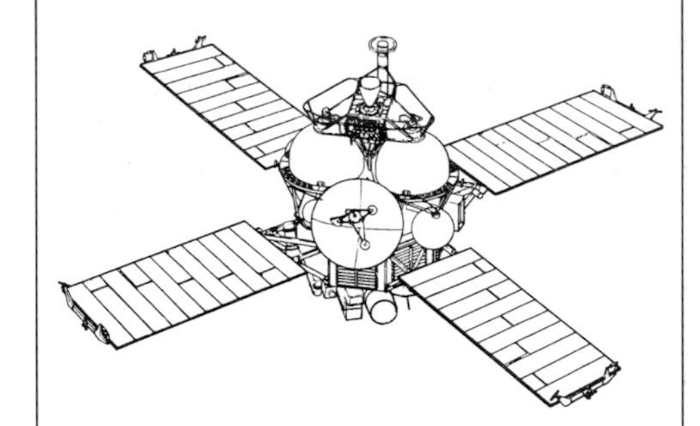

Mariner 9 Mission to Mars

Summary

This spacecraft was the first to orbit the planet Mars. In fact, *Mariner 9* was the first spacecraft to orbit a world other than the Earth or Moon. During its operational lifetime, *Mariner 9* transmitted more than 7,000 photographs of the Martian surface back to Earth. Some of the important discoveries from this mission include giant volcanoes and canyons.

Dimensions

Width	6.9 meters from end of one solar panel to end of other
Mass	1,030 kilograms

Mission Dates

Launch	30 May 1971
Arrival at Mars	13 November 1971

RULE OF THUMB

Spaceflight engineers use C_3 as a measure of how much energy it takes to send a spacecraft to another planet. Higher C_3 numbers require more rocket propellant to achieve than lower C_3 values. In general, distant planets take more C_3 to reach than closer planets. For example, a mission to Mars requires a C_3 of about 10. On the other hand, reaching Jupiter (four times farther from the Sun than Mars) requires a C_3 of close to 80.

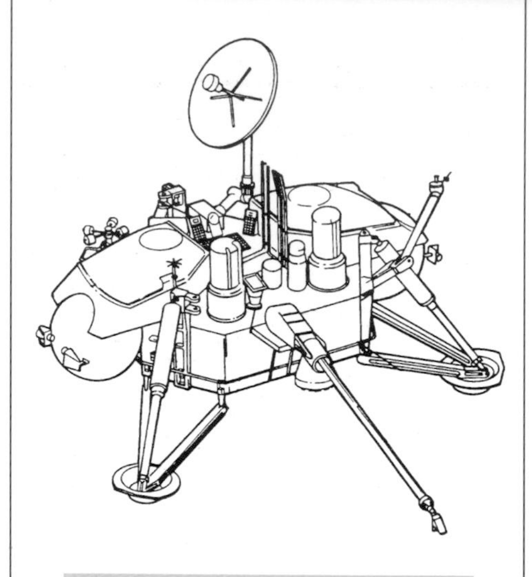

Viking Missions to Mars

Summary
These two spacecraft used a combination of rocket engines and a parachute to make the first successful soft landing on Mars. There, they took numerous photographs of the surface, weather readings, and performed a chemical analysis on the soil to test for the presence of life. Both landers transmitted data back to Earth for several years.

Dimensions

Shape	Hexagonal
Length	3.7 meters from end to end
Height	2.1 meters to top of antenna
Mass	576 kilograms

Mission Dates

Launch	20 August 1975 *(Viking 1)*
	9 September 1975 *(Viking 2)*
Mars Landing	20 July 1976 *(Viking 1)*
	3 September 1976 *(Viking 2)*

RULE OF THUMB

Interplanetary spacecraft of today are often called *probes* because they do not carry astronauts. NASA engineers also use the term *probe* to describe a specific type of interplanetary spacecraft that is carried to a planet by another interplanetary spacecraft, and then dropped into the atmosphere of that planet to perform a chemical analysis of the gasses.

beyond the destination planet's orbit also work, but require more energy to create during the Earth escape burn.

One of the key characteristics of the interplanetary Hohmann transfer is that the spacecraft travels 180° (half an orbit) around the Sun before encountering the destination planet. In other words, arrival occurs on the opposite side of the Solar System from launch. Therefore, if Earth escape occurs at perihelion, as in the case of a journey to an outer planet, then the encounter will occur at aphelion. Conversely, if Earth escape occurs at aphelion, as in the case of a journey to an inner planet, then the encounter will occur at perihelion. Figure 51 on page 251 and Figure 52 on page 252 shows how this concept works.

Although interplanetary Hohmann transfers minimize the propellant required to reach another planet, the energy savings comes at the expense of an increased time of flight. Think about the problem of reaching a distant planet as comparable to climbing a hill. Walking up the hill requires less energy than running up but takes a longer time to reach the top. Analogously, a rocket can send a spacecraft to another planet in a faster time than the Hohmann transfer time by supplying more than the minimum amount of C_3 during the Earth escape burn. The key characteristic of a faster interplanetary transfer is that the solar orbit extends beyond the orbit of the destination planet. Consequently, the spacecraft travels less than 180° during the transfer and will reach the planet before aphelion for a trip to an outer planet, or will reach the planet before perihelion for a trip to an inner planet. Figure 51 on page 251 and Figure 52 on page 252 also shows how this concept works.

Increasing the C_3 of the Earth escape burn to decrease the flight time to the destination planet also comes at an expense. The reason is that rockets almost always leave the launchpad with their propellant tanks filled. Since no room exists in the tanks for more propellant, achieving the extra energy needed for an increase in C_3 requires reducing the total mass of the spacecraft. Think about this trade-off in terms of two runners of equal strength trying to run up a hill. The one carrying the most amount of weight will reach the top last. Analogously, interplanetary Hohmann transfers may take longer than other transfers, but they allow the most amount of mass to be sent to the destination planet.

After escaping the Earth, a spacecraft coasts in an orbit around the Sun under the predominant influence of solar gravity until reaching the orbit of the destination planet. The emphasis here is on the word *coast*. Contrary to popular belief, interplanetary spacecraft do not burn their rocket engines all the way to the destination planet. Remember that Earth-orbiting satellites are in a perpetual state of free fall around the Earth. They require no propulsion to remain in orbit because gravity provides the power. The same concept applies to spacecraft and planets in orbit around the Sun. In this case, free fall occurs around the Sun instead of around the Earth.

	Mercury	Venus	Mars	Jupiter	Saturn	Uranus	Neptune	Pluto
Number of NASA Missions Flown	1	6	7	4	3	1	1	0
Name of First NASA Mission	*Mariner 10*	*Mariner 2*	*Mariner 4*	*Pioneer 10*	*Pioneer 11*	*Voyager 2*	*Voyager 2*	n/a
Date First Mission Arrived	29 Mar 1974	14 Dec 1962	14 July 1965	3 Dec 1973	1 Sep 1979	24 Jan 1986	25 Aug 1989	n/a
Next Scheduled Mission	n/a	n/a	*Polar Lander* *Climate Orbiter*	TBD *(Europa Orbiter?)*	*Cassini*	n/a	n/a	TBD
Min. C_3 to Reach Planet Direct	55.8	6.4	8.8	77.3	105.8	127.2	135.8	139.5
Velocity for Earth Escape	5.51 km/s	3.49 km/s	3.62 km/s	6.30 km/s	7.28 km/s	7.98 km/s	8.25 km/s	8.36 km/s
Ideal Hohmann Transfer Time	106 days	146 days	259 days	2.73 years	6.04 years	16.02 years	30.63 years	45.30 years
Max. Mass on Delta 2 Rocket	414 kg	1,127 kg	1,073 kg	264 kg	135 kg	71 kg	50 kg	42 kg
Max. Mass on Atlas 2AS Rocket	1,053 kg	2,366 kg	2,269 kg	748 kg	483 kg	347 kg	303 kg	285 kg
Max. Mass on Titan 4 Rocket	3,758 kg	8,621 kg	8,292 kg	2,478 kg	1,225 kg	520 kg	280 kg	183 kg
List of NASA Missions Flown	*Mariner 10*	*Mariner 2 & 5* *Mariner 10* *Pioneer Orbiter* *Pioneer Probe* *Magellan*	*Mariner 4 & 6* *Mariner 7 & 9* *Viking 1 & 2* *Pathfinder* *Global Surveyor*	*Pioneer 10* *Pioneer 11* *Voyager 1* *Voyager 2* *Galileo*	*Pioneer 11* *Voyager 1* *Voyager 2*	*Voyager 2*	*Voyager 2*	n/a

Despite the fact that interplanetary spacecraft require no propulsion to maintain an orbit around the Sun, they typically fire their small rocket engines several times on the way to the destination planet. Spaceflight engineers call these burns *trajectory correction maneuvers*, or TCMs for short. These burns allow ground controllers to correct for slight errors in a spacecraft's velocity. Often, the large rocket that performs the Earth escape burn causes these small errors. The reason is that large rockets provide so much power that achieving the precise velocity required to reach another planet is difficult. However, spaceflight does not work like driving down the freeway, where exceeding the speed limit by a small amount will not be noticed by the officer with the radar gun. Even a velocity error as small as 1 meter per second (2.2 miles per hour) requires a TCM because after a two-year journey through space, that error can result in more than a million-kilometer miss at the destination planet.

How do spaceflight engineers determine a spacecraft's velocity? The answer comes from the same effect that makes a police siren appear to take on higher than normal pitched tones when approaching, and lower than normal pitched tones when receding. Physicists call this effect a *Doppler shift*. Spaceflight engineers on Earth responsible for navigating a spacecraft to a distant planet use large antennas to record electronic "blips" emitted by the spacecraft's radio. The antennas feed the signal into computers. In turn, the computers use Doppler equations to analyze the changes in "pitch" of the radio signal as compared to its normal "tone." Using this method, navigators can determine changes in a spacecraft's velocity on the order of millimeters per second.

Arrival Choices

One of two things can happen at the point where the spacecraft encounters the destination planet. If the spacecraft does nothing, then the gravity of the planet will deflect the craft's motion from the

Above:

The table above provides a quick way to look up information about sending a spacecraft to another planet. An entry of "n/a" for "Next Scheduled Mission" indicates that NASA currently has no definite plans to explore that planet as of June 1999. Values for "Minimum Velocity for Earth Escape" indicate how much speed a spacecraft in low Earth orbit (300 km altitude) must attain in order to escape the Earth and reach the listed planet. Keep in mind that a spacecraft in a 300-km orbit is already moving at 7,726 meters per second. For more information about the rockets listed in this table, refer back to Chapter 1.

RULE OF THUMB

Successfully navigating a spacecraft to another planet requires the same amount of accuracy as throwing a baseball from Los Angles to New York and having the baseball fly through a specific (preplanned) window in the Empire State Building.

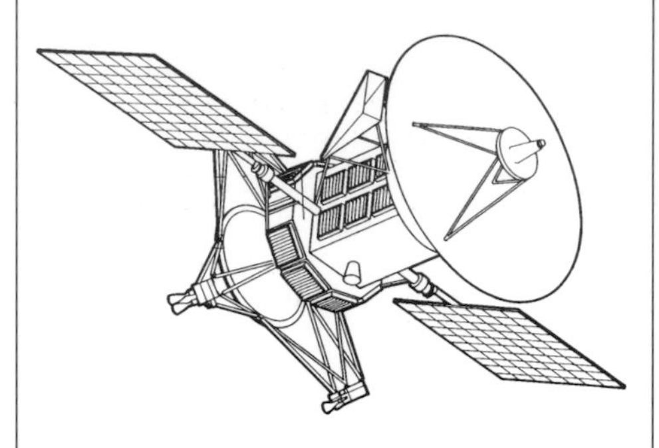

Magellan Mission to Venus

Summary

Magellan was sent to map the surface of Venus from orbit. However, the spacecraft had no camera. The reason is that Venus' surface is always covered with a thick layer of clouds that obscures the surface. *Magellan* bounced radar signals through the clouds and listened to the echoes from the surface. NASA scientists used the radar data that *Magellan* sent back to generate computer maps of the surface.

Dimensions

Width	6.4 meters from tip of antenna to rocket engines
Mass	3,235 kilograms

Mission Dates

Launch	5 May 1989
Venus Arrival	10 August 1990

RULE OF THUMB

After a spacecraft escapes the Earth, it enters an orbit around the Sun that will take it to the destination planet if the engineers planned the orbit correctly. Upon arrival at the destination planet, the spacecraft can either slow down to enter an orbit around the planet, or use the planet's gravity to sling the craft into a different orbit around the Sun.

original transfer orbit into a different orbit around the Sun. Spaceflight engineers call this type of scenario a *hyperbolic flyby*. The other option involves entering an orbit around the planet by slowing the spacecraft down and allowing the planet to "capture" the craft from the Sun. In general, a spacecraft on a direct solar transfer orbit from the Earth to another planet moves too fast to be automatically captured by the destination planet's gravity. Slowing down to orbit a planet requires the use of an amount of propellant comparable to the amount that sent the spacecraft on the escape trajectory away from the Earth.

The choice of sending a spacecraft to simply fly by the target planet, or to orbit the planet represents one of the most important decisions that mission planners must make. A flyby scenario allows the mission planning team to design a spacecraft lighter than that required on a mission where the goal involves orbiting the target planet. In turn, a lighter spacecraft allows NASA to use a lighter, less expensive rocket for launch, or allows the project scientists to attach more scientific instruments to the spacecraft. The savings in mass comes from the fact that the spacecraft need not carry the large amounts of propellant necessary to slow down so that the target planet can "capture" the spacecraft. However, depending on the specific scenario, a flyby spacecraft will zoom by the target planet in the time span of several hours to several weeks. All of the photography and gathering of scientific data must occur in that short time span. As a contrast, spacecraft sent to orbit a planet usually spend several months to several years gathering scientific data.

Gravity Assist

On a flyby mission, the hyperbolic flyby phase begins when a spacecraft approaches the orbit of the destination planet and passes into its sphere of influence. The spacecraft moves in the direction of an imaginary line called the arrival asymptote at the beginning of the planetary encounter. This line marks the spacecraft's natural direction of motion along its solar transfer orbit from the Earth. During the encounter, the planet's gravity accelerates the spacecraft inward and curves the flight path away from the arrival asymptote. Then, it slings the craft around the planet, and then away from the planet in a direction different from the approach. The net effect is that the hyperbolic flyby encounter changes the spacecraft's direction of motion relative to the Sun. As a consequence, it also changes the craft's velocity and solar orbit.

A parameter called the *turn angle* measures the angle between the arrival and departure asymptotes. In other words, the turn angle measures the change in direction of a spacecraft's motion due to the hyperbolic flyby of a planet. Larger planets typically change a spacecraft's direction of motion more than smaller planets because of their larger gravitational pull. Other factors that affect the turn angle include the distance between the arrival asymptote and the planet, and the speed of the craft at the beginning of the encoun-

Figure 53: *How to Use a Gravity Assist to Save Propellant* ◄

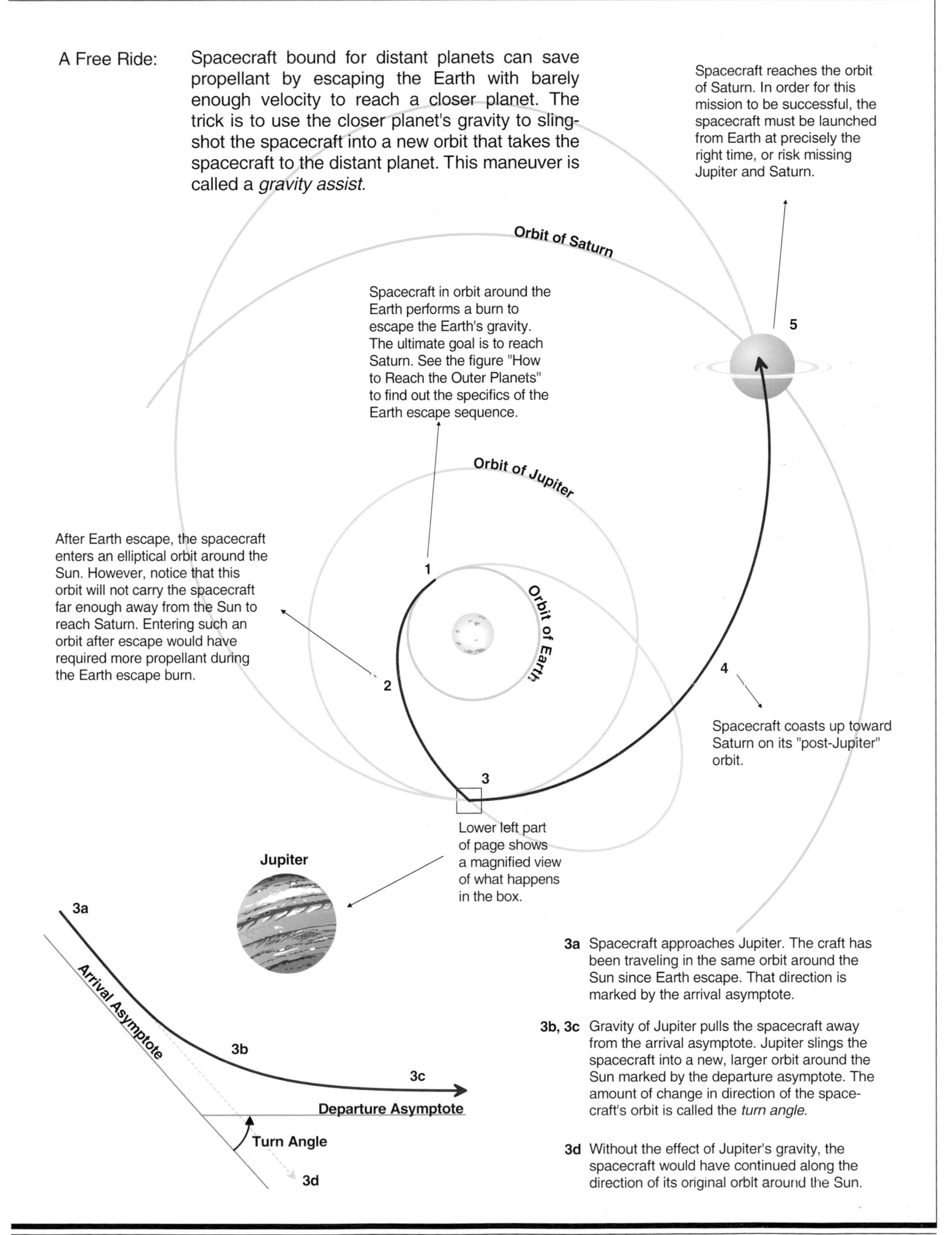

► Figure 54: *The Grand Tour*

NASA sent these two spacecraft to make a flyby of the outer planets of the solar system. *Voyager 1* flew by Jupiter and used its gravity to reach Saturn. *Voyager 2* flew by Jupiter and then used gravity assists to reach the planets Saturn, Uranus, and Neptune. That mission marked the first flyby of Uranus and Neptune in history. Many consider these two *Voyager* flights as the most successful interplanetary missions in history. Most of the information that scientists know about the outer planets today were made possible by these two spacecraft. Both *Voyagers* will fly away from the Sun forever. NASA hopes to maintain contact with them until 2019.

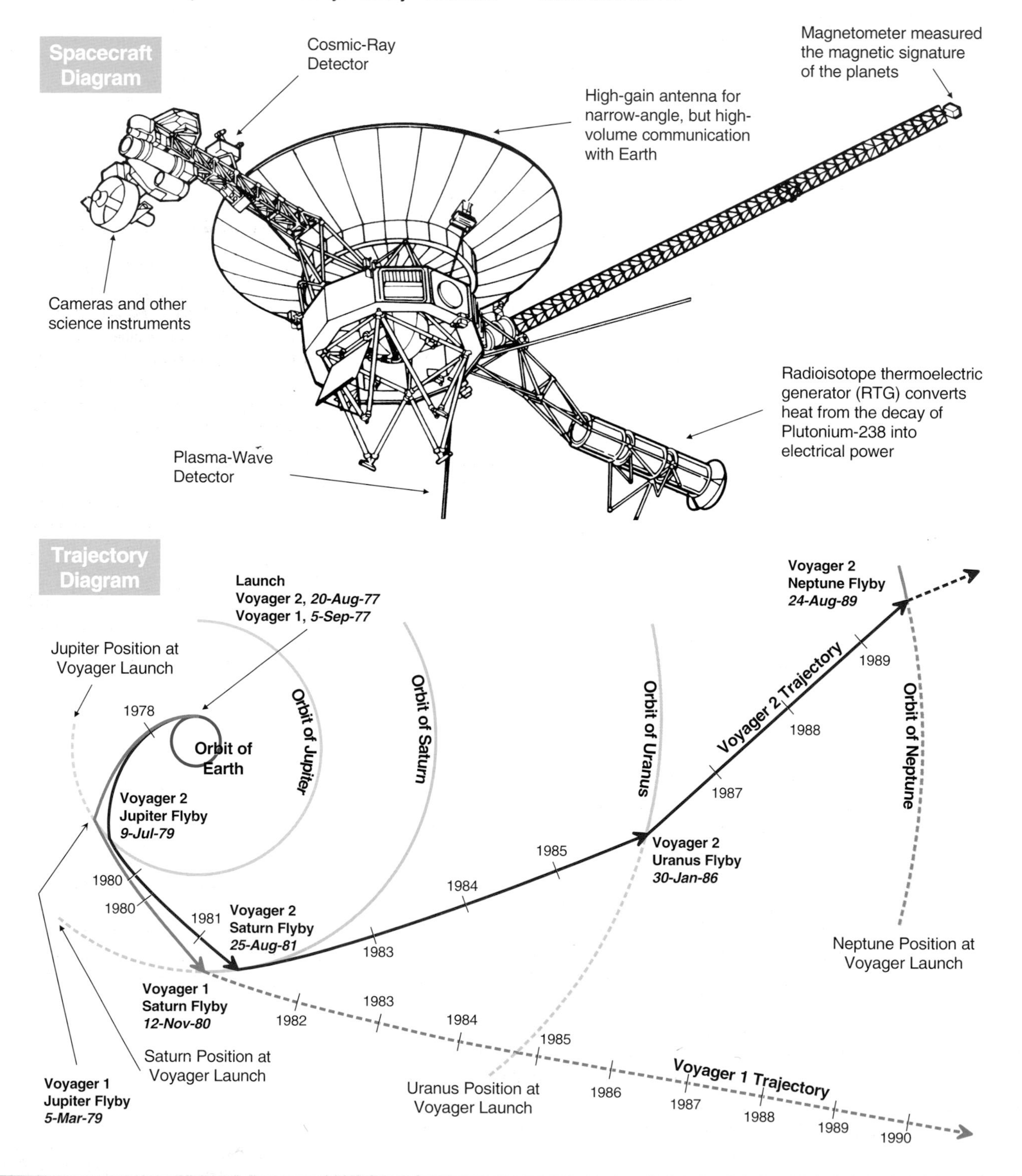

ter. In general, a slow, close flyby of a planet results in more turn angle than a fast, distant flyby. Giant planets like Jupiter and Saturn can turn a spacecraft by 90° or more.

On a simple mission where the goal involves gathering data about a planet during a quick hyperbolic flyby, spaceflight engineers typically do not care about the spacecraft's post-flyby solar orbit. However, with careful planning, a hyperbolic flyby of one planet can alter a spacecraft's trajectory to reach another planet. The advantage of this scheme is that a spacecraft can reach a more distant planet with the use of a smaller amount of rocket propellant than normal (see Figure 53 on page 257). Essentially, this technique works by launching the spacecraft with barely enough velocity to reach a planet closer to Earth than the final destination planet. Then, the gravity of the closer planet slings the spacecraft toward its final, more distant destination. The savings in propellant occurs because it takes less propellant and energy to send a spacecraft on a trajectory to a closer planet than to a more distant planet. Gravity from the closer planet provides the rest of the energy.

You might be wondering where this extra energy comes from? Also, how does gravity sling a spacecraft away from a planet when it is supposed to pull things inward? As explained earlier, the key involves understanding that the laws of physics dictate that spacecraft flying around the Sun typically move too fast to be automatically captured by a planet's gravity into an orbit around the planet. However, during a hyperbolic flyby, a spacecraft temporarily passes through a planet's sphere of gravitational influence. When the spacecraft exits this imaginary sphere, its new velocity will equal its original velocity, plus the planet's velocity around the Sun. Think of this idea as similar to walking through an airport and stepping onto a moving conveyer-belt-style sidewalk. Ever notice what happens when you step off? If you keeping walking at the same speed while on the moving sidewalk, your speed the moment you step off will equal your original speed, plus the speed of the sidewalk.

Interestingly enough, the gravity assist technique was used for the first time on a mission to an inner planet, not an outer planet. In 1974, *Mariner 10* flew to Mercury. During the planning stages, a detailed computer analysis showed that the powerful Titan 3C rocket could place *Mariner 10* on a direct trajectory to Mercury. The analysis also showed that it was possible to launch *Mariner* on a cheaper, less powerful Atlas-Centaur rocket. Such a scheme involved launching *Mariner* with barely enough velocity to reach Venus (located closer to the Earth than Mercury) and then using Venus' gravity to sling the spacecraft inward toward Mercury. By choosing the latter option, NASA saved a lot of money. The only problem was that *Mariner* had to fly by Venus within a 400-kilometer (250-mile) "corridor" after a journey of more than 160,000,000 kilometers (100,000,000 miles). If *Mariner* had missed the corridor at Venus, it would have eventually missed Mercury. Fortunately, the guidance was extremely accurate,

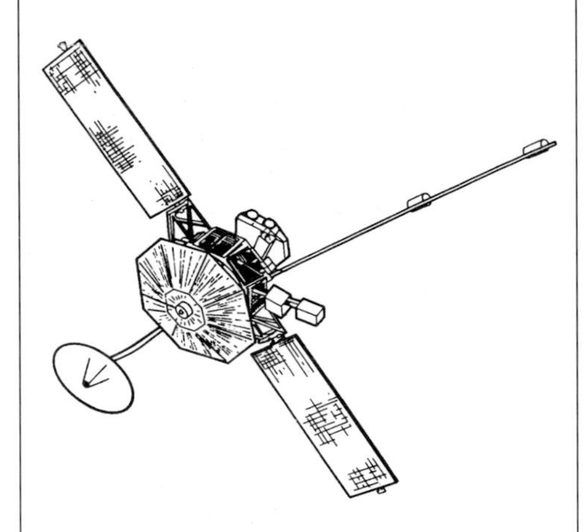

Mariner 10 Mission to Mercury

Summary

This spacecraft was launched toward Venus. There, it used Venus' gravity as a slingshot to reach Mercury. In doing so, *Mariner 10* was the first spacecraft to use another planet's gravity to alter its trajectory, and the first to fly by Mercury. *Mariner 10* revealed Mercury as a barren world, much like our Moon. The spacecraft also transmitted the first television pictures of Venus.

Dimensions

Width	6.8 meters from end of one solar panel to end of other
Mass	503 kilograms

Mission Dates

Launch	3 November 1973
Venus Flyby	5 February 1974
Mercury Flyby	29 March 1974

Scientific Fact!

Jupiter is so large that under the right conditions, a hyperbolic flyby of it will cause a spacecraft to fly away from the Sun forever. In contrast, the inner planets are much too small to provide this large a gravity assist. In everyday conversation, engineers rarely use the term *hyperbolic flyby*. Instead, they prefer to use the term *gravity assist* or *swingby*.

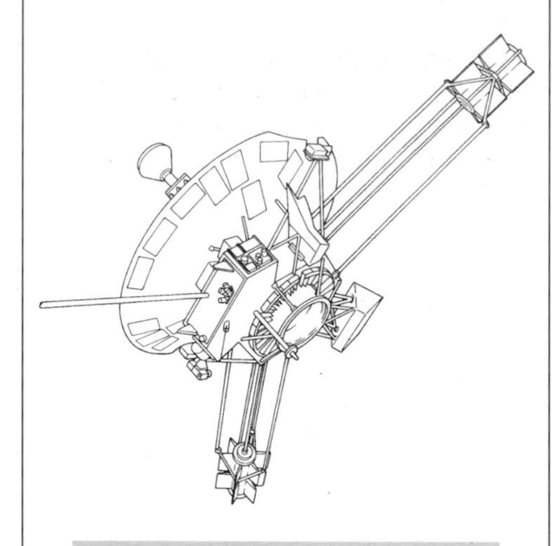

Pioneer Missions to Jupiter

Summary
These two spacecraft were NASA's first outer-planets spacecraft. *Pioneer 10* became the first human-made object to fly by Jupiter in December 1973. Then, Jupiter's gravity sling-shot the spacecraft on a trajectory that took it past Pluto. *Pioneer 11* flew by Jupiter in December 1974, used Jupiter's gravity to sling-shot it to a September 1979 flyby of Saturn, and then used Saturn's gravity to slingshot it past the orbit of Pluto. Both are currently on trajectories that will take them away from the Sun forever. NASA hopes to maintain contact with the two *Pioneers* until 1999 or later.

Dimensions

Size	Large antenna dish is 2.75 meters in diameter
Mass	258 kilograms

Mission Dates

Launch	2 March 1972 *(Pioneer 10)* 5 April 1973 *(Pioneer 11)*
Jupiter Flyby	3 December 1973 *(Pioneer 10)* 2 December 1974 *(Pioneer 11)*
Saturn Flyby	1 September 1979 *(Pioneer 11)*

RULE OF THUMB

A direct trajectory is not a straight-line path. Remember, orbits that spacecraft use to move between planets in the Solar System take the shape of ellipses. Direct trajectories are orbits that connect two planets such that a space-craft moving on that orbit will not encounter another planet along the way.

and *Mariner 10* flew by Venus within 43 kilometers (27 miles) of its preplanned position.

One of the most dramatic uses of the gravity assist technique occurred on a mission that began in 1977. Mathematicians discovered in the 1960s that the four giant outer planets "line up" every 176 years. They also discovered that the next alignment was scheduled to occur in the 1980s. In such a configuration, a spacecraft can visit all four outer planets in one mission without using an exorbitant amount of propellant. NASA launched two spacecraft to take advantage of this unique opportunity. *Voyager 1* was launched with barely enough velocity to reach Jupiter. Then, *Voyager 1* used Jupiter's gravity to sling it toward Saturn. *Voyager 2* was also launched toward Jupiter. It used Jupiter's gravity to reach Saturn, Saturn's gravity to reach Uranus, and Uranus' gravity to reach Neptune. The hyperbolic flybys added so much velocity to the *Voyagers* that both spacecraft have already passed the orbit of Pluto and will eventually escape the Sun's gravity.

Both *Voyager* missions proved to be an unqualified success for NASA. The scientific data and photographs transmitted back to Earth by the two *Voyager* spacecraft have allowed scientists to learn more about the four giant planets in one decade than all of the previously recorded knowledge of those planets throughout history. In fact, most of the information presented in this book about the outer planets came from the scientific data transmitted back to Earth by the two *Voyager* spacecraft.

By far, the most complicated and ingenious use of a gravity-assist trajectory began on 18 October 1989 with the launch of the Jupiter-bound *Galileo* spacecraft. NASA originally planned for the Space Shuttle to carry *Galileo* and a powerful Centaur rocket into Earth orbit. From there, the liquid-hydrogen-fueled Centaur was supposed to propel *Galileo* on a direct trajectory to Jupiter. After the accident that destroyed the Space Shuttle *Challenger* and killed seven astronauts in 1986, NASA reviewed all of its shuttle safety rules and declared that the Centaur was too dangerous for the shuttle to carry. This decision forced the *Galileo* mission design team to find a way to reach Jupiter using the much less powerful, but shuttle-approved Inertial Upper Stage (IUS).

The new solution involved using Space Shuttle *Atlantis* to carry *Galileo* and an IUS rocket into low Earth orbit. From there, as ludicrous as it sounds, the IUS propelled *Galileo* inward toward Venus instead of outward toward Jupiter. This new plan called for *Galileo* to orbit the Sun three times in the vicinity of Earth and Venus. On each orbit, *Galileo* used a gravity assist from one of the two planets to travel farther away from the Sun than on the previous orbit. The first gravity assist, from Venus, put *Galileo* on an orbit that took it slightly beyond the Earth's orbit around the Sun. On *Galileo's* next pass around the Sun, a gravity assist from Earth put the spacecraft on an orbit that reached roughly halfway to Jupiter. Then, on the third and last pass around the Sun, another gravity assist from the Earth

Figure 55: *Using Orbit Pumping to Reach Jupiter* ◄

This large NASA spacecraft arrived at Jupiter in December 1995 and spent over to two Earth years orbiting the giant planet and studying its family of more than 16 moons. *Galileo* also dropped a probe into Jupiter's turbulent atmosphere in order to perform a chemical analysis. Unfortunately, the main high-gain antenna did not work properly. At Jupiter, *Galileo* was forced to transmit its scientific data back to Earth at a much slower rate over the low-gain antenna. Although less data was returned to Earth than planned, the quality of the science was not impacted.

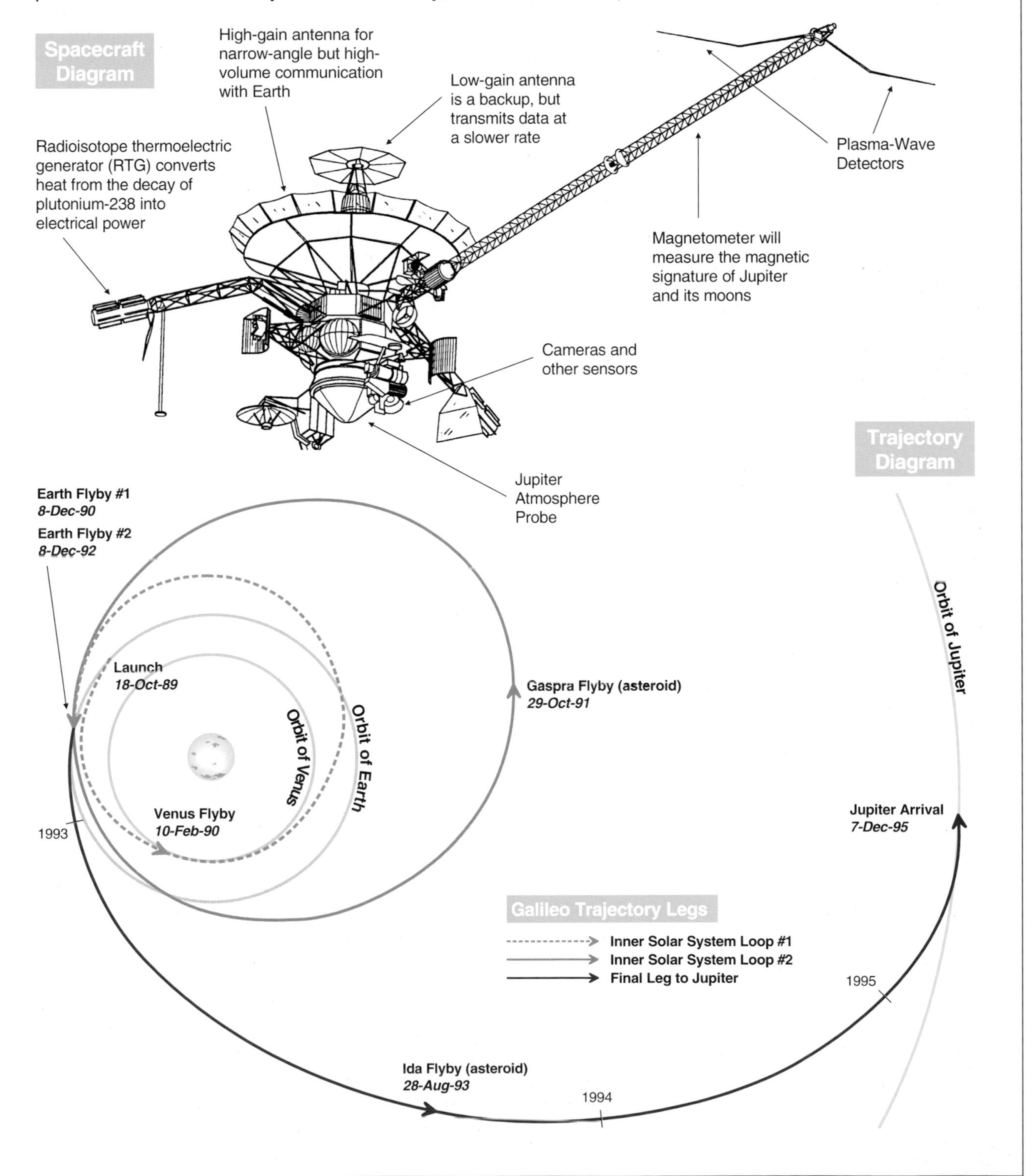

put *Galileo* on a direct trajectory to Jupiter. *Galileo* required three gravity assists because both the Earth and Venus lack the gravitational strength required to sling a spacecraft to Jupiter on one try. The spacecraft received its last gravity assist on 8 December 1992 and arrived there and entered orbit in December 1995.

NASA engineers refer to the *Galileo* trajectory as a *Venus-to-Earth-to-Earth* gravity assist or VEEGA. Think of the multiple gravity assist trajectory as comparable to the case of a mother trying to push her child higher into the air on a playground swing. One way to send the child to the top of the swing would be to use a lot of energy in one giant push. She can also choose to push her child gently at first. In this case, the child will not reach the top on the first try, and the swing will come back down. Now, if she pushes the child gently from behind when the swing is at the bottom, the child will rise higher into the air than before. After a few pushes at the bottom, the child will reach the top. VEEGA works like the latter case. For this reason, some engineers refer to the technique of multiple gravity assists as *orbit pumping*. Another NASA spacecraft, called *Cassini*, is using an orbit pumping Venus-to-Venus-to-Earth-to-Jupiter gravity assist trajectory (VVEJGA) to reach Saturn. *Cassini* launched on 15 October 1997 with a Saturn arrival scheduled for June 2004.

Scientific Fact!

Where does the electricity come from to power an interplanetary spacecraft's instruments? No battery can last long enough to provide power for a several-year mission. Instead, spacecraft to Mars, Venus, and Mercury use solar panels to generate electricity. However, the strength of the Sun's rays at Jupiter's orbit and beyond is too weak for solar panels. A spacecraft bound for one of the outer planets must depend on radioactive Plutonium-238 for power. A device called a radioisotope thermoelectric generator (RTG) converts heat from the decay of the plutonium into electricity.

Not as Easy as It Looks

NASA's launch records since 1960 include more than fifteen robotic spacecraft dispatched to explore the other planets of the Solar System. These automated machines have traveled through a combined total of more than 32 billion kilometers (20 billion miles) of space. To put this number in perspective, consider that a person flying on a standard passenger jet would need roughly 5,000 uninterrupted years of flight time to travel the same distance. Amazingly enough, only one interplanetary mission, excluding failures due to the rocket exploding during launch, failed and returned no scientific data. That spacecraft, called *Mars Observer*, stopped communicating on 21 August 1993 for reasons still unknown. Every other spacecraft transmitted a vast amount of valuable photographs and scientific data back to Earth.

RULE OF THUMB

Despite the best of planning, unforeseen and unpredictable events can destroy an interplanetary spacecraft. For example, a collision with a tiny pebble in orbit around the Sun would instantly shatter the spacecraft because of the extremely high velocities involved in the collision. Such an occurrence is unlikely, but still within the realm of possibility.

Unfortunately, NASA's amazing success record tends to hide the true hardships required to make an interplanetary mission successful. Consider the fact that most missions last for at least several years. Some missions may take over a decade to complete. If a component on the spacecraft breaks, no means exist to send a crew out to the far reaches of the Solar System to perform a repair. Consequently, most spacecraft must carry two of every critical part, such as the control computer, radio transmitters for communicating with the Earth, electric power generating devices, and rocket propellant tanks. This redundancy keeps any single failure from destroying the spacecraft's ability to operate properly. However, many critical items lack redundancy. For example, most interplanetary spacecraft carry only one

Figure 56: *Last of the Grand Missions* ◄

This latest NASA interplanetary mission is currently on its way to Saturn. The goal is to send the *Cassini* spacecraft to orbit the ringed-planet for four years. During that time, the spacecraft will study the planet and its family of more than 21 moons. *Cassini* will also drop a small probe to land on the moon Titan. Scientists are interested in this large moon because it is the only known moon in the solar system with a sizable atmosphere. *Cassini* will use gravity assists from Venus, Earth, and Jupiter to gain the necessary velocity to reach Saturn. Arrival is scheduled for July 2004.

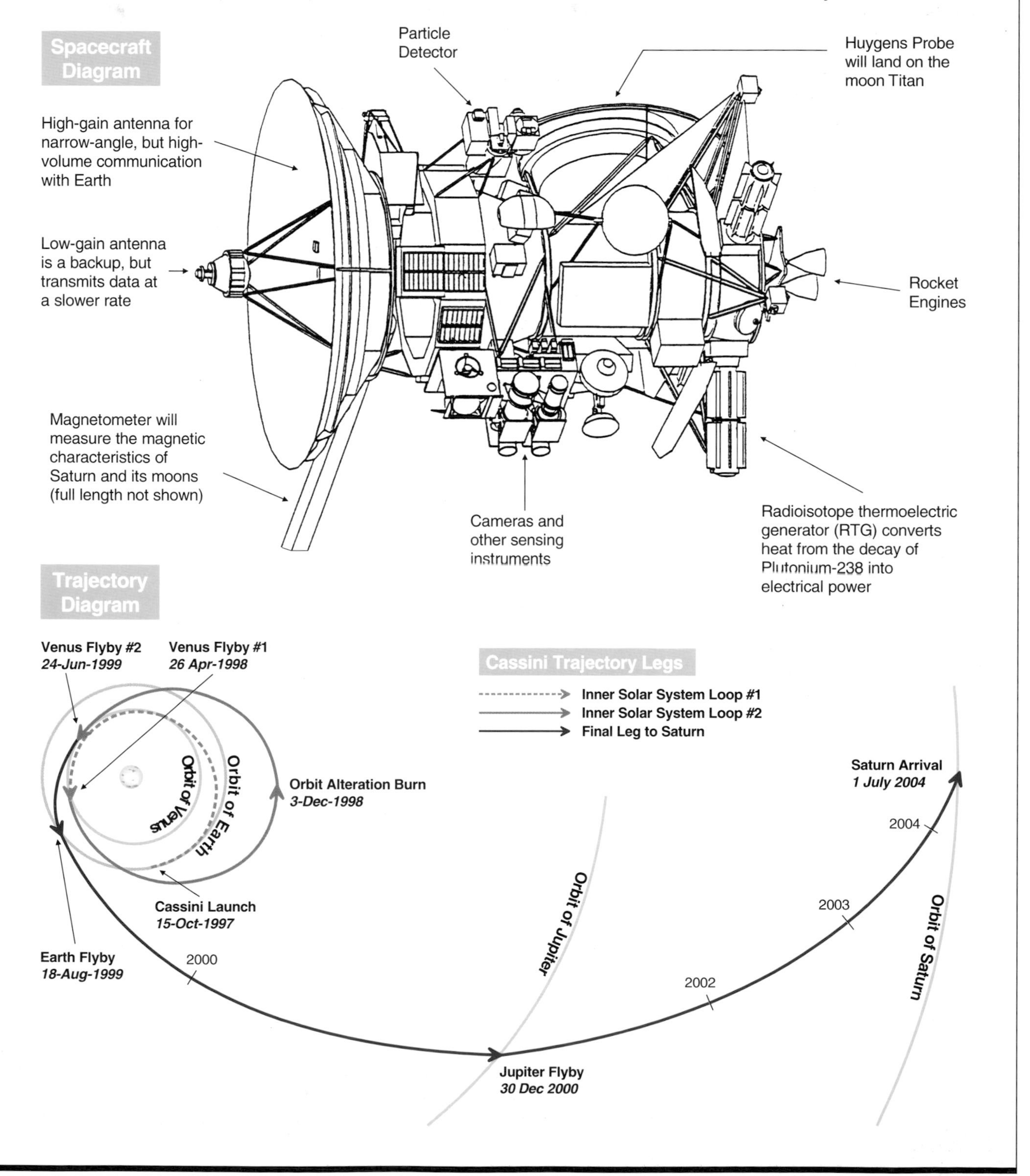

CONTEL
BOEING
JPL
NASA
EXIT
EMERGENCY EXIT ONLY

large antenna dish for high-volume communications with the ground because a two-dish spacecraft design would weigh too much.

Communications with the Earth from deep space also presents a difficult challenge. All of the pictures and scientific data must be transmitted back by radio because interplanetary spacecraft rarely return to Earth. If such a task sounds easy, consider the fact that most commercial radio and television stations broadcast with thousands of watts of power to reach listeners and viewers within a rough 100-kilometer (62-mile) radius of the station. On the other hand, spacecraft lack the ability to generate huge amounts of electrical power and must broadcast their signals back to Earth with about 20 watts of power or less. That 20-watt signal, with three times less power than required to light an average light bulb, must cross millions, and sometimes billions, of kilometers of space. By the time the radio signal reaches the Earth, the power level measures significantly less than that at the time of transmission. For example, *Voyager 2's* signal from the distant planet Neptune registered a power level of less than one millionth of one billionth of a watt (0.000000000000001 watts) at the surface of the Earth. Trying to receive a transmission from an interplanetary spacecraft amounts to standing on a beach and attempting to listen through the crashing waves to hear a whispering voice hundreds of kilometers offshore.

NASA uses the Deep Space Network (DSN) to track interplanetary spacecraft and to receive their transmissions. This communications system consists of the Goldstone Complex in the Southern California desert, the Madrid Complex in Spain, and the Canberra Complex in Australia. Each complex houses one monster antenna dish with a 70-meter (230-foot) diameter, and several other dishes with 34-meter and 26-meter (112- and 85-foot) diameters. All of these antennas look like gigantic versions of the backyard satellite-television antenna dishes owned by an increasing number of households around the world. The DSN antennas essentially function as "giant electronic ears," allowing them to listen to the faint signals sent by interplanetary spacecraft from the distant parts of the Solar System. Still, NASA engineers often compare the process of tracking a deep-space spacecraft to the proverbial problem of finding a needle in a haystack. ❑

Above:
NASA uses giant antennas at three sites around the world to track interplanetary spacecraft. This photograph shows antenna DSS-26 at the Goldstone tracking site near Barstow, California. The diameter of this antenna is 34 meters.

Opposite Page:
3 August 1989 – Technicians at the Kennedy Space Center work to prepare the *Galileo* spacecraft for launch. It was carried into space by Space Shuttle *Atlantis* on 18 October 1989. *Galileo's* ultimate destination is Jupiter. There, *Galileo* will spend 23 months orbiting the giant planet and will return detailed scientific data about the planet and its extensive system of moons. In addition, *Galileo* will drop a probe into Jupiter's atmosphere that will perform a chemical analysis.

Invading Mars

Chapter 7: Scientific Armada to the Red Planet

For many centuries, the existence of Mars mystified and terrified a countless number of generations of people on Earth. Long ago, astronomers in the Roman Empire named Mars after their god of war because they superstitiously associated the red-looking planet with fear and destruction. More recently, false sightings of canals and green vegetation on the Martian surface in the late 1800s led to widespread rumors of an advanced race of aliens inhabiting the planet. So compelling was the idea of "little green men" living on Mars that Orson Welles was able to panic America in 1938 with his dramatic, but false news broadcast of invaders from the Red Planet.

Americans did not know any better at the time, but Welles envisioned the invasion backward. While no Martian has ever seen Earth, NASA sent eight robotic spacecraft to study the Red Planet between 1965 and 1976 and is now in the process of reviving the spirit of Mars exploration. This "scientific armada" was led by the *Mars Global Surveyor* and the *Mars Pathfinder* spacecraft that arrived during the summer of 1997. *Surveyor* is currently conducting the most comprehensive scientific study of Martian surface features, atmosphere, thermal properties, weather patterns, and magnetic field in history. While *Surveyor* remained in orbit, *Pathfinder* landed and deployed a six-wheeled robotic micro-rover to drive around and collect surface data. NASA's plans involve sending spacecraft to study Mars every 25 months for the next decade.

Sometime in the early twenty-first century, human explorers will land on Mars, following in the footsteps of their robotic predecessors. By studying Mars, the next target for human exploration in space, NASA scientists hope to gain insight into the formation and evolution of Earth and the inner Solar System. The question is not, "Can we do it?" Clearly, the technology already exists. The question is "When will we do it?" More than 500 years after Columbus explored the New World, NASA is ready to explore a new world of its own. This chapter will take a look at some of these plans.

Opposite Page:
September 1997 – *Sojourner*, a tiny robotic rover, explores the Martian surface at an area called Ares Vallis. The hill near the horizon is approximately 35 meters (115 feet) tall and is located about 1 kilometer (0.6 miles) away. This photograph was taken by the camera onboard the *Mars Pathfinder* lander.

7.1 Dispelling Old Myths

On the eve of Halloween in 1938, more than six million listeners tuning in to the New York CBS radio studio and all of its affiliates around the country heard the shocking announcement. "Ladies and gentlemen," the supposed news broadcaster said, "we interrupt our program of dance music to bring you a special radio bulletin from the Intercontinental Radio News." With that announcement, Orson Welles began the single most stunning program ever broadcast over radio. Throughout the evening, Welles interspersed bulletins

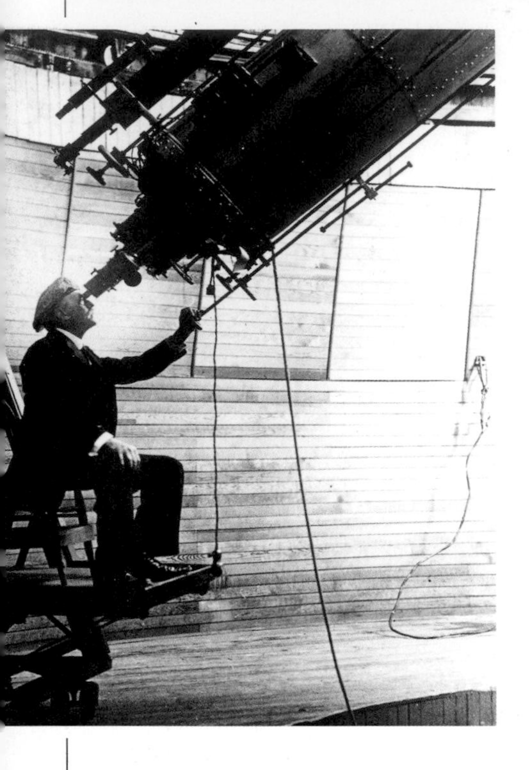

Above:

American astronomer Percival Lowell studies Mars through his 24-inch refractor telescope on a high mesa near Flagstaff, Arizona. Lowell was arguably the most famous of all the "Martian astronomers." In 1894, he reported sighting more than 184 canals on the planet.

(Photograph provided by Lowell Observatory in Flagstaff, Arizona)

Historical Fact

Skeptics of Lowell and his Martian observations argued that the canals had to measure an unreasonable width of over 30 miles in order to be seen from Earth. Lowell was ready with an explanation. He claimed that the sightings were actually areas of irrigated crop farms situated next to the waterways.

of a reported Martian invasion of New Jersey with the normal evening program of dance music. The Martians were invading the United States with their flashing ray guns and poison smoke to set fire to the countryside, or so people thought.

At the beginning, Welles announced that the evening's program was a radio dramatization of a version of science fiction writer H.G. Wells' *War of the Worlds*. Many who tuned in after this initial disclaimer were convinced that an actual invasion was taking place. Thousands fled their homes, prayed in churches over the coming end of the world, and many national guard troops reported for mobilization to defend the Earth. Residents of Rhode Island demanded that authorities shut down all electrical power stations with the hope that the invaders would overlook their tiny state. Some even took action in their own hands by leaning out their windows with shotguns in order to repel the arriving Martians. Mars, the red planet in the sky, had touched an entire generation of Americans. Today, many uninformed schoolchildren and some adults still equate Martians with hostile creatures from space.

Canals?

The idea of intelligent creatures inhabiting Mars was not new to Americans in 1938. Between 1877 and 1888, an Italian astronomer named Schiaparelli made sketch drawings of Mars based on crude telescopic observations. His drawings showed several dark areas and a vast network of what he called *canali*. In Italian, *canali* means *grooves*. However, the word can also mean *canal* and was often translated that way. The crux of the problem was with the translation. Many people assumed that if these canals existed, then somebody or something must have built them. Furthermore, some people interpreted Schiaparelli's observation of dark areas on Mars as proof of the existence of seas.

In 1894, Percival Lowell confirmed the existence of these so-called canals from his observatory in Flagstaff, Arizona. He spotted close to 200 canals, ice caps at the poles, and greenish-looking dark areas on the planet that seemed to spread in the summer and recede in the winter. The popular hypothesis of the time conjectured that the melting of the polar ice caps in the summer channeled water into the canals. This vast network of irrigation canals spread water around the planet to produce a seasonal revival of vegetation. Some textbooks even went so far as to claim that the existence of canals and vegetation provided concrete proof that intelligent life must exist on Mars.

By 1960, observations made from telescopes more powerful than Lowell's allowed scientists to conclude that Mars lacked oceans and contained much less moisture than the Earth. Their best estimates placed the surface atmospheric pressure at slightly less than 10% that of Earth's. However, many thought that a limited amount of surface water might exist in some places along with moisture in the soil. The moisture hinted at the possible existence of simple plant life on Mars. After all, what else could have explained the cyclic change of the Martian surface color that was synchronized with the seasons and the advance and retreat of the polar ice caps on the planet?

Mariner and the Ghoul

All of the questions about Mars remained unanswered until the beginning of the space age in the 1960s. At that time, NASA and the Soviet Union raced each other to send small, automated spacecraft to fly by the Red Planet, photograph it, and gather scientific data. The Soviets sent several spacecraft toward Mars, but they lost communication with them halfway there. Many joked that a mythical beast, called the "Great Galactic Ghoul," lurked halfway between Earth and Mars to guard the secrets of the Red Planet by gobbling up all spacecraft that tried to pass.

A NASA spacecraft named *Mariner 4* made it past the Ghoul in 1964. On 15 July 1965, *Mariner 4* became the first to fly past Mars at close range. In total, *Mariner* took twenty-two pictures of Mars and transmitted them back to NASA's Jet Propulsion Laboratory (JPL) for analysis. The photographs showed that Mars somewhat resembled the Moon in that large craters littered the surface of the planet. There was no trace of the infamous canals or plant life that had dominated scientific theory for the past half century.

Mariner 4 also transmitted extremely valuable scientific data. For example, temperature sensors that scanned the southern polar ice cap revealed a composition of frozen carbon dioxide ice (popularly known as dry ice), not water ice. In addition, a "radio-occultation" experiment provided data regarding the atmosphere. This clever scheme depended on the fact that the planet blocked all radio contact between Earth and the spacecraft at closest approach. However, radio signals transmitted by the spacecraft passed through the Martian atmosphere for a few moments before the spacecraft passed behind the planet, and for a few moments after the spacecraft emerged from behind the planet. By analyzing the strength of the radio signals as they faded and reappeared, scientists at JPL concluded that the atmospheric pressure at the surface of Mars measured less than 1% that of Earth's, not 10% as previously thought.

The *Mariner 4* radio-occultation experiment showed that the atmosphere on Mars is so thin that any water on the surface will instantly evaporate. Furthermore, unlike the Earth's atmosphere, the thin layer of gas surrounding Mars offers little protection against the constant bombardment of harmful ultraviolet radiation from the Sun. Although large living creatures and complex plant life cannot exist on Mars today, scientists cannot rule out the possibility that bacterial life might exist in scattered oases, or that life may have existed billions of years ago.

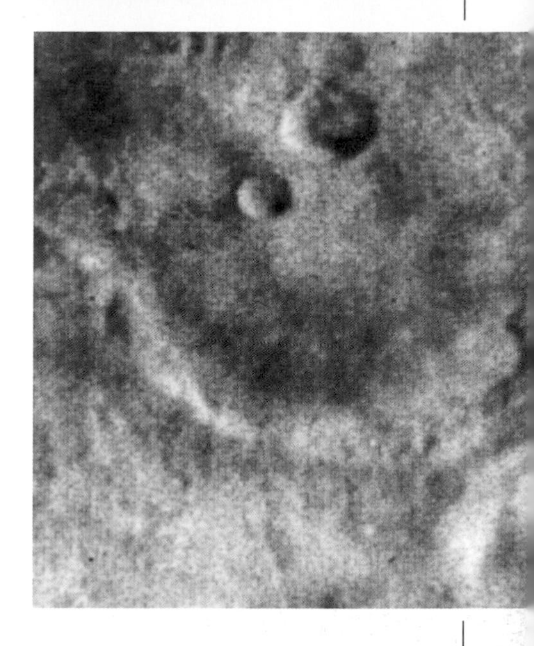

Above:
The first photographs of Mars ever taken from a spacecraft at close range greatly disappointed scientists. Instead of canals and vegetation, they saw a dead landscape littered with craters, similar to the Moon's surface. A spacecraft named *Mariner 4* was the first to fly by Mars. This was one of the twenty-two images it recorded on 14 July 1965. Due to *Mariner's* primitive communications system, each image took nearly twelve hours to be completely transmitted to Earth.

Scientific Fact!

By now, scientists are certain that no wild animals or hostile aliens carrying laser weapons inhabit Mars. However, some believe that Mars may be home to primitive bacterial life or may have supported the development of life sometime in its past. Such a find would be one of the most important in the history of science.

Above:
Much of the Martian northern hemisphere looks like a desert littered with rocks. Most of the rocks in this photograph are about 50 centimeters (20 inches) across. The rock near the upper left-hand corner, named "Big Joe," is about 2.5 meters (8 feet) across and about 8 meters (26 feet) from the camera. The metal rod in the middle of the picture holds wind and temperature sensors. This photograph was taken by the *Viking 1* lander at Chryse Planitia on 8 August 1978, 730 days after landing.

In 1969, two more identical Mariner spacecraft followed the *Mariner 4* flyby of Mars. They arrived within a few days of each other and only a week after the end of the historic *Apollo 11* Moon landing mission. Each spacecraft carried more instruments than their *Mariner 4* predecessor and observed roughly 20% of the Martian surface area, mostly in the southern hemisphere. Again, the photographs transmitted back to Earth revealed a Moon-like surface, devoid of any interesting features. By sheer coincidence, NASA's first three Mars missions missed all of the fascinating terrain now known to exist.

On the next Mars mission, a spacecraft named *Mariner 9* left its flyby trajectory and went into orbit around the planet. It spent eleven months between November 1971 and October 1972 photo-

graphing and mapping Mars. The 7,329 images that *Mariner 9* transmitted back to Earth revolutionized scientific thinking about the Red Planet. Some of the Mars images revealed surface details as small as 100 meters (330 feet). On the other end of the size spectrum, *Mariner 9* discovered giant volcanoes that dwarf Mount Everest in size, an extremely deep canyon that stretches for a length equal to that of the continental United States, and channels probably cut by torrential floods millions of years ago. The photographs allowed scientists to create maps of Mars more accurate than many maps of the Earth fifty years prior.

Reaching Utopia and the Search for Life

NASA delegated the question of life on Mars to two automated spacecraft named *Viking 1* and *Viking 2*. Both arrived in Mars orbit during the summer of 1976. After arrival, each spacecraft proceeded to send a lander craft to touch down on the surface. The delicate landers used small rockets to slow down from orbital speeds to a gentle landing. *Viking 1* touched down at a location called Chryse Planitia on 20 July, and its counterpart from *Viking 2* landed at Utopia Planitia on 3 September. Cameras on each spacecraft lander transmitted photographs that indicated a desert-like Martian surface filled with rocks and reddish orange-colored sand dunes. The initial mission plans called for the landers to return data back to Earth for 90 days. Their performances exceeded all expectations as the *Viking 1* and *Viking 2* landers operated for four and two years, respectively.

The answer to the "big" question involved the use of the lander's tiny but sophisticated biological laboratory. First, a mechanical robot arm scooped dirt from the Martian soil and placed the sample in the test chamber inside the lander. Then, three tests were performed. One of them, called the Gas Exchange Experiment (GEX), looked for evidence of biological processes by adding moisture and a broth of vitamins, amino acids, purines, and organic acids to the soil. When the automated equipment added moisture to the soil, it gave off oxygen. However, nothing happened when the broth was added.

The second experiment, called *Labelled Release* (LR), worked like GEX except for the fact that the broth contained radioactive carbon-14. Scientists speculated that any organisms present in the soil would ingest the broth and give off radioactive carbon dioxide as a byproduct. The instruments did detect such a release, but the flow stopped in the middle of the experiment. Organisms on Earth would have continued to produce radioactive carbon dioxide throughout the entire experiment, and for some time afterward.

The design of the last test, called the *Pyrolytic Release Experiment* (PR), avoided the assumption that Martian microbes function like their counterparts on Earth. Instead, the designers of the experiment postulated that the principal constituent of the Martian atmosphere, carbon dioxide (CO_2), must somehow be involved with the metabolic processes of any organisms present in the soil. The auto-

American Mars Missions

Mariner 4
Launch 28 Nov 1964, Arrival 14 July 1965
First flyby of Mars and first close-up images

Mariner 6
Launch 24 Feb 1969, Arrival 31 July 1969
Photographic flyby mission of equatorial area

Mariner 7
Launch 27 March 1969, Arrival 5 Aug 1969
Photographic flyby mission of southern areas

Mariner 9
Launch 30 May 1971, Arrival 13 Nov 1971
First spacecraft to orbit and map Mars

Viking 1
Launch 20 Aug 1975, Arrival 19 June 1976
Dual lander and orbiter mission, lander set down at Chryse Planitia on 20 July 1976

Viking 2
Launch 9 Sep 1975, Arrival 7 Aug 1976
Dual lander and orbiter mission, lander set down at Utopia Planitia on 3 Sep 1976

Mars Observer
Launch 25 Sep 1992, Lost 22 Aug 1993
Orbiter mission lost three days before arrival

Mars Global Surveyor
Launch 7 Nov 1996, Arrival 11 Sep 1997
Orbiter mission to replace *Mars Observer*, most detailed Mars survey in history

Mars Pathfinder
Launch 4 Dec 1996, Arrival 4 July 1997
Lander mission, deployed miniature rover to drive around on the surface to gather data

Mars Climate Orbiter
Launch 11 Dec 1998, Arrival 23 Sep 1999
Spacecraft lost during orbit insertion

Mars Polar Lander
Launch 3 Jan 1999, Arrival 3 Dec 1999
Study of the southern polar region

Human Mars Mission
Not yet accomplished, not yet officially scheduled on NASA's calendar, probably will not happen before 2018

Mars Pathfinder Quick Facts

Launch Date	
4 December 1996	
Launch Vehicle	
Delta 2, from Pad 17 at Cape Canaveral, FL	
Trajectory	
Type 1, landing date of 4 July 1997	
Spacecraft Mass	
At Launch (includes propellant)	800 kilograms
At Mars Entry	463 kilograms
Lander (includes rover)	264 kilograms
Rover Only	12 kilograms
Rover Chassis	
Rocker-bogie suspension	
Rover Transmission	
All terrain six-wheel drive, 2000:1 gear ratio	
Rover Dimensions	
Length	63 cm
Width	48 cm
Height	28 cm
Wheel Base (diameter)	13 cm
EPA Mileage Estimate	
10 watts of power from solar panel will allow rover to drive at about one meter per minute	
Preferred Option Package	
Laser range-guides for navigation, UHF modem for communications with the lander, internal 80C85 computer control with half a megabyte of RAM, radioactive heater units for environmental control	
Safety Features	
Airbags included	
Roadside Assistance Plan	
None available at Mars	
Manufacturer's Standard Warranty	
Rover must work for at least 7 Martian days	
MSRP	
$25 million for the rover, $150 million for the lander, does not include $60 million destination charge (for the Delta 2 rocket)	

mated laboratory tested this hypothesis by surrounding the soil sample with radioactive carbon dioxide and shining a xenon lamp on it to simulate sunlight. Later, the gas was flushed from the chamber, and the soil sample was baked at intense temperatures to vaporize any microbes present in the sample. Such an action would have released radioactive carbon dioxide from the soil if tiny organisms lived there. Unfortunately, the results were inconclusive.

Scientists working on the Viking project arrived at nonbiological explanations for results from the three experiments. They conjectured that the reactions were the result of chemical oxidation reactions between the iron-rich soil, water, and nutrients. Although the experimental results indicated that life probably did not exist at the test sites, most scientists concede that a biological explanation will continue to be a possibility until some future test provides conclusive proof that the results came from nonbiological processes. In addition, the two Viking landers conducted these three tests at only one location apiece. Other sites may yield more definitive clues. Future missions that involve more sophisticated scientific equipment, automated landers on wheels that can drive to different sites, and astronauts with the flexibility to change scientific procedures as they see fit will yield more conclusive results.

7.2 In the Footsteps of Vikings

After the two Viking missions, not a single NASA spacecraft studied Mars for the next twenty years. That drought ended on Independence Day in 1997 when a tiny tetrahedron-shaped probe called *Pathfinder* arrived high above the Martian sky after a seven-month journey from Earth. On that day, NASA engineers successfully crash-landed this spacecraft on the surface of Mars. Yes, the operative word in this mission is "crash." Instead of using traditional retro-rockets to slow the spacecraft lander from space to a gentle touchdown, *Pathfinder* plunged directly into the Martian atmosphere at fantastic speeds. After safely crashing onto the surface, *Pathfinder* released a six-wheeled miniature robot rover (about 0.60 meters, or 2 feet, long) that drove around and gathered scientific data about the Martian terrain, soil, and rocks.

A Crash Landing

In order to survive the extremely rugged landing, *Pathfinder* employed several different devices. The first, a blunt, saucer-shaped aeroshell heat shield covered the bottom. This shield prevented the spacecraft lander from melting due to the intense heat generated by entry into the Martian atmosphere at velocities approaching 7.6 kilometers (4.7 miles) per hour. As *Pathfinder* sliced through upper reaches of the Martian atmosphere during its descent toward the planet's surface, the collisions with the molecules of carbon dioxide

Above:

July 1996 –Two engineers at NASA's Jet Propulsion Laboratory prepare the *Mars Pathfinder* lander for shipment to the launch site at Cape Canaveral. In the photograph, the rover is sitting on one of the three petals of the lander's pyramid-shaped shell. *Pathfinder* flew to Mars encased in this shell. Upon arrival, the petals opened to expose the solar cells that lined the inside wall of each petal. These solar cells provided electricity to operate the *Pathfinder's* electronics.

gas in the atmosphere slowed the lander to slightly less than twice the speed of sound.

At this point, a parachute was deployed to slow the lander's decent to a velocity of about 235 kilometers (146 miles) per hour. Unfortunately, a parachute large enough to provide for a soft landing would have weighed too much because the thickness of the Martian atmosphere measures much less than Earth's. The ingenious solution to solve the large parachute problem involved using safety airbags similar to those found in ordinary automobiles, but much larger. Immediately before landing, four of these gigantic airbags inflated to encase *Pathfinder* in a protective bubble about 5 meters (16 feet) in diameter. Touchdown occurred at 9:57 Pacific Daylight Time on the Fourth of July with an impact velocity of about 65 kilometers (40 miles) per hour.

Upon impact, the airbag-encased lander bounced about 15 meters (49 feet) into the air. For the next 150 seconds, *Pathfinder* bounced at least another 15 times and rolled about 1 kilometer (0.62

Top Photograph:

5 July 1997 - The *Sojourner* rover waits for its next command as it sits next to a rock that the *Pathfinder* scientists nicknamed "Barnacle Bill." On the rover, the metallic-colored mast pointing into the air is a radio antenna that was used by *Sojourner* to communicate with the *Pathfinder* lander. The dark-colored grid on top of the rover contain solar cells that were used to provide power for *Sojourner's* electronics and wheel motors.

Bottom Photograph:

12 August 1997 - This image of the *Pathfinder* lander was taken by *Sojourner* on the thirty-ninth day of the mission. The crumpled material under the body of the lander are the airbags that protected *Pathfinder* when it crashed into the surface of Mars on Independence Day. On the lander's body, the American flag and Jet Propulsion Laboratory logo are clearly visible. In addition, the image also shows *Pathfinder's* camera sitting on top of the wire-frame style structure immediately above the flag.

miles) from the initial impact point. After arriving at a complete stop, the lander deflated and retracted the airbags. Then, the three metallic sides of the pyramid-shaped lander opened like petals on a flower to expose the science instruments and tiny six-wheeled rover to the Martian environment. Although the motors that opened the petals produced enough power to force *Pathfinder* upright regardless of its orientation after the rolling stopped, the lander was fortunate to come to rest right side up.

Exploring Mars by Remote Control

Several hours after touchdown, *Pathfinder* began to transmit the first images from its landing site in Ares Vallis, a location slightly north of the Martian equator. The images showed a dusty, rolling plain littered with a countless number of rocks and boulders. Scientists chose this site for study because evidence gathered from orbit during prior missions indicated that torrential floods long ago in Martian history deposited much of the rocks in that region. Studying the Ares Vallis area contributed toward understanding the Martian geological and climatic history.

Although *Pathfinder* carried a camera that produced several breathtaking panoramic images of the landing site and meteorology sensors to gather weather data, the rover conducted most of the exploration activities. This six-wheeled miniature vehicle was named in honor of antislavery activist Sojourner Truth. The word *miniature* describes the *Sojourner* rover accurately, as it weighed only 11.5 kilograms (25 pounds) and measured 630 millimeters (2 feet) in length. In order to see, this rover carried two tiny cameras and measured distances to potential obstacles using five small laser rangers.

Because of the desert-like, rocky surface at Ares Vallis, NASA engineers designed *Sojourner* with enough traction to climb over small rocks and 30° slopes in dry sand. Both front and rear wheels were able to turn independently to allow the rover to turn in place. Despite these spectacular maneuvering capabilities, the tiny vehicle inched across the Martian surface at a snail's pace of about one meter per minute. The reason for this slow pace was that the onboard solar panels provided a maximum of 15 watts of power. In fact, the panels only provided enough power for driving when the Sun shone directly overhead, between 10:00 in the morning and 2:00 in the afternoon local Mars time. Backup batteries took over to power the electronics during the night and on cloudy days.

Mission controllers at the NASA Jet Propulsion Laboratory in Pasadena, California, directed *Sojourner*'s activities by remote control. Depending on the position of the planets, radio signals can take as long as forty minutes to travel from Earth to Mars and back. Consequently, the operators were not able to drive the rover in "real time." Instead, pictures transmitted from the lander's cameras allowed mission analysts to determine the tiny vehicle's next destination and the safest route of travel. After receiving a destination command from

Earth, relayed through the lander, the rover drove autonomously with a limited ability to avoid unexpected obstacles. At the designated destination, *Sojourner* stopped to allow the operators a chance to select the next destination.

Although *Sojourner's* radio link to *Pathfinder* was designed to transmit at distances up to a kilometer (slightly more than half a mile), the science teams found ample rocks to study in close proximity to the lander. As a result, *Sojourner* never ventured more than 10 meters (33 feet) away from the landing site. The data gathered during these short traverses consisted primarily of pictures of the Martian surface and read-outs from the onboard alpha, proton, and X-ray spectrometer (APXS). This device slowly bombarded rocks with atomic particles to determine their chemical signature.

After eighty-three Martian days of collecting data on the surface of the Red Planet, mission controllers at the Jet Propulsion Laboratory lost contact with *Pathfinder*. Several of the engineers who built and operated the lander theorize that the battery that powered *Pathfinder* during the Martian night may have failed in a way that crippled the electrical system. Because *Sojourner* received its commands from Earth through a relay antenna on *Pathfinder*, the tiny rover's mission also came to an end. Despite the loss of contact on 27 September 1997, the mission proved to be an unqualified success as the lander functioned for nearly three times its original design life of thirty days.

A Bargain-Basement Discount

Despite the unique scientific opportunity, NASA classified *Pathfinder* as an "engineering demonstration" mission. Some of the

Above:

4 July 1997– After landing on Mars, the camera on the *Pathfinder* lander took more than sixty images in order to create a panoramic mosaic of the Ares Vallis landing site. When the data reached Earth, sophisticated computer software was employed to construct this larger image from the smaller individual frames. The empty boxes near the bottom of the image indicate areas containing frames that had not yet been transmitted to Earth. In the mosaic, the rover appears curved and the horizon looks broken because the software has not yet corrected for an optical effect called *parallax.*

Historical Fact

On Independence Day in 1997, *Pathfinder* became the eighth American spacecraft to successfully reach Mars, but the first in over two decades. The Russians have never implemented a successful Mars mission despite more than three decades of trying.

Above:

July 1997 – This image of the *Mars Pathfinder* landing site at Ares Vallis is the left half of a 360° panoramic image taken by the lander's camera (see opposite page for the right half). At the bottom center of the image, the triangular-shaped petal is one of three solar panels that generated electricity for the lander. Airbags used to cushion *Pathfinder* during its "crash landing" can be seen as deflated, crumpled material under the solar panel. The two gold-colored planks extending from the solar panel are ramps that the rover used to drive off of the lander and onto the Martian soil. In the distance approximately 1 kilometer (0.6 miles) away, two small hills rise above the horizon. Scientists nicknamed these hills "The Twin Peaks."

primary objectives involved testing the inexpensive, but rugged and untried landing techniques, and proving the concept of using microrover vehicles for exploration. The lessons learned by designing and operating *Sojourner* will be crucial toward the development of an exciting new mission called Mars Sample Return. Early in the twenty-first century, this mission will utilize advanced rovers to gather rock samples that will be returned to Earth for analysis. These rovers will weigh almost five times more than tiny *Sojourner*.

Just as important as the technology demonstration aspect, *Pathfinder* also served as a fiscal demonstration. Typically, the total bill for a lander mission to another planet can add up to more than $1 billion dollars. NASA spent about $150 million for the *Pathfinder* lander, $25 million for the *Sojourner* rover, and $60 million for the Delta 2 rocket that launched the mission. Prior to 1996, NASA used these inexpensive Delta boosters only for launching small communications satellites into Earth orbit. The more powerful Space Shuttle or Titan rockets typically used for interplanetary launches cost much more. In total, $225 million represents a "bargain-basement" price in the business of space exploration.

Scientific Fact!

A day on Mars lasts for 24 hours, 39 minutes. NASA engineers and scientists refer to this amount of time as a *sol*. By coincidence, the length of a sol and Earth day are almost identical.

7.3 A Martian Survey

NASA's *Pathfinder* represented just one of the two American-launched spacecraft that arrived at Mars in 1997. On September 11th, the 1,060-kilogram (2,337-pound) *Mars Global Surveyor* spacecraft swung into orbit around the Red Planet after a journey from Florida of nearly 10 months. Until March 2001, *Surveyor* and its five extremely sophisticated scientific instruments will continuously gather data to map the Red Planet. The goal is to collect an unprecedented amount of scientific information regarding the Martian sur-

face features, topography, mineral distribution, and magnetic properties. At the end of the mission, the total data transmitted to Earth from *Surveyor* will be enough to fill more than 100 CD ROMs.

Aerobraking

Initially upon arrival, geometrical constraints forced *Surveyor* to whirl around the Red Planet in an highly elliptical orbit taking 45 hours to complete a single revolution. At the low point (periapsis) in the orbit, the spacecraft's several-hundred-kilometer altitude would have allowed the instruments to see the surface clearly. However, the highly elliptical nature of the orbit took *Surveyor* out to 54,000 kilometers (33,554 miles) at the high point (apoapsis). This altitude was much too high above the surface for the spacecraft's instruments to gather useful scientific data. In order to carry out a worthwhile mapping mission, controllers back on Earth were forced to lower the orbit to a circular path no higher than several hundred kilometers in altitude.

The traditional method for shrinking the size of a spacecraft's orbit involves slowing down by burning a significant amount of propellant using rocket engines. In the business of orbital mechanics, if a spacecraft slows down at the low point in the orbit, its high point will be lower than on the previous orbit. Unfortunately, the small and inexpensive Delta 2 rocket that launched *Surveyor* from Earth lacked the power to lift both the spacecraft and the propellant needed to hit the brakes in Mars orbit. Without a clever solution, NASA would have been forced to purchase a more powerful, but more expensive, rocket.

Fortunately, spaceflight engineers at NASA's Jet Propulsion Laboratory in Pasadena, California, developed an unique method to

Above:

July 1997 - This image of the *Mars Pathfinder* landing site at Ares Vallis is the right half of a 360° panoramic image taken by the lander's camera (see opposite page for the left half). Near the left edge of the image, the *Sojourner* rover is sitting next to a waist-high-size rock. Scientists working on this mission nicknamed this rock "Yogi."

Historical Fact

In the spring of 1995, NASA and the Planetary Society sponsored an essay contest for America's schoolchildren to name the *Mars Pathfinder* rover. The contest rules called for the rover to be named after a famous woman, either real or fictional. Valerie Ambroise of Bridgeport, Connecticut, won the contest with her essay on Sojourner Truth, an American abolitionist and feminist. Sojourner Truth was an African-American citizen born into slavery in 1797. After gaining freedom in 1827, she devoted her life to preaching against slavery and for the rights of women. The *Pathfinder* rover was named *Sojourner* in her honor.

Above:

Unfortunately, there are no news cameras in orbit around Mars to photograph the *Mars Global Surveyor* spacecraft in action. This computer illustration depicts what the spacecraft would look like to an observer at Mars. The wing-like appendages on the sides of the spacecraft are two solar panels that provide nearly 1,000 watts of power for *Surveyor's* electronics. The circular structure at the top of the boom sticking up from the spacecraft is a high-gain antenna capable of transmitting data to Earth at a rate of 85,333 bits per second. From tip to tip, *Surveyor* measures almost 10 meters (33 feet) long.

Opposite Page:

March 1996 - Test engineers at the Lockheed Martian Astronautics facility in Denver, Colorado, prepare the *Surveyor* spacecraft for a critical thermal and vacuum test. During the test, all of the air was pumped out of the chamber, and the spacecraft was subjected to heat lamps to simulate the Sun. This test was performed to verify the spacecraft's ability to withstand the space environment. Lockheed Martian was the company selected by NASA to design and build *Surveyor*.

save propellant by using the Martian atmosphere to lower the orbit. They called this new idea *aerobraking*. Essentially, this solution works in the same way as how children lower the height at which they swing on a playground swingset by dragging their feet through the sand as they swing pass the bottom.

In order to initiate aerobraking, ground controllers fired the spacecraft's main rocket engine to lower the low point of *Surveyor*'s orbit into the upper fringes of the Martian atmosphere. Every time *Surveyor* passed through the atmosphere at the low point on the orbit, collisions with the air molecules caused the spacecraft to slow down by a slight amount. Consequently, *Surveyor* did not climb as high as on the previous orbit. This technique was designed to lower the orbit from highly elliptical to nearly circular by causing the spacecraft to slowly spiral inward toward Mars.

Although using the Martian atmosphere to slow down represented an elegant solution to a seemingly unsolvable problem, aerobraking was dangerous and presented many significant challenges. When *Surveyor* slowed from bumping into air molecules at orbital velocities, the energy from the collisions dissipated as heat. Flight controllers faced the challenge of finding a safety corridor through the atmosphere high enough to prevent the spacecraft from burning up like a meteor, but not too high as to prevent not bumping into enough air molecules to slow down appreciably.

In addition, *Surveyor* also faced an unexpected aerobraking challenge. Shortly after launch, the structure holding one of the two solar panels to the spacecraft's body cracked. Several weeks after the start of aerobraking activities, flight controllers discovered that the Martian air pushing on the spacecraft during passes through the atmosphere caused the cracked panel to bend backward. In order to keep that solar panel from breaking, the flight team was forced to design a plan that employed flying through the atmosphere at a much gentler pace. The end result was that the reduction of the orbit size down to several hundred kilometers in altitude was delayed from February 1998 to February 1999.

An Orbit Made for Mapping

The successful completion of aerobraking placed *Surveyor* in an orbit that circles the Red Planet every 118 minutes at an average altitude of 378 kilometers (235 miles). In this *mapping orbit*, the spacecraft travels northward on the daylight side of Mars and southward on the night side. Trajectory analysts working on the project designed the orbit so that the Sun always appears in the same part of the sky every time the spacecraft flies over the equator. Specifically, the local Mars time on the ground under the spacecraft is always 2:00 in the afternoon. This *Sun-synchronous* characteristic ensures that the shadows and lighting conditions in all of the images taken from *Surveyor* will match, giving scientists a common frame of reference from which to analyze all of the data.

Surveyor Spacecraft Statistics

Launch Date	
7 November 1996	
Launch Vehicle	
Delta 2, from Pad 17A at Cape Canaveral, FL	
Trajectory	
Type 2, Mars on arrival 11 September 1997	
Mass	
Spacecraft	600 kg
Propellant	380 kg
Science Instruments	75 kg
Total	1,060 kg
Communications	
Command Rate	12.5 per second
Uplink Data Rate	500 bits per second
Downlink Data Rate	85,333 bits per second
Downlink Power	22 watts
Propulsion	
Fuel	Monomethyl Hydrazine
Oxidizer	Nitrogen Tetroxide
Main Engine	596 Newtons of thrust
Attitude Control Jets	4.45 Newtons of thrust
Power	
Two solar arrays, each with two solar panels, will provide a maximum of about 1,000 watts of power. This value drops to about 660 watts when Mars is farthest from the Sun	
Data Storage	
Solid-state recorders (functions like computer memory) can store 3.0 gigabits (375 Megabytes), equivalent to 26 hours of data	
Science Instruments	
Camera, thermal emission spectrometer, laser altimeter, mars relay, radio science, magnetometer, electron reflectometer	
Design Life	
About five years in Mars orbit	

Surveyor's orbit lies tilted at nearly a right angle to the Martian equator and passes over both the north and south polar regions of the planet. This polar orientation was chosen to allow *Surveyor's* scientific instruments to eventually image the entire surface area of Mars as the planet rotates under the orbit. In fact, the orbit allows the spacecraft to repeatedly map the entire planet in detail once every 7.2 Earth days. At the conclusion of the mission in 2001, *Surveyor* will have repeated this cycle of global observation numerous times over a period of one Martian year (two Earth years). A year's worth of observations will provide scientists with an opportunity to compare the effects of weekly and seasonal variations in Mars' weather.

Tools for Surveying

Compared to the *Sojourner* rover that required constant attention from ground controllers on Earth, *Surveyor* will essentially fly in autopilot mode throughout the mapping mission. At all times, the spacecraft's onboard computer systems keep the electronic eyes of its sensitive imaging camera focused on the Martian surface. This device, called the *Mars Orbiter Camera*, photographs objects in either color or black and white mode. Its powerful, high-resolution telephoto lens can resolve Martian rocks and other objects as small as 1.4 meters (4.6 feet) across from orbit.

In contrast to the telephoto mode, the camera's wide-angle mode uses a fish-eye type lens to generate spectacular panoramic images that span almost from horizon to horizon. These images are similar to weather photos commonly shown on the evening news. Scientists will play sets of these images like a motion "flip-book" or a film. This ability will allow them to see the life history of Martian weather phenomena such as dust storms, cloud formations, the growth and contraction of the polar ice caps, and surface features that get blown by the wind.

Another instrument working in parallel with the camera constantly bombards the Martian surface with a laser. Contrary to popular myth, the laser was not designed to fend off hostile Martians waiting to ambush the spacecraft. Instead, the laser fires ten bursts of light at the ground every second. As the spacecraft flies over valleys, mountains, canyons, hills, and craters, its altitude above the ground changes. By measuring the time required for the beam to leave the spacecraft, reflect off the ground, and return to the collecting mirror on the instrument, scientists are able to determine *Surveyor's* altitude above the surface to within several meters. By combining the laser data with high-resolution camera images, NASA will be able to reconstruct a detailed topographical atlas of the planet. Such maps will help scientists to better understand the geological forces that shaped Mars and help future mission planners to determine possible landing sites for astronauts.

Although scientists will depend heavily on visual images for their studies, a third instrument on *Surveyor* is yielding interesting

clues about Mars by imaging the Red Planet in regions of the energy spectrum that humans cannot see. This instrument, called the *Thermal Emission Spectrometer*, scans the Martian surface in the infrared range. On the color spectrum, infrared represents the zone that would take on an extremely deep red hue if the human eye could see it. In practical terms, infrared colors manifest themselves as heat energy.

Most compounds radiate a unique thermal signature. The data radioed back to Earth from the spectrometer allows NASA scientists to analyze heat emissions from the Martian surface, providing them with an opportunity to determine the general mineral composition of patches of ground as small as 9.0 square kilometers (3.5 square miles) in area. Scientists will gather this type of data over many days

Above:

This photograph shows a view of a small section of Argye Planitia (bottom left-hand corner), as seen from the *Viking* orbiter in 1976. Argye is a thousand-kilometer-wide impact basin formed billions of years ago when a huge asteroid collided with Mars. Although much of the terrain around the basin is rough and jagged, the floor of the basin is relatively smooth. Keep in mind that the basin is much larger than the four craters seen in this photograph.The thin Martian atmosphere appears as a thin, slightly transparent streak of haze above the planet.

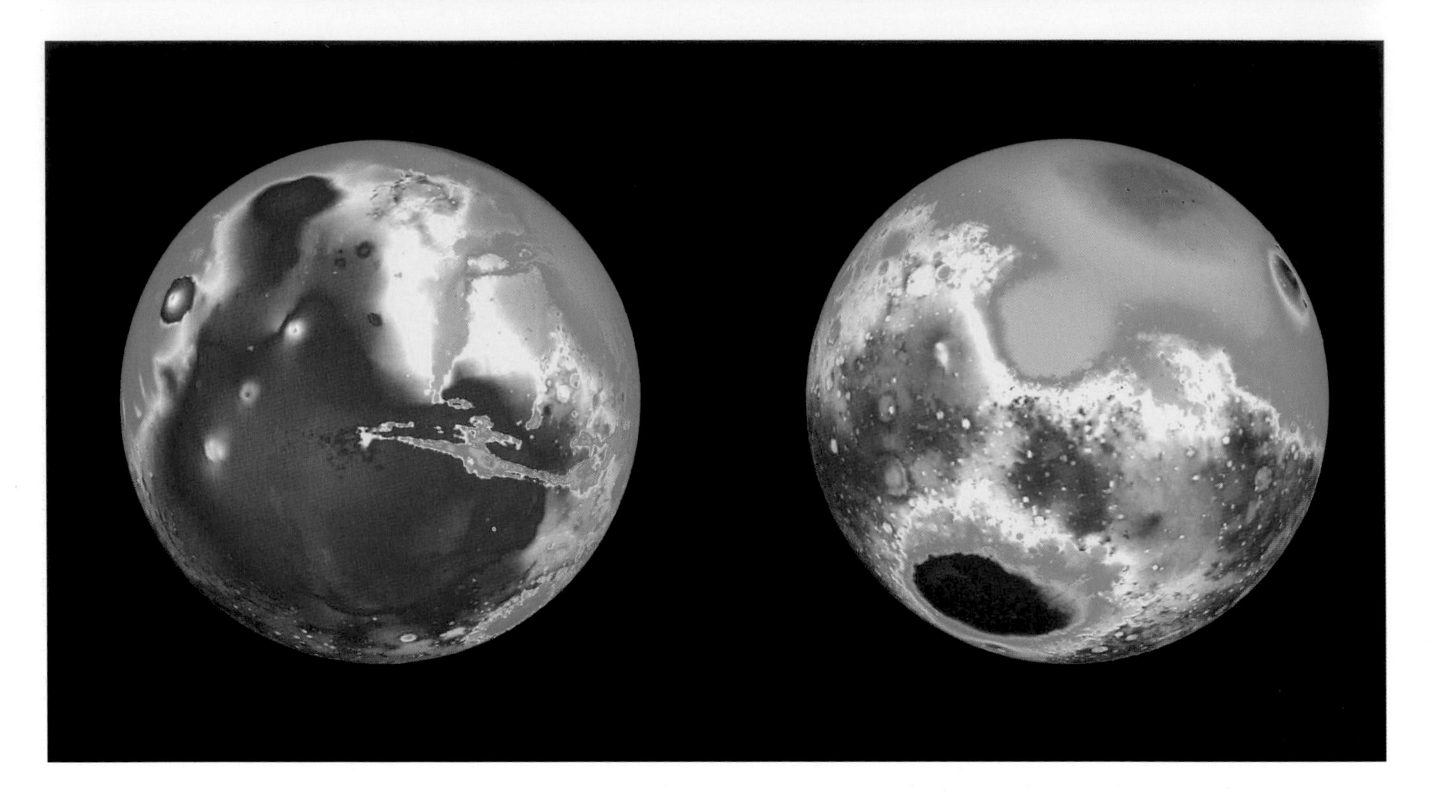

Above:

The two colored globes in this image represent portions of the first Martian topographic map generated using laser altimeter data from the *Mars Global Surveyor* spacecraft. Red colors indicate terrain at high altitudes, while blue colors indicate low elevations.

The left globe shows the northern hemisphere of Mars. Four red-white spots near the left edge of the map represent giant volcanoes with summits three times higher than Mount Everest on Earth. The green-blue streak toward the right center is a canyon system longer than the length of the United States.

The right globe shows the southern hemisphere of Mars. A large blue area near the bottom of this globe indicates an extremely deep basin nearly 2,100 km (1,305 miles). This area was formed long ago when a gigantic asteroid impacted Mars.

The laser altimeter's science team is led by Dr. David Smith and Dr. Maria Zuber of NASA's Goddard Space Flight Center in Greenbelt, Maryland.

in order to conduct a general, planet-wide geological survey of Mars without ever setting foot on the planet. In addition, thermal data helps scientists to study the Martian atmosphere, clouds, and weather. Many also hope that the spectrometer data will yield clues into the location of deposits containing water or carbonate minerals. Although no liquid water exists on Mars today, clay deposit areas might have been shorelines long ago in Martian history, and would serve as prime locations for future astronauts to explore and search for fossil remains of primitive Martian life.

The fourth instrument, called the *magnetometer*, is attempting to measure the poorly understood global magnetic properties of Mars. On Earth, a large central core of churning molten iron acts to create an extremely powerful magnetic field surrounding the planet. In comparison, the magnetometer data shows that the Martian magnetic field is almost nonexistent except for a few local disturbances. By studying this data, scientists hope to learn about the interior composition of Mars. Such a study will yield insight into the history of the geophysical forces that shaped the surface of Mars, and how they differ from those on Earth.

All of the data collected by *Surveyor's* four instruments arrive at Earth by way of radio signals transmitted from the spacecraft. A special group of scientists will use sophisticated computers to analyze these radio signals. They are not interested in the data contained in the signals, but in the electrical strength and "tone" of the transmissions. By analyzing tiny variations in the tone as *Surveyor* is bumped by minor variations in the strength of the Martian gravity,

scientists will be able to determine the planet's shape to a very high degree of accuracy. This data will also be combined with data from the laser altimeter to improve on the accuracy of the topographical maps.

In addition, this team of "radio scientists" is also conducting occultation experiments. When *Surveyor* flies over the back side of Mars with respect to the Earth, the planet will block the spacecraft's radio signal from reaching Earth. Spaceflight engineers call this event an *occultation*. For a few moments before the spacecraft flies behind the planet, and for a few moments after it reemerges, the radio signal passes through the thin Martian atmosphere on its way to Earth. By analyzing the strength of the signal as it fades and re-appears, scientists are able to determine the atmospheric pressure at a specific location on Mars. When combined with data from the spectrometer, scientists will gain a greater understanding of the Martian atmosphere than ever before.

7.4 Reaching the Red Planet

After examining data and photographs sent back from automated probes in Mars orbit and from miniature robotic landers on the surface, NASA scientists will want to study the Red Planet in person, and at close range. Such a mission would provide an unprecedented opportunity for scientific discovery not otherwise possible with robotic spacecraft. Although funding for American astronauts to visit Mars has yet to be allocated by Congress and firm dates for the flight have not yet been determined, NASA space mission design engineers have already worked out possible scenarios for the first piloted Mars mission.

Hohmann Transfers

Much of the initial planning for the first piloted Mars mission will focus on how to reach the Red Planet. A direct, linear flight to Mars from the surface of the Earth is impossible from an energy perspective because in spaceflight, there is no such thing as a straight-line path. Remember that all planets and spacecraft travel in long, curved paths around the Sun that take the shape of circles and ellipses. To reach Mars, a spacecraft's rocket engines need to impart enough energy to escape the Earth's gravitational field. After Earth escape, it enters an orbit around the Sun and coasts to Mars under the predominant influence of the Sun's gravity.

Most spacecraft bound for Mars utilize a Hohmann-style transfer. This type of trajectory provides the most energy-efficient method to transfer a spacecraft between two different orbits that do not intersect. An ideal Hohmann transfer orbit between Earth and Mars takes the form of an ellipse that barely touches the Earth's orbit at the closest point to the Sun and barely touches the orbit of Mars at

Historical Fact

In December 1997, Paramount Pictures and 20th Century Fox released the Oscar-winning film *Titanic*. This extremely successful movie, staring Leonardo DiCaprio and Kate Winslet, cost more than $200 million to produce and distribute according to reports from the entertainment industry. As a comparison, NASA's budget allocated spending only $154 million to develop the sophisticated *Mars Global Surveyor* spacecraft.

RULE OF THUMB

How far away is Mars? On the average, Mars lies about 225 million km from Earth. How far is that? If the Earth were the size of a tennis ball, then Mars would be the size of a golf ball 1.3 km (0.8 miles) away. Although Mars is one of the closest planets to the Earth, it is still far enough away that rockets need to accelerate Mars-bound spacecraft to a velocity 34 times faster than a speeding bullet in order to reach the Red Planet.

A Round-Trip Ticket

If all goes according to plan, real Martians may land on Earth in late October 2008, nearly 70 years to the day that Orson Welles terrified America with his fictional radio broadcast of an invasion from the Red Planet. Contrary to theories proposed by disciples of the television series *The X-Files*, the landing will not represent the culmination of a diabolical conspiracy hatched by the government to overthrow humanity. Instead, the possible arrival of Martians is part of NASA's historical effort to collect samples from Mars and deliver them to Earth. Scientists are hopeful that these samples may contain tiny Martian organisms or evidence of their past existence.

Not surprisingly, the immense interest regarding the mystery of Martian life is not limited to labcoat-cloaked scientists toiling in basement laboratories among their chemicals and microscopes. Anyone who has bothered to gaze into a star-filled night sky will attest that the question, "Are we alone in the universe?" is as intrinsic to philosophical debate as scientific inquest. The discovery of life elsewhere than Earth might suggest that life is a natural consequence of the way the universe works. Such a finding may indicate that life can develop anywhere in the universe given the right conditions and has probably done so a countless number of times. A negative finding is equally important.

At best, the resolution of such questions comes with great difficulty. On the Earth, life arose rather rapidly from a cosmic perspective. Less than 500 million years after the end of the torrential space-debris bombardment that accompanied the formation of the Earth more than 4 billion years ago, flourishing ecosystems of bacteria had developed and then left fossilized clues of their existence in rocks for scientists to decipher. Like the Earth of the past, ancient Mars was also warm and wet at the same time that life made its first appearance at home. If conditions were comparable on these two celestial neighbors, it is entirely possible that life developed in an analogous fashion on the Red Planet prior to leaving their traces in Martian rocks.

If pondering these philosophical concepts seems formidable, consider the fact that NASA's strategy to gather rock samples on Mars and deliver them to Earth for comprehensive scrutiny is equally as challenging. The plans currently call for launches in 2003 and 2005 with a proposed sample delivery in October 2008. In addition,

NASA is currently in the midst of negotiations for the French space agency CNES to be a major partner in this ambitious endeavor. Engineers working on this project call it "Mars Sample Return," or MSR for short.

Planners at the NASA Jet Propulsion Laboratory have determined that the most efficient method to conduct this mission is *Mars orbit rendezvous*, analogous to the ingenious lunar orbit rendezvous scheme used for Apollo in the 1960s. Specifically, MSR will employ two NASA-provided landers of nearly identical design and one French orbiter transfer vehicle carrying a NASA payload of rendezvous sensors, mechanics to capture objects in Mars orbit, and an Earth return capsule. The high-level concept is that the landers will launch surface samples into Mars orbit, and the French orbiter will retrieve the samples in orbit and then carry them back to Earth.

The Mars Sample Return odyssey will begin in May 2003 with the launch of one of the two landers from Cape Canaveral, Florida. At liftoff, this gigantic Martian explorer will weigh in at more than 1,800 kilograms (3,968 pounds) and will rank as the heaviest robotic spacecraft NASA has ever designed to land on an extraterrestrial body. However, unlike the revolutionary *Pathfinder* lander that crash-landed with airbags in July 1997, the twice-as-heavy MSR lander will utilize traditional retro-rockets to slow to a safe touchdown.

Thanks to a celestial coincidence, the planetary positions of Earth and Mars dictate that the 2003 lander will arrive at the Red Planet in late December of that year, almost exactly one century to the day that the Wright brothers inaugurated the aviation age at Kitty Hawk, North Carolina. After touchdown, the lander will deploy a six-wheeled rover resembling the *Sojourner* rover from the *Pathfinder* mission, but several times larger. This all-terrain descendant of the successful 1997 mission will survey the local terrain, extract core samples from selected rocks using a miniature drill, and deliver the cores back to a sample canister mounted to the top of a small, three-stage rocket attached to the deck of the lander. At the conclusion of the surface mission, this rocket will bast-off and insert the sample canister into a low Mars orbit. The canister will wait in orbit until it is retrieved by the French orbiter in early 2007.

Part two of the Mars Sample Return mission will commence in August 2005 as the French orbiter and second NASA-provided lander depart Earth. Current proposals call for the two spacecraft to be launched together on an Ariane 5 rocket from French Guiana in the mosquito-infested jungles lining the Atlantic coast of South America. The French designed this heavy-lift launch vehicle primarily to tap into the lucrative business of boosting communications satellites into geosynchronous orbit. Their offer to donate this rocket for the Mars Sample Return campaign will save NASA nearly $100 million.

Upon arrival at Mars eleven months later, the 2005 lander will touch down and perform a surface mission nearly identical in concept to the one that was executed during the 2003 opportunity. Similar to its predecessor, the 2005 lander will also launch a sample canister into a low Mars orbit at the conclusion of the surface mission. Although NASA scientists have not selected the landing sites, the 2003 and 2005 sites are expected to be different to enhance the diversity of the collected samples.

The orbiter's arrival at Mars in August 2006 will be highlighted by a fiery plunge into the Martian atmosphere. In the past, such a shocking scenario would only have occurred as the result of a catastrophic error. In this case, the plunge will be intentional, as the French are planning to exercise an untried and daring technique called *aerocapture* to enter Martian orbit. This scheme works by allowing the orbiter to fly through the atmosphere immediately upon arrival. The enormous amount of air resistance generated by this process will slow the spacecraft and allow it to be captured into orbit around Mars.

After surviving the difficult task of remote-control piloting the orbiter through the Martian atmosphere at fantastic speeds, the flight team will turn its voracious appetite for seemingly impossible tasks toward the equivalent of finding the proverbial needle in a haystack. The challenge will be to find and capture the tiny 15-centimeter (6-inch) sample canister launched into orbit by the 2003 lander. What will make this task difficult is that the canister will be adrift among the billions of kilometers of orbital paths around Mars.

In order to prevent this search from turning into an exercise in futility, the orbiter will be aided by a tiny radio beacon on the canister. This radio will emit electronic beeps that will allow the French spacecraft to home in on the canister's approximate position. Once in close proximity, the orbiter will use a laser for precision guidance, leading to capture. Upon completion, this process will be repeated to capture the canister launched by the 2005 lander.

All of the orbital search and capture operations must be completed by the time Earth and Mars align for the return trip home in July 2007. At that time, the orbiter will fire its main rocket engine for the trans-Earth injection burn to escape Mars and achieve a return trajectory. As the spacecraft approaches Earth toward the end of October 2008, it will release a small, blunt-shaped capsule that will contain the sample canisters. This capsule will protect and isolate the samples during its entry into the Earth's atmosphere. Although a landing site has not been selected, NASA is considering military bases in the western United States.

Amazingly, this ambitious project is just one component of a much larger, sustained effort to study the Red Planet. Starting in 1994, NASA has worked at a furious pace to build and launch small but capable spacecraft at every biannual Earth-Mars celestial alignment. The first in this scientific armada, the *Mars Global Surveyor* orbiter and *Mars Pathfinder* lander, blasted off in 1996. They were quickly followed by the *Mars Climate Orbiter* and *Mars Polar Lander* in late 1998. More of the Mars exploration fleet is currently under construction and will follow early in the twenty-first century. This rapid-fire style of exploration is a boon to scientists who are used to waiting decades between mission opportunities. ❑

Opposite Page:

December 2003 - The rover drives off of the Mars Sample Return lander on its way to extract core samples from rocks (computer rendition).

Below:

March 2004 - After a ninety-day surface mission, the Mars Sample Return lander launches its rock samples into Mars orbit (computer rendition).

the farthest point from the Sun. Spacecraft using this type of trajectory take 259 days to reach Mars and move through precisely 180° (half an orbit) around the Sun (see Figure 57 on page 287).

The first step in understanding how Hohmann transfers to Mars work is to remember that objects in orbit around the Sun move at a rate that decreases with increasing distance from the Sun. For example, Mars moves around the Sun with an average speed slower than Earth because it orbits the Sun farther out than the Earth. On the other hand, consider a spacecraft on a Hohmann transfer orbit between Earth and Mars. Such a spacecraft moves with an average speed less than that of Earth's, but greater than that of Mars.

A spacecraft on a Hohmann transfer from Earth to Mars must leave Earth at a time when Earth is behind Mars relative to the orbital positions of the two planets around the Sun. The reason is that on the average, the spacecraft moves faster on the transfer orbit than Mars moves in its orbit. Leaving while the Earth is behind works analogously to giving Mars a head start. Specifically, the spacecraft must leave at a time when Mars is ahead of the Earth by about 44°.

Where did this figure come from? A spacecraft on a Hohmann transfer to Mars takes 259 days to reach the planet. During that time, the spacecraft moves through 180° around the Sun. Mars takes 687 days to complete one orbit (360°) around the Sun. At that rate, Mars moves through 0.524° every day, or 136° in 259 days. Since the spacecraft moves 44° around the Sun more than Mars during the time of the Hohmann transfer, it must start 44° behind Mars.

Launch Windows

It is impossible to launch a spacecraft to another planet at any arbitrary time. Interplanetary mission success critically depends on precise timing. To understand the difficulty of planning such a mission, imagine two people driving from opposite ends of the United States with the goal of meeting in Chicago. However, neither of them can stop at any time, and both must arrive at the same time. They must figure out how fast each drives, and then take those speeds into account when calculating the departure and arrival times. Interplanetary mission planners face a similar task. They must calculate departure times so that the spacecraft arrives at the destination point at the same time as the destination planet. Spaceflight engineers refer to the calendar dates that satisfy the previous constraint as the *launch window* or *launch period*. The window is considered to be *open* during the dates within the timeframe and *closed* at all other times.

Phase angle measures the relative difference between the positions of Earth and Mars with respect to the Sun. The exact phase angle required for an interplanetary transfer determines the date of launch that will require the least amount of propellant to escape Earth and reach Mars. In the case of an ideal Hohmann transfer, the phase angle is 44°. Nonoptimal launches that occur before or after

Comparison of Earth and Mars

Earth	Mars
Solar System Rank	
3rd from the Sun	4th from the Sun
5th out of 9 in size	7th out of 9 in size
Diameter	
12,786 km	6,786 km
Distance from Sun	
149,600,000 km	227,940,000 km
Length of Year	
365.25 Earth days	687 Earth days
Length of Day	
23 hours, 56 minutes	24 hours, 37 minutes
Mass of Planet	
5.972×10^{24} kg	11% of Earth's Mass
Density (Water = 1)	
5.52	3.95
Gravity (Earth = 1)	
100%	38%
Temperature Range	
-70°C to 40°C	-120°C to 25°C
Tilt of Equator	
23.5° from vertical	25.2° from vertical
Number of Moons	
One large	Two extremely small
Atmosphere	
Nitrogen and oxygen	Carbon dioxide (CO_2)
Air Pressure	
1.00 at sea level	about 1% of Earth
Surface Water	
Covers 70% of planet	None
Polar Ice Caps	
All water ice	Mixture CO_2 and water
Life	
Lots and lots	To be determined

Scientific Fact!

Going to Mars on a Hohmann transfer requires that the spacecraft leave Earth when Mars is about 44° ahead of the Earth with respect to the two planets' orbital positions around the Sun. The reason for the head start is that the spacecraft's average velocity on the Hohmann transfer is faster than Mars.

Figure 57: *What's Your Type?* ◀

A Hohmann transfer provides the most energy-efficient way to move a spacecraft from Earth to Mars or vice versa. In theory, the transfer orbit takes the shape of an ellipse that barely touches the orbits of both Earth and Mars. Spacecraft on Hohmann transfers to Mars move around the Sun exactly 180° in about 259 days. Due to the fact that Mars' orbit is not perfectly circular and the orbits of Earth and Mars lie slightly tilted to each other, optimal trajectories to Mars are almost never pure Hohmann. They take the form of either Type-1 or Type-2 paths. Type-1 requires a trip around the Sun of slightly less than 180°, while the Type-2 requires slightly more than 180°.

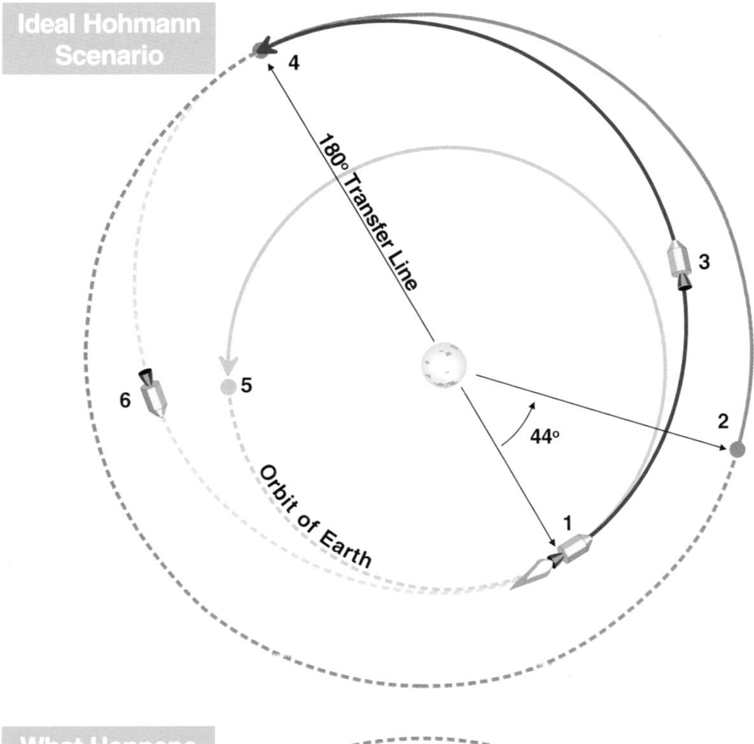

1 Earth position at launch. Spacecraft performs burn to escape the Earth's gravity.

2 Mars position at launch. Notice that Mars is 44° ahead of the Earth. Mars must have a head start because it moves around the Sun slower than the spacecraft.

3 Spacecraft coasts to Mars on a Hohmann ellipse. The orbit has a perihelion at Earth's orbit and aphelion at Mars' orbit.

4 About 259 days after launch, spacecraft reaches the same point in space as Mars.

5 Earth position at Mars arrival.

6 If the spacecraft does not enter Mars orbit, it will fly right past Mars and back down the other half of the Hohmann ellipse.

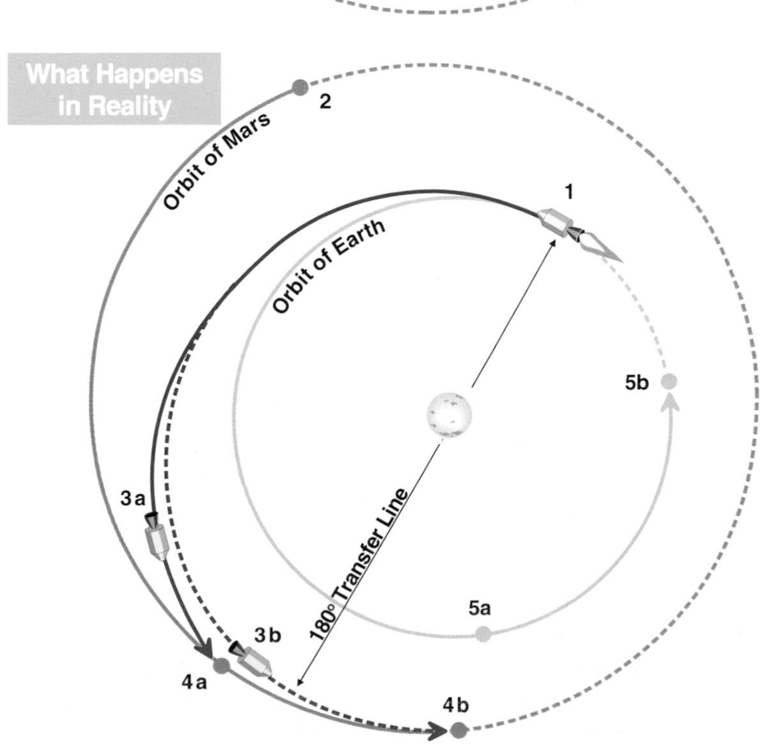

1 Earth position at launch. Spacecraft performs burn to escape the Earth's gravity.

2 Mars position at launch. Notice that Mars is ahead of the Earth. Mars must have a head start because it moves around the Sun slower than the spacecraft.

3 Spacecraft coasts to Mars using a Type-1 or a Type-2 trajectory (3a and 3b, respectively).

4 Spacecraft arrives at the same point in space as Mars. The trip takes roughly 200 days for the Type 1 and about 300 days for the Type 2 (4a and 4b, respectively).

5 Earth position at Mars arrival for Type-1 and 2 trajectories (5a and 5b, respectively).

this optimal time are possible, but require more propellant for Earth escape.

The total amount of propellant available to accelerate a spacecraft to escape Earth and reach Mars affects the length of the launch window. In general, increasing the booster rocket's propellant supply extends the launch window both forward and backward in time from the optimal target date. However, increasing the amount of propellant almost always comes at the expense of reducing the total spacecraft mass sent to Mars. Spaceflight engineers typically chose a total spacecraft mass low enough to allow for a launch window of between two and three weeks. This length of time is necessary because the laws of orbital mechanics do not tolerate unexpected delays, human tardiness, or bad weather at the rocket's launch site.

A spacecraft that misses its launch window must wait until Earth and Mars realign properly. This time period, called the *synodic period,* lasts between twenty-five and twenty-six months. Every synodic period lasts for a different time than the previous period because the orbits of the Earth and Mars are elliptical. If the orbits of both planets were perfect circles, every synodic period would last for the same amount of time. The pattern of synodic periods repeats about every thirty-two years (every fifteen periods). Some of the launch opportunities within the thirty-two-year repeat cycle are more favorable from a minimal propellant usage perspective because they allow a spacecraft to intercept Mars at its closest point (perihelion) to the Sun. These opportunities repeat approximately every 15 to 17 years and almost exactly every 284 years.

Mars Launch Windows

Launch	Arrival
1996 Opportunity	
5 December 1996	4 July 1997
7 November 1996	11 September 1997
1998 Opportunity	
16 January 1999	30 July 1999
7 February 1999	31 December 1999
2001 Opportunity	
19 March 2001	10 September 2001
16 April 2001	27 January 2002
2003 Opportunity	
7 June 2003	25 December 2003
10 May 2003	29 December 2003
2005 Opportunity	
10 August 2005	22 February 2006
2 September 2005	8 October 2006

The first set of dates in each row represent the mission dates for a Type-1 trajectory, while the second set represent the dates for a Type-2 trajectory. Launch and arrival dates were chosen to minimize the energy needed to depart the Earth with the exception of the 1996 opportunity. Those dates reflect the launch and arrival dates for the *Mars Pathfinder* and *Mars Global Surveyor* missions.

RULE OF THUMB

The Mars lander spacecraft flown by the astronauts on America's first piloted mission will carry an all-terrain vehicle for them to drive in while they explore the Martian surface. Many design options exist. The simplest designs will look like open-air jeeps. More sophisticated "rovers" will function like miniature campers. The inside of these will contain a pressurized oxygen atmosphere type of environment. This way, astronauts will be able to take multi-day trips away from the main landing site with the luxury of being able to take their spacesuits off at night.

What's Your Type?

Unfortunately, ideal Hohmann transfers between Earth and Mars do not exist in real life because the theory assumes that the two planets orbit the Sun in the same plane. In reality, the orbits of Earth and Mars lie slightly tilted from one another by an angle slightly less than 2°. This tilt differential alters the mathematical solution to the Earth-Mars transfer problem and results in two sets of optimal trajectories to Mars from a minimal propellant usage perspective. Spaceflight engineers refer to these solutions as Type-1 and Type-2 transfer trajectories (see Figure 57 on page 287). The main difference between these two options is the length of time required to reach Mars and the amount of propellant that a spacecraft must expend to slow down to enter Martian orbit at the end of the transfer.

A spacecraft on a Type-1 trajectory takes slightly less time to reach Mars than in the ideal Hohmann case, while a Type-2 trajectory takes slightly more time. For example, flight times for the two types average in the range of 200 days and 300 days, respectively. Although the Type-1 trajectories allow spacecraft to arrive at Mars quicker, they typically require more propellant to slow down and enter Martian orbit at the end of the transfer than when on a Type-2 trajectory.

The choice of which trajectory type to utilize for a Mars mission often depends on the spacecraft mass. Lighter spacecraft, or those sent to gather data on a simple flyby of Mars, normally fly on Type-1 trajectories. Conversely, heavier spacecraft, or those needing to slow down after reaching Mars in order to enter orbit, usually depart Earth on Type-2 trajectories. For example, this pattern held in 1996 when NASA launched the *Mars Global Surveyor* orbiter on a Type-2 transfer in November. In December of the same year, NASA launched *Pathfinder*, a Mars lander lighter than *Surveyor*, on a Type-1 transfer.

Mission Scenarios for Astronauts

Piloted spacecraft will weigh many times more than their robotic predecessors. Therefore, sending astronauts on Type-1 or -2 trajectories looks attractive because these flight paths provide the optimal way to reach Mars from a minimal propellant usage perspective. Unfortunately, they also require a total completion time much too long for the first piloted mission. The problem lies not in the transit time of 200 to 300 days needed to fly to Mars or return to Earth, but in the wait at Mars. If astronauts exclusively use Type-1 and -2 trajectories to reach the Red Planet, they will be forced to wait between 300 to 550 days after Mars arrival for the opening of the return-to-Earth launch window. A minimal propellant journey could therefore last more than 1,000 days (slightly more than three years). Currently, the longest stay in space by a human lasted for about 400 days, and the entire mission took place in Earth orbit.

Use of trajectories to Mars that require more propellant than optimal can allow a piloted spacecraft to catch the Red Planet in a position favorable for an almost immediate Earth return. Some scenarios allow for a return less than sixty days after arriving at Mars. However, use of more propellant will also reduce the total mount of mass dispatched to Mars. In turn, lowering the mass may degrade the quality of science to be carried out on the Martian surface because the astronauts will not be able to bring along as much equipment.

One proposed plan to compensate for the less efficient trajectory involves flying toward Venus in the direction of the Sun instead of flying away from the Sun toward Mars. As the spacecraft flies toward the Sun, it will gain the necessary velocity to reach Mars. Unfortunately, it will also be headed in the wrong direction. By flying past Venus, the astronauts can use the gravity of the planet to swing the spacecraft around and head it toward Mars in what spaceflight engineers call a *gravity assist maneuver*. However, no solution exists without inherent disadvantages. The temperature range near Venus from the Sun is about four times hotter than at Mars. Consequently, payload space will need to be sacrificed for a cooling system useless at Mars.

In either the optimal propellant or Venus assist scenario, the spacecraft must slow down to orbit Mars or risk flying past it. The

Mars Launch Energies

Launch	C_3 Value (see below)
1996 Opportunity	
30 November 1996	9.00 (1,068 kg)
21 November 1996	8.93 (1,070 kg)
1998 Opportunity	
16 January 1999	9.01 (1,068 kg)
7 February 1999	8.44 (1,081 kg)
2001 Opportunity	
19 March 2001	8.63 (1,077 kg)
16 April 2001	7.85 (1,094 kg)
2003 Opportunity	
7 June 2003	8.81 (1,073 kg)
10 May 2003	12.56 (993 kg)
2005 Opportunity	
10 August 2005	15.88 (928 kg)
2 September 2005	15.45 (936 kg)

The first number listed in the second column is the C_3 corresponding to the listed launch date. The number next to the C_3 value represents the maximum mass that a Delta 2 rocket can send to Mars for the listed C_3 value. As in the table on the previous page, the first date in each row corresponds to the launch date for a Type-1 trajectory, and the second date is for a Type-2 trajectory.

Scientific Fact!

Keep in mind that not all launch opportunities to Mars are created equal. Some years are much better than others in that they require less C_3. Remember, C_3 is a measure of how much energy it takes to send a spacecraft to another planet. A higher C_3 value always requires a lighter spacecraft because the rocket must work harder. As shown in the table above, the 2001 launch window is an extremely favorable opportunity, but the 2005 window is a very poor one.

► Figure 58: *Ideas for a Piloted Mars Mission*

Like the Apollo missions to the Moon in the 1960s, the first piloted Mars mission will employ two different spacecraft. Some of the astronauts will climb into the lander to explore the surface while the remainder of the crew will stay in orbit around Mars. The ideas presented on this page were developed by NASA in 1988.

Inner Cruise to Venus	166 Days
Outer Cruise to Mars	192 Days
Stay Time on Mars	40 Days
Return Cruise to Earth	192 Days
Total Mission Duration	590 Days

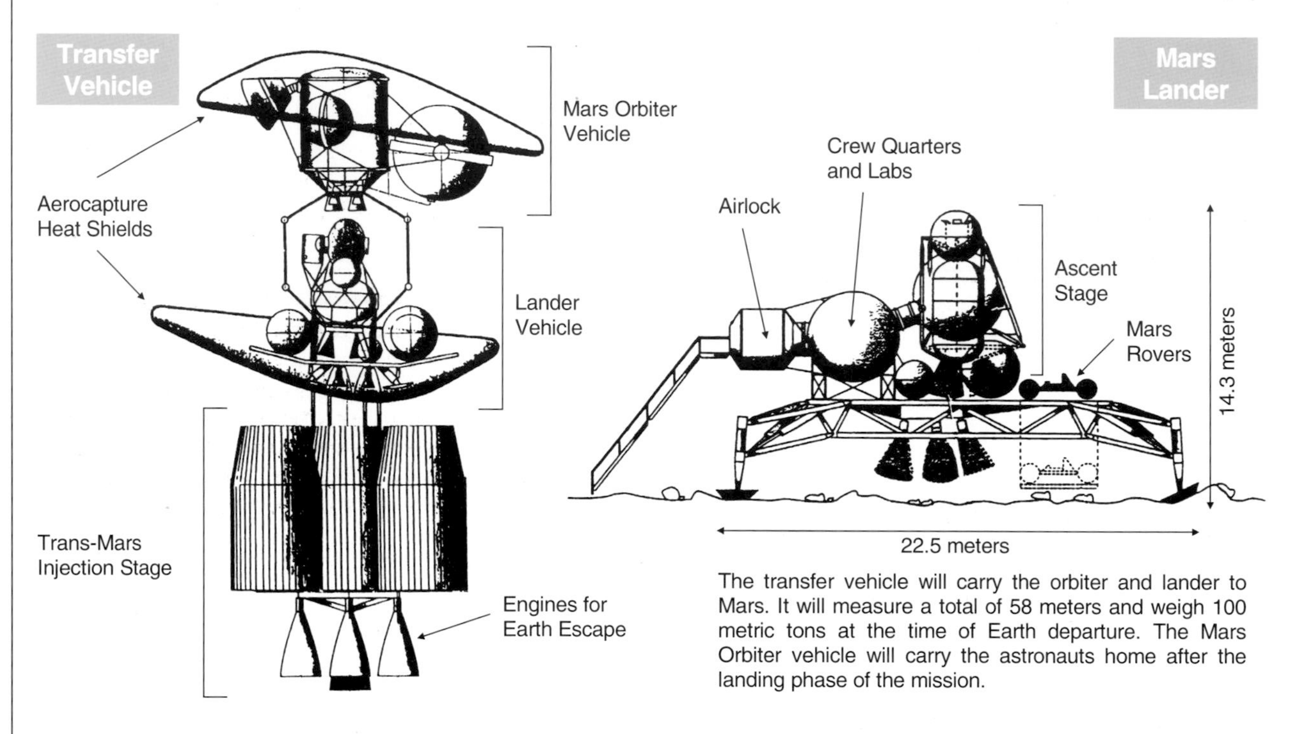

The transfer vehicle will carry the orbiter and lander to Mars. It will measure a total of 58 meters and weigh 100 metric tons at the time of Earth departure. The Mars Orbiter vehicle will carry the astronauts home after the landing phase of the mission.

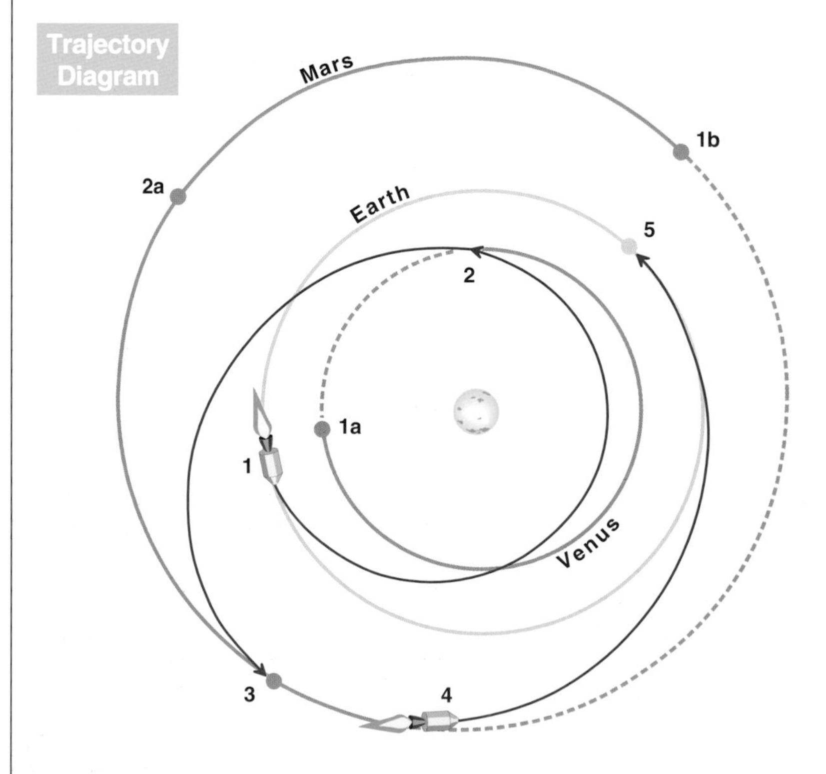

1 *Day 0* - ***Departure from Earth***
Spacecraft must perform a burn to escape the Earth's gravity.

1a Venus Position at Earth Launch
1b Mars Position at Earth Launch

2 Day 166 - ***Venus Flyby***
Gravity assist from Venus slings spacecraft toward Mars for "free."

2a Mars Position at Venus Flyby

3 *Day 358* - ***Mars Orbit Insertion***
Spacecraft must perform a slow-down burn to enter Mars orbit. If the burn is not performed, the astronauts will fly right past Mars.

4 *Day 398* - ***Departure from Mars***
Leaving Mars for Earth requires a burn to escape Mars' gravity.

5 *Day 590* - ***Return to Earth***
Spacecraft must use its rockets to slow down or it will fly right past Earth.

same condition holds upon return to the Earth. Currently, use of rocket engines to kill the speed at closest approach to the destination is the accepted method. However, valuable payload mass must be sacrificed for additional propellant. A new and risky technique called *aerocapture* may provide an alternate solution. This new method involves intentionally targeting the spacecraft to pass through the destination planet's atmosphere at the point of closest approach. Plunging into the atmosphere will allow for air friction and drag to slow the craft enough for orbital capture.

Unfortunately, many extremely challenging engineering problems exist in developing aerocapture devices. A saucer-shaped shield to protect the bottom of the spacecraft from intense heating must be developed. Many control problems also exist. If the spacecraft plunges through the atmosphere too steeply, then the frictional heating will burn through the spacecraft and incinerate the crew and cargo. If the entry path is too shallow, then the craft will skip off the atmosphere like a flat stone on water.

Spacecraft Design

Nobody has yet designed the spacecraft for the first piloted Mars mission. Despite this fact, some NASA mission design experts predict that the mission will probably use two separate components. The first, an unpiloted cargo vehicle, will carry all the surface supplies, surface habitat, and possibly propellant for the return trip home. NASA plans call for this cargo to leave for Mars on a minimal propellant trajectory in advance of the astronauts. Only after the cargo safely arrives will the astronauts depart Earth. Their spacecraft, called the *piloted component*, will consist of an interplanetary transfer vehicle and the Martian lander.

Astronauts will direct the cargo vehicle to land by remote control after verifying the safety of the landing site. Half of the astronaut crew will also proceed to the surface in the lander vehicle, while the remaining half will remain in orbit around Mars in the interplanetary transfer vehicle. Astronauts in the lander will not have any help from mission control on Earth during the landing phase as radio signals can take up to twenty minutes to reach Mars.

Like the primitive lunar module used to land astronauts on the Moon during the Apollo program, the Martian lander will probably be comprised of both a lower stage and an upper stage. Engines on the lower stage will slow the lander toward a soft touchdown, while the upper stage will contain the crew accommodations. The upper stage will also ferry the astronauts back to the transfer vehicle for the trip home. Current plans call for the astronauts to live in the surface habitat delivered by the unpiloted cargo vehicle for their initial thirty-day stay. In the event that this habitat is unusable, the lander's design will provide at least a week's worth of emergency provisions, allowing the astronauts time to conduct the most important scientific experiments on their agenda.

RULE OF THUMB

Typically, a spacecraft that arrives at Mars must wait about 550 days before Earth and Mars realign for the trip home. At that time, the astronauts will be able to use their rocket engines to leave Mars and return home.

However, NASA may chose a "free-return" option for the first piloted Mars mission. In this scenario, the astronauts will arrive in such a way that they can choose to either slow down and enter Martian orbit or allow the gravity of Mars to automatically bend the spacecraft's trajectory and send it home without the use of rockets. Such an option will allow the astronauts the flexibility of either landing on Mars or immediately beginning the return trip home in the event of an emergency occurring before Mars arrival. If they choose to land, the free-return option will no longer apply, and they will need their rockets to return home. Typically, they will be able to begin the return trip after only 40 to 60 days at Mars instead of 550.

Possible Dates for the First Piloted Mars Mission

Event	Date
2015 Opportunity	
Earth Launch	22 May 2015
Venus Flyby	19 October 2015
Mars Flyby (no landing)	21 April 2016
Earth Arrival (free return)	24 February 2017
Mars Arrival (w/ landing)	21 April 2016
Mars Launch	31 May 2016
Earth Return	26 January 2017
2017 Opportunity	
Earth Launch	28 March 2017
Venus Flyby	10 September 2017
Mars Flyby (no landing)	21 March 2018
Earth Arrival (free return)	19 July 2018
Mars Arrival (w/ landing)	21 March 2018
Mars Launch	30 April 2018
Earth Return	8 November 2018

All landings assume a forty-day stay time at Mars

Sites of Interest to Visit on Your Next Trip to Mars

Olympus Mons (#1)
Tallest Martian volcano and largest in the Solar System. At an altitude of 29 km at the peak, Olympus Mons measures more than three times higher than Mount Everest on Earth.

Tharsis Volcanos (#2)
Giant volcanos located to the east of Olympus. They are not as tall, but peaks are at the same altitude because they sit on top of a tall plateau. All are taller than Mount Everest.

Valles Marineris (#3)
Giant canyon that stretches for a distance longer than the United States. Canyon depth is much deeper than the Grand Canyon.

Ares Vallis (#4)
Site of torrential floods long ago in Martian history. Study of rocks in this area may yield clues into the mystery of Mars' missing water.

Argyre (#5) and Hellas Planitia (#6)
Located in the rough terrain of the ancient, heavily cratered southern hemisphere. These large basin-like plains are thousands of kilometers across and were formed by collisions with huge meteors billions of years ago.

Chryse (#7) and Utopia Planitia (#8)
Historical site where the *Viking 1* (Chryse) and *Viking 2* (Utopia) landers touched down in the summer of 1976.

Polar Ice Caps (#9)
Many layers of dust trapped between sheets of ice will yield clues into the history of Mars' climate and past atmospheric conditions.

RULE OF THUMB

When visiting Mars, try to avoid planning your trip during the summer in the southern hemisphere. During this season, global dust storms tend to sweep across the planet with wind speeds of up to 320 kilometers (200 miles) per hour. However, since the Martian atmosphere is so thin, the winds do not pack as much punch as they would on Earth.

Surface Operations

Astronauts on the surface of Mars will be well equipped to resolve many unanswered questions. To discover whether organic material exists beneath the surface, they could systematically drill core sections and analyze them with biochemical experiments. If the landing site is close to one of the Viking or Pathfinder landers, the astronauts will take pieces of it back so that engineers on Earth can assess how well the materials withstood the Martian environment. Geological investigations will be the other major concern. At least one member of the landing party will be a geologist. Explosive underground charges that produce mini-Marsquakes will yield data on the internal structure of the Martian crust and core. Information about the development of the Martian atmosphere, where rocks formed and at what pressures, and the history of water on the planet will be gained by examining the chemical and mineral composition of Martian rocks. Understanding the history of water on the surface of the planet will also reveal clues regarding the possible formation of organic substances or life.

Balloons will be released into the atmosphere to monitor wind conditions and weather patterns. Automatic science stations will be left on the surface to provide data long after the astronauts return home. These instruments will include seismographs, heat probes, barometers, and a mass spectrometer to measure the chemical composition of the air. Understanding Mars will allow us to better understand the history of the Earth because current conditions on Mars are similar to conditions on Earth during the ice ages.

7.5 A Tourist's Guide to Mars

Visitors to the Red Planet, whether human or their robotic servants, will find plenty to explore. Although the diameter of Mars measures only half that of Earth's, Mars contains as much land surface area because most of Earth's surface lies under the ocean. Even if NASA's early missions to Mars dispelled old myths of vast artificial canal networks or ancient civilizations, they revealed a world of scientific fascination inviting future study. On Mars, the largest volcanoes in the Solar System tower above the local landscape, rolling sand dunes cover lands larger than Africa, global dust storms howl at hurricane speeds, and deep chasms stretch for distances longer than the continental United States.

Despite the fascinating geology, much of the scientific mystery revolves around the subject of missing water. On a planet that makes the Sahara look lush in comparison, scientists have found evidence of extensive flood valleys and dry riverbeds. If water once existed in vast quantities, then simple forms of bacterial life may have evolved. In fact, life may still exist in watery oasis-like areas unknown to humans. Only continued exploration will reveal the answers.

Giant Volcanoes

Mars is home to Olympus Mons, the tallest known volcano in the Solar System. This Martian mountain soars to an altitude of about 27 kilometers (close to 90,000 feet) above the local terrain. In comparison, the summit of Mount Everest on Earth only reaches about 8.9 kilometers (5.5 miles) above sea level. Olympus Mons also dwarfs the 10-kilometer (6.2-mile) tall Mauna Kea volcano in Hawaii, the largest mountain on Earth when measured from the floor of the ocean to the summit. In addition, the size of Olympus Mons is as impressive as its height. From edge to edge, the base spans a distance of roughly 650 kilometers (404 miles), enough to cover an area the size of Arizona.

Nobody knows for certain how a volcano three times larger than Earth's tallest mountain formed on a planet only half as large. On Earth, volcanoes form near the edges of huge tectonic plates that

Above:

This image of the giant Martian volcano, Olympus Mons, was prepared from photographs taken from *Viking 1* on 31 July 1976. The crater at the top of the volcano measures about 80 kilometers (50 miles) across. In the photograph, a thick layer of clouds at an altitude of 19 kilometers (12 miles) hides a majority of the lower sections of Olympus. The top of the mountain is at an altitude of 27 kilometers (90,000 feet) above the Martian lowlands. This altitude exceeds the summit height of Mount Everest by a factor of three.

Visitor's Map of Mars

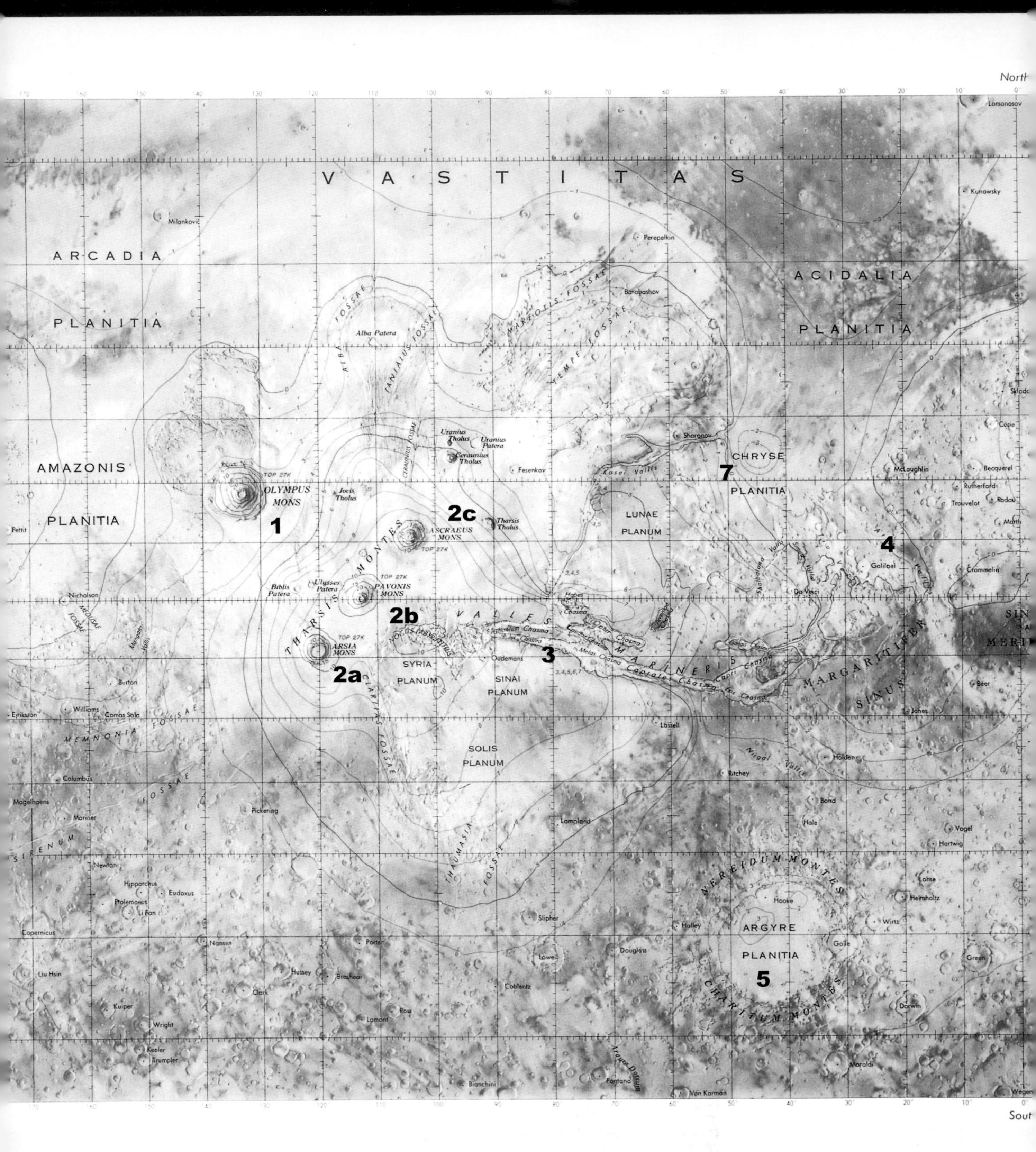

The map of Mars on these two pages covers the entire range of longitudes, but only spans the range of latitudes from 65° South to 65° North. Because Mars has no seas, altitudes as displayed by the contour lines (expressed in kilometers) cannot be referenced with respect to "sea level." Instead, all altitudes are given with respect to an average altitude called the *Mars Datum Surface*. This map was created by the United States Geological Survey for NASA using photographs and data transmitted back to Earth by the Viking missions during the late 1970s. The individual locations numbered on the map correspond to the list of "Interesting Sites to Visit on Your Next Trip" three pages prior to this one.

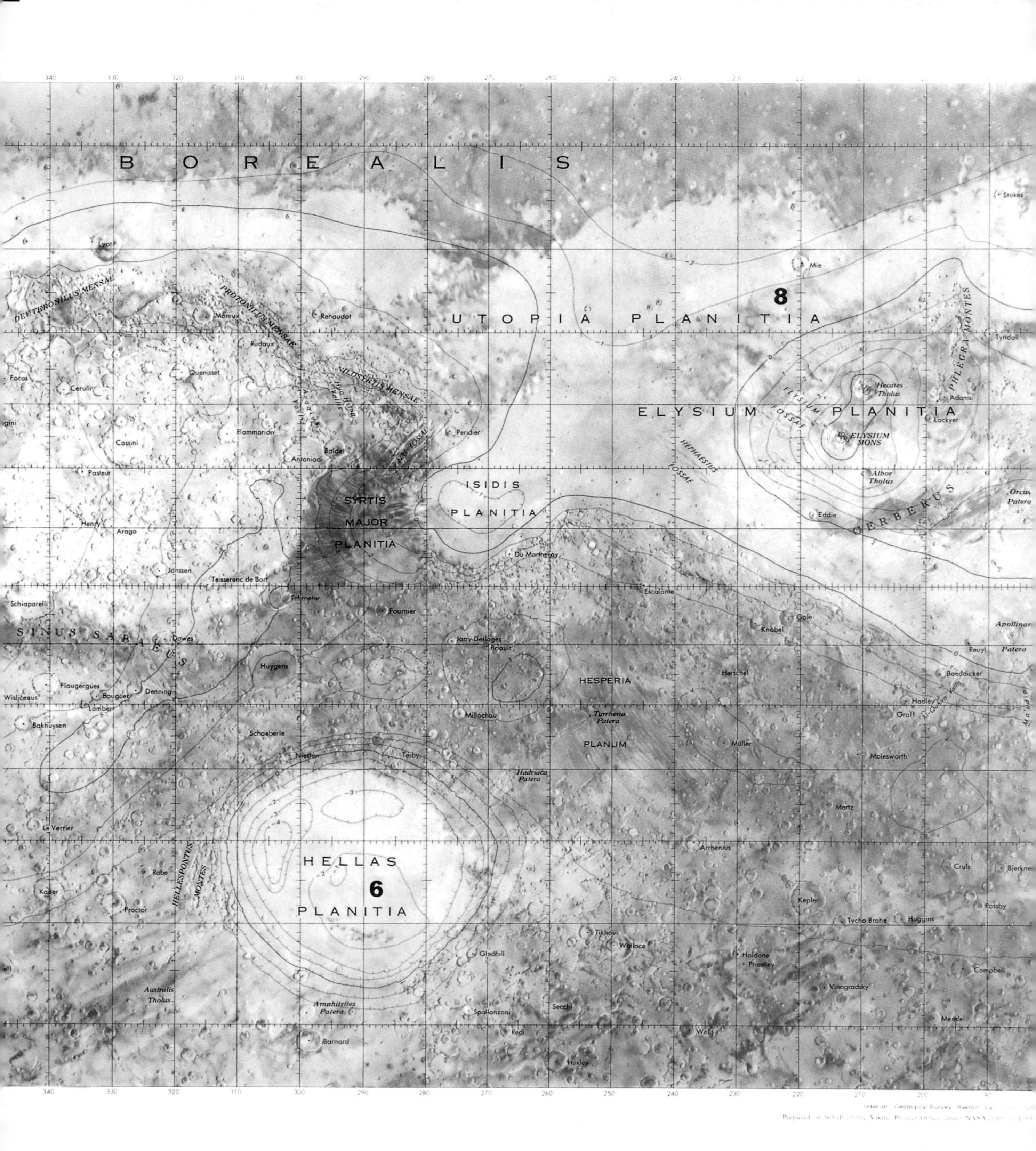

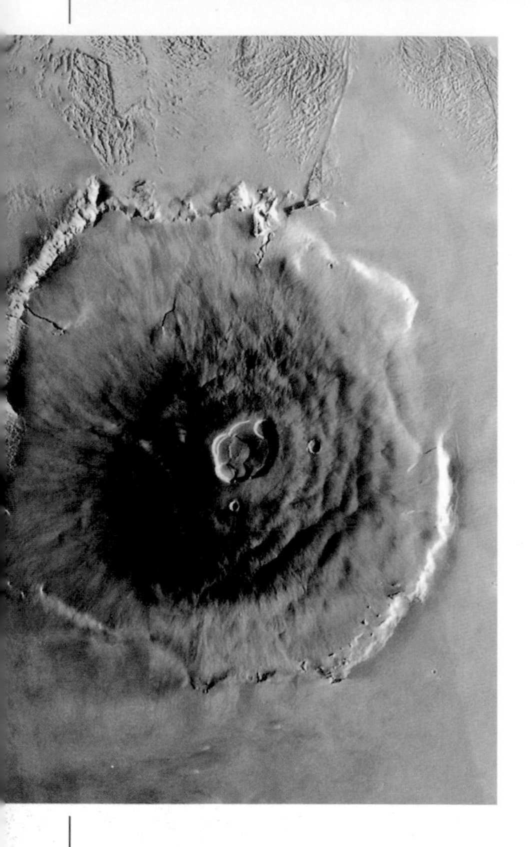

Above:

Olympus Mons, tallest volcano in the Solar System, also ranks as one of the largest. At its widest point, this gigantic mountain measures about 600 kilometers across, enough to cover a surface area the size of Arizona. Notice the cliffs that skirt the entire circumference of the mountain's base. They are about six kilometers tall and would dominate the horizon as seen by astronauts on the surface. As seen in this photograph taken from the *Viking* orbiter, the lack of impact craters on the slopes of Olympus indicates that the volcano is relatively young from a geological perspective.

cover the entire surface of our planet like a giant jigsaw puzzle. These plates are in constant motion and slide past each other at average rate of about one centimeter every year. This motion creates earthquakes and causes the continents to change position relative to each other over millions of years. Over a long time, the movement of the plates will carry a volcano away from its source of underground magma, and the volcano will no longer be able to erupt and add to its height by pouring new lava over flows from previous eruptions.

The net result of plate movement is that an Earth volcano will die out before it has a chance to build up to tremendous heights as found on Mars. Data returned from past Mars missions suggests the existence of a rigid crust that either lacked tectonic plates or lacked significant plate motion. Consequently, scientists think that Martian volcanoes tend to "stay put" near their magma source and grow to extreme heights over hundreds of millions of years. In comparison, Earth volcanoes become extinct after only several million years at most. By studying the mineral composition of the enormous lava deposits near Olympus Mons, scientists will learn more about the Martian interior and how it differs from the inside of Earth. Also, by using geological techniques to determine the age of the lava flows, scientists will be able to determine when Olympus Mons stopped erupting. This data will yield clues into when and why the interior of Mars cooled down.

Although the size and height of Olympus Mons is impressive when compared to mountains on Earth, it is certainly not unique with respect to the Martian landscape. A set of three other volcanoes lies slightly southeast of Olympus and forms a line across the Martian equator running from the northeast to southwest. These volcanoes, named Ascraeus Mons, Pavonis Mons, and Arsia Mons, measure a mere 10 kilometers (6.2 miles) from base to tip, but sit on an elevated plateau called the Tharsis Bulge. Consequently, the summits of the Tharsis volcanoes lie at an altitude comparable to that of Olympus Mons.

The Grand Canyon of Mars

The formation of the four giant volcanoes on Mars may mystify scientists, but some theorize that the consequences appear prominently in the form an extremely large and long canyon system immediately to the southwest. This system, known as Valles Marineris to Martian cartographers, stretches for a length in excess of 4,000 kilometers (2,485 miles), a distance roughly equal to that of the continental United States. Individual subcanyons within Valles Marineris measure up to 200 kilometers (124 miles) wide and about 7 kilometers (4.3 miles) deep. In comparison, the Grand Canyon in Arizona measures only 450 kilometers (280 miles) long, 2 kilometers (1.2 miles) deep, and 30 kilometers (19 miles) wide at most. Essentially, the Grand Canyon would barely be a scratch in the Valles Marineris canyon system.

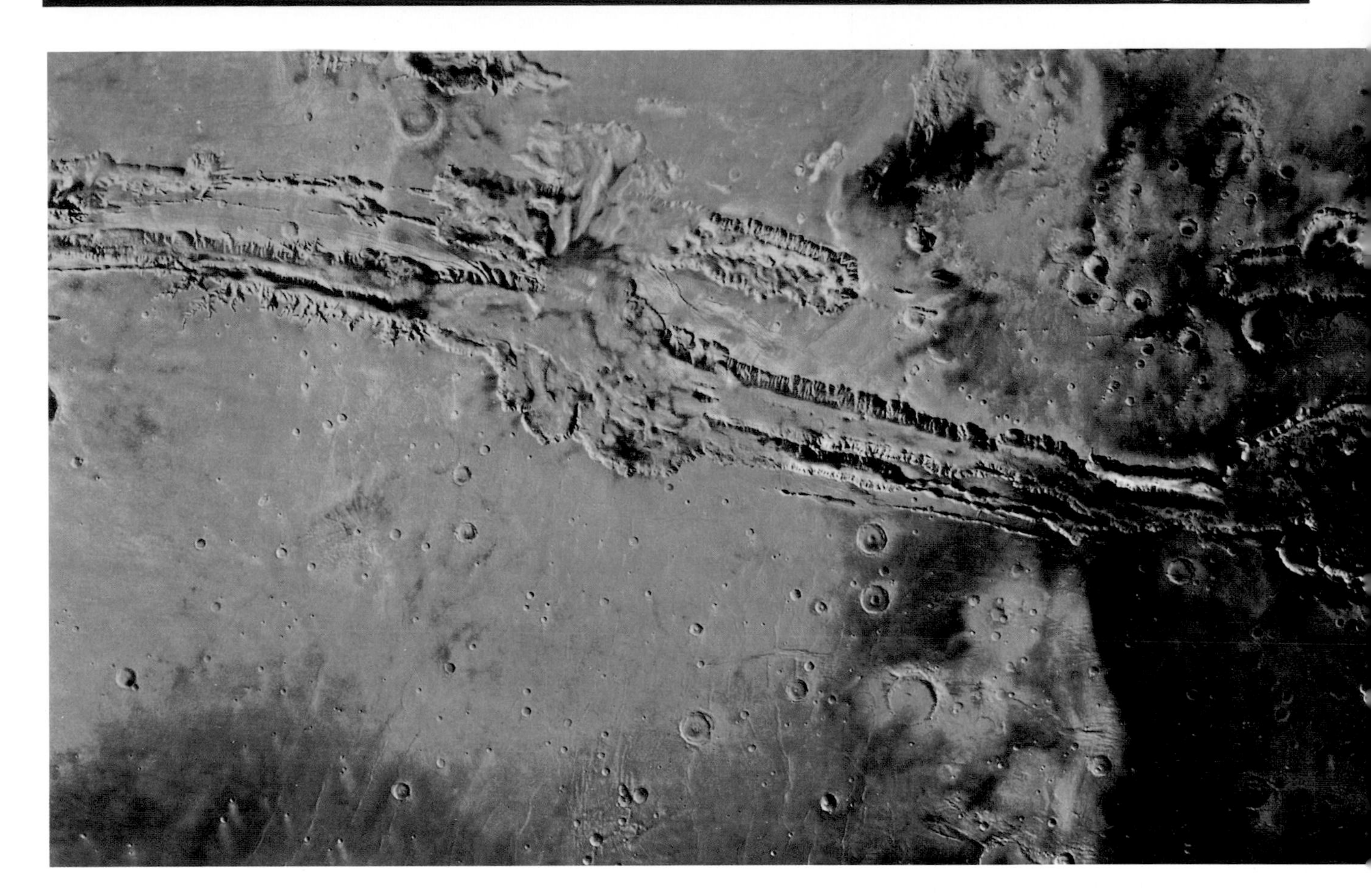

The proximity of Valles Marineris to the four giant volcanoes and the Tharsis Bulge plateau may suggest that water played an insignificant role in the formation of this canyon. Some geologists believe that the release of tremendous amounts of underground magma needed to feed the gigantic volcanoes gradually created an empty pocket or sinkhole underneath the present location of the canyon. Consequently, the surface probably collapsed under its own weight and created the gigantic network of canyons. A study of this canyon area will yield many valuable clues into the geological history of Mars. In addition, even though geologists speculate that Valles Marineris was not formed by water erosion, sedimentary layers near the bottom seem to have been deposited by running water. Future astronauts sent to explore Mars may search these sedimentary layers in an attempt to find fossil evidence of previously existing microbiotic life.

Landing in the canyon, either with astronauts or robotic probes, will pose difficulties. A landing in the extreme western section of the canyon, called Noctis Labyrinthus, is definitely out of the question because that area contains a maze of interconnected, narrow canyons. However, middle areas may provide a stable area to land. The choice of an exact landing site will depend on several factors. For starters, the smoothness of the canyon floor varies from location to location. In areas where the canyon is wider, the floors are flatter. On

Above:

Even when observed from hundreds of kilometers above the surface, the giant canyon Valles Marineris still looks huge. In order to understand the enormous size of the canyon, consider the fact that the distance from edge to edge in the photograph measures longer than the distance between San Francisco and New York. The vertical faces of Valles Marineris' walls are nearly seven kilometers high in places. This photographic mosaic was generated using images returned in the late 1970s from the two *Viking* spacecraft sent to orbit the Red Planet.

Above:
As seen in the photographic mosaic, the canyon floor of Valles Marineris is filled with debris that has fallen from the cliff faces during enormous avalanches. Scientists do not know whether the avalanches were caused by quakes or melting subsurface ice that caused the land to collapse. The canyon walls in this photograph measure about five kilometers high.

the other hand, the narrower areas will be more interesting to explore from a geological perspective, but the floor in those areas tends to contain more rocks, hills, and craters that could potentially damage the lander. Also, a landing in the canyon will restrict surface exploration of the planet because the Mars rover vehicles will probably not be able to climb the steep walls. Finally, an unexpected Mars-quake could trigger devastating landslides.

The Enigma of Water

Did liquid water ever exist on Mars? Today, the atmospheric pressure on the Red Planet is much too low to support water. Any liquid water on the surface would instantly evaporate. However, a big mystery exists because NASA's early Mars probes discovered quite a few large, long valleys that look like dry river channels meandering over the surface of Mars. Many of these channels start in a region of chaotic terrain immediately to the east of Valles Marineris and empty out in the Martian plains hundreds of kilometers to the north. This chaotic terrain at the source appears heavily cracked with many shallow depressions and is littered with many jumbled boulders. Incidentally, these channels are not the canals supposedly seen by pre-spaceage astronomers. Despite their large size, they are too small to have been seen by the limited resolving power of late nineteenth- and early twentieth-century telescopes.

Most scientists and geologists who study Mars speculate that these valleys were cut by torrential floods long ago in Martian history. Teardrop-shaped islands downstream, similar to islands found on floodplains on Earth, support this theory. In addition, unlike typical river systems on Earth, these valleys possess few tributaries and show evidence of scouring from much debris violently swept downstream. For this reason, geologists refer to the river valleys as *outflow channels*. Another bit of evidence that supports the torrential flood theory is that the large flood valleys appear similar to the giant scablands in the state of Washington. That area formed millions of years ago when one of the containing ice walls of ancient Lake Missoula broke. The water from the lake covered much of Idaho and Montana prior to scouring eastern Washington in a matter of a few days.

The amount of water that flowed through these large Martian channels may have exceeded 10,000 times the average discharge rate of the Amazon River into the Atlantic Ocean. Some of the larger flood valleys measure more than 100 kilometers (62 miles) wide in places. Currently, one plausible theory as to the source of the water involves a substantial-sized pack of subterranean ice melted by volcanic heat. When the ice melted, the rock and dirt above it collapsed to cause the release of a catastrophic amount of floodwater that cut the valleys. The collapse of the rock and dirt created the chaotic terrain at the source of the flood channels.

Mars also contains many medium-size channels in the neighborhood of 20 kilometers (12 miles) wide and up to several kilome-

Scientific Fact!

Despite the fact that Mars is only half as large as the Earth in diameter, Mars contains almost as much land surface area as Earth. The reason is that most of Earth's surface is covered by oceans.

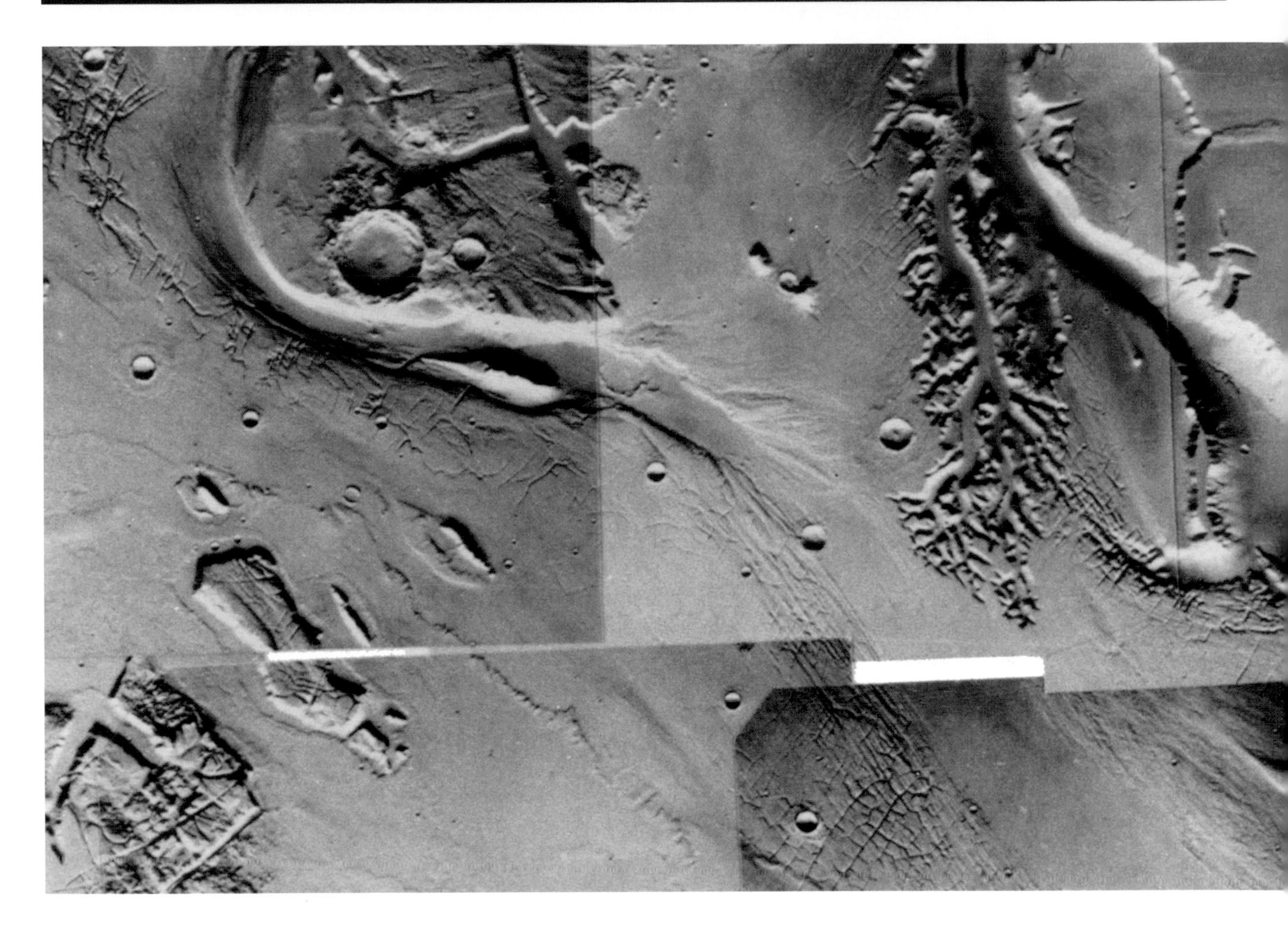

ters deep. These channels are more numerous and tend to meander much more than the large flood valleys, are often joined by many tributaries, and increase in size downstream. Their shape and geological characteristics appear similar to streams created by minor, intermittent flooding in deserts on Earth.

Unfortunately, nobody knows what happened to the water on Mars. The relatively weak gravitational pull of Mars allows light gasses to easily escape into space. Data returned from the Viking missions showed signs of the chemical constituents of water, hydrogen and oxygen gas being lost to space at the rate of thousands of liters every day. Mathematical models based on this data theorize that Mars may have once harbored enough water to cover the entire surface of the planet up to one kilometer deep.

Recent evidence suggests that much of the Martian water that did not escape into space may currently lie partially frozen beneath its surface. Scientists arrived at this conclusion after studying meteorites that landed on Earth after being blasted away by asteroids that collided with Mars millions of years ago. In particular, they examined the ratio between hydrogen and a heavier form of hydro-

Above:

Large outflow channels carved into the ground at Kasei Vallis (north of Valles Marineris) mark the site of a catastrophic flood long ago in Martian history. The teardrop-shaped sandbar island in the middle of the channel, along with the silt deposits and scour marks at the bottom provide indisputable evidence that large amounts of water once flowed on Mars. In order to gain an idea of the size of the flood, consider the fact that the channel is over a kilometer deep, and the crater near the bend in the channel measures 25 kilometers across. This photograph was taken by the *Viking 2* spacecraft in 1976, many millions of years after the water dried out.

Above:

Although scientists classify 40% of the Martian surface as plains, these areas are anything but smooth. As shown in this *Viking 2* photograph of the Utopia Planitia area, most Martian plains are littered with boulders. In the image, the horizon lies about 3.2 kilometers (2 miles) from the camera. The horizon appears slanted about 8° from the horizontal because the spacecraft set down unevenly with one of its landing footpads on a small rock.

Opposite Page:

As on the Earth, sheets of ice cover the surface of Mars at each pole. This photograph of the Martian south pole is a mosaic of 18 photographs taken by *Viking 2* on 28 September 1977. At the time the photographs were taken, it was summer in the southern hemisphere of Mars. Most of the ice sheet had melted already. The remaining ice pack measures about 400 kilometers across and is composed almost entirely of frozen carbon dioxide, popularly called "dry ice." Some of the surface features near the top of the mosaic were etched by violent winds. The United States Geological Survey in Flagstaff, Arizona, prepared this mosaic.

gen known as deuterium. Common hydrogen is more likely to escape into space because it weighs less than deuterium.

Over time, the proportion of deuterium in the atmosphere will increase unless the hydrogen that escapes into space is regularly replenished from water on the surface. However, one of the meteorites was dated at three million years old and contained hydrogen and deuterium in the same ratio found on Mars today. This finding points to the fact that Mars must contain large amounts of water under the surface that is regularly exchanged with the atmosphere. Scientists believe that life on Earth began in the oceans. They theorize that if similar bodies of water existed on other planets, life may have had a chance to develop.

Plain Old Plains

Although gargantuan volcanoes, great canyons, and giant flood valleys dominate the Martian surface, cartographers characterize about 40% of the Red Planet's surface as plains. Technically, they refer to the plains by the Latin term *planitia*. Some plains in the southern hemisphere, such as Argyre Planitia and Hellas Planitia, are the smooth floors of huge, ancient impact basins formed when large asteroids collided with Mars long ago. They measure more than a thousand kilometers across and several kilometers deep. However, a vast majority of the land in the south looks like scarred, rough terrain littered with thousands of craters. They formed during a period of cataclysmic bombardment from meteors during the first billion years of the planet's existence.

Most of the Martian plains lie in the northern hemisphere. In contrast to the relatively smooth floors of the southern impact basins, the planitias of the north take on anything but a smooth appearance. These plains consist largely of volcanic lava flows, flood deposits, or rolling sand dunes and scattered boulders. Currently, scientists do not understand the reasons for this hemispheric discontinuity.

Frozen Club Soda

As on Earth, a layer of ice covers the surface of Mars at both poles. However, a major difference exists. The Martian polar ice caps consist primarily of frozen carbon dioxide, popularly known as dry ice. Think of it as similar to frozen club soda. During the winter months, a substantial amount of carbon dioxide from the atmosphere "freezes out," accumulates at the winter pole, and causes the ice caps to expand toward lower latitudes. In the summer, the ice melts and releases the carbon dioxide back into the atmosphere. This release in the summer is a great force in Martian weather because the atmospheric density blooms by as much as 30%. One interesting fact is that the northern residual ice cap, the ice that remains in the summer after the expanded portion melts, consists mostly of water ice. Conse-

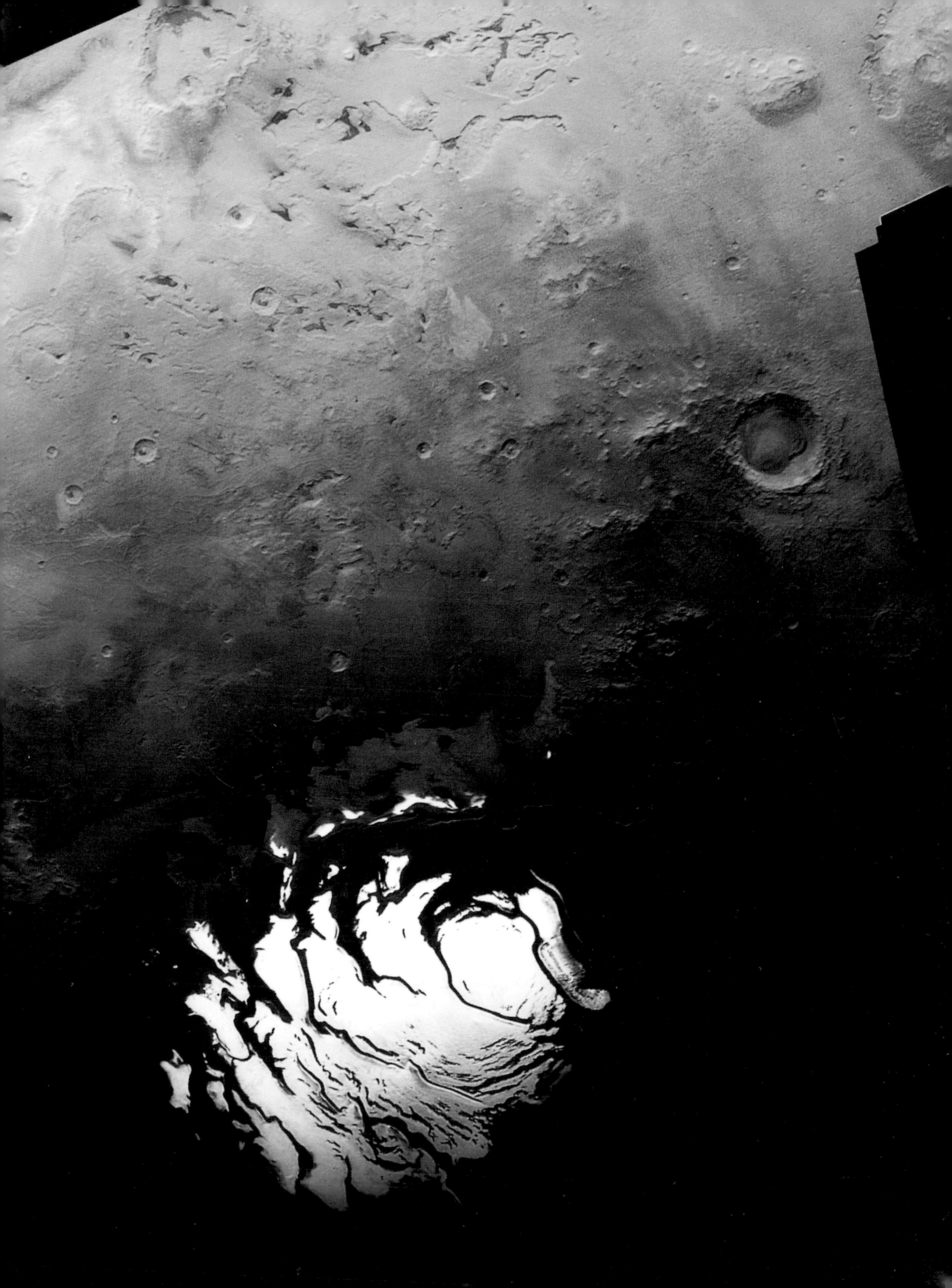

Below:

This 16-photograph mosaic of the Lunae Planum area of Mars (north of Valles Marineris) was taken by *Viking 2* on 18 November 1976. The picture covers an area approximately 2,000 kilometers (1,240 miles) long by 500 kilometers (311 miles) wide. The channels toward the bottom were cut by torrential floods long ago in Martian history.

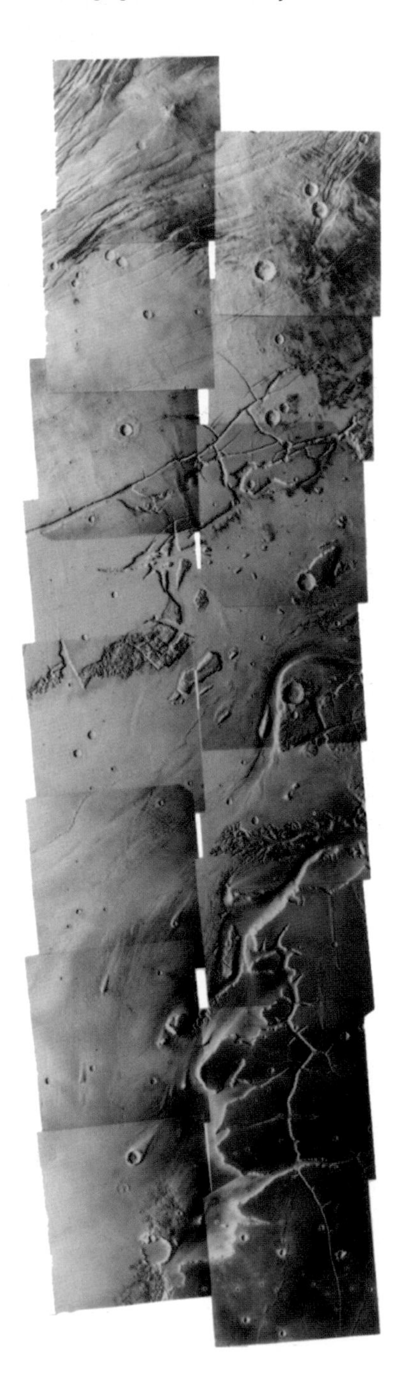

quently, the presence of water in the soil near the Martian north pole may support microbiotic life in the soil.

The polar ice caps also store "snapshots" in time from a geological perspective. Over a period of millions of years, the ice has been gradually eroded by the wind, covered with a layer of soil, and re-formed over the soil. Many of these ice-soil-ice layers exist within the polar ice caps. Different layers of a core sample drilled from the ice could be examined to determine the overall Martian climate and the atmospheric thickness at different periods during the planet's history. The importance of such a determination is that many scientists believe that the atmosphere of Mars was once dense enough to support running water on the surface. Also, by comparing the core sample results with tests conducted on sediment samples from Earth, scientists could determine whether cold periods on Earth and Mars occurred simultaneously.

How's the Weather?

Mars may rank among the planets as the most Earth-like, but temperatures still average well below freezing. Typical thermometer readings at the equator during the daylight hours fall near the -50° C (-58° F) mark. During a rare, summer heat wave, the temperature might climb as high as a balmy 20° C (70° F) at noon. At night, the thin Martian atmosphere retains only a tiny fraction of the heat absorbed during the day. Consequently, daily temperature variations fluctuate over a far wider range than on Earth. Equatorial temperatures on the dark side of Mars may plummet down to -100° C (-148° F) or lower, far colder than it ever gets in Antarctica.

What is the chance of rain? On Mars, the weather forecast will always call for cold and dry air with a possibility of winds, but no chance of precipitation. What little amount of water known to exist above ground, outside the northern polar ice cap, takes the form of vapor in the atmosphere. In total, if all of the vapor in the Martian atmosphere instantly turned into rain, enough water would fall to cover the entire surface with only a paltry 0.0001 to 0.01 centimeters of precipitation. As a comparison, that figure measures two to three centimeters (about an inch) for the Earth. However, since water cannot exist in a liquid state on Mars today due to the extremely low barometric pressure, rain can never fall. Despite this fact, the atmospheric water often freezes out and appears on rocks as frost in the regions near the Martian poles.

Will it be cloudy or clear? Despite the relative lack of moisture in the Martian sky, clouds seem to form frequently in the atmosphere just as on Earth. Some of these formations appear puffy and form in the equatorial regions early in the summer afternoon near the slopes of the four giant volcanoes. Others types called *wave clouds*, look like ripple patterns formed on ponds by wind. They often appear on the leeward (downwind) side of large surface features such as craters and mountains. At lower elevations, early morning white fog commonly appears in canyons such as Valles Marineris or deep depression basins like Hellas Planitia.

An Empire of Dust

At least once every Martian year, violent dust storms raging as fast as hurricane speed winds erupt. Apparently, these winds start in the southern summer at Martian perihelion. When Mars comes closest to the Sun, the southern hemisphere heats up and the south polar ice cap begins to melt. The heat variation between the relatively warm, newly exposed ground and the remaining freezing ice stirs fierce winds that inject large amounts of dust into the air. Sometimes, the temperature differential between the relatively warm summer south and freezing winter north causes the local disturbances to spread to global proportions. These global dust storms engulf the entire planet and blot out the Sun, sometimes for weeks at a time. Curiously, these dust storms do not occur during the opposite season. Due to the eccentricity of the Martian orbit, the Red Planet finds itself 46 million kilometers (29 million miles) farther from the Sun at northern summer. Consequently, northern summers are not as warm as the ones in the south.

Almost every land feature on Mars shows signs of extreme scouring and erosion from the winds that race around the planet. However, they do not strike with as much force as the winds on Earth due to the thin Martian air. Nevertheless, every rock face, mountain, and crater wall is well worn from eons of bombardment from blizzards of dust. In the late 1800s and early 1900s, wind erosion played tricks on observers by depositing streaks of light-colored dust on darker terrain. Astronomers of the past who looked through their telescopes mistook the streaks for growths of vegetation and networks of canals. Because the dust storms were seasonal, the supposed growth of vegetation also appeared seasonal.

Why Mars?

There are those who argue that the United States should not continue to explore Mars. For example, in the summer of 1995 at the NASA Jet Propulsion Laboratory open house, a few concerned citizens voiced their opinion that money should be used for federal programs more directly beneficial to the average American citizen. They said, "It's none of the average citizen's business what goes on in outer space." Some in Congress have echoed the same sentiment. A representative once stated that he could not justify spending money to find evidence of life on Mars when he knew that there were rats in Harlem apartments, overflowing sewers, hunger, and national health problems to solve. To some extent, these critics have an important message. In the process of exploring space, Americans must not forget about the Earth, because the ultimate goal of space exploration is to expand knowledge and ultimately improve humanity.

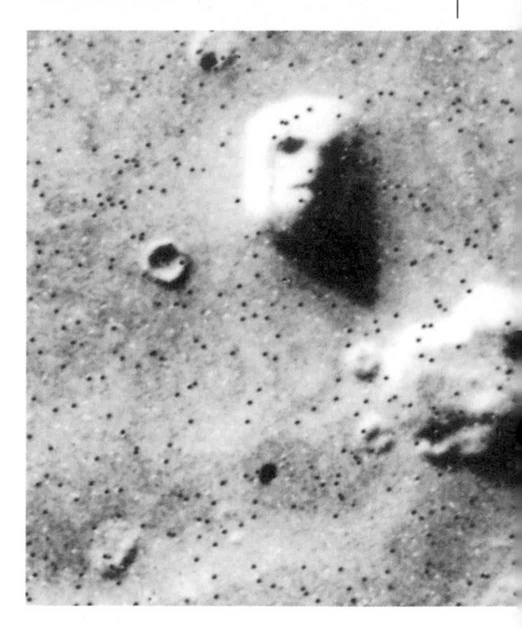

Above:
Although NASA scientists are certain that no intelligent life exists on Mars, wind erosion and tricks of lighting might lead some to believe otherwise. Humans tend to "see things" they want to believe due to paradigms intrinsic to the human mind. Most people as children imagine clouds to look like everyday objects. However, nobody would suggest that there are intelligent creatures in the sky carving pictures in the clouds.

Consider the fact that problems have always existed on Earth ever since the dawn of civilization. In the year 1492, Christopher Columbus' generation experienced significant problems of their own. Street gutters served as their sewers, rats ran rampant through houses, and communities suffered from epidemics like the plague. Citizens of Europe may not have battled drive-by shootings, lung cancer, or the Ebola virus, but their problems were every bit as tough as those of today. However, if Columbus had waited until all the problems of his time were solved before heading out for the seas, the wood used to build his ships would now be rotting at the docks in Lisbon, Portugal.

History has shown time and time again that those who lead in exploration always lead the world. Years ago, world leadership revolved around ideological concepts such as imperialism, conquest, and colonization. Today, leadership is about science and technology. By studying Mars, Americans will gain greater insight into the history and formation of Earth. On our planet, were water flows, life follows. Eons in the past, Mars resembled the Earth with respect to temperate climates and flowing water on the surface. Now, the Red Planet lies dry as a desert, trapped in ice-age conditions. Did life ever exist on Mars? Will Earth suffer the same fate? Where did we come from, and where are we headed? If Americans care about the future of humanity, then these questions are important and deserve to be answered.

Spaceflight costs Americans about $10 billion dollars every year. Of that $10 billion, Mars exploration costs less than $300 million dollars annually. In total, the NASA budget amounts to less than one penny of every tax dollar. However, that penny is well spent, as it represents an investment into this country's technological, and therefore economical, success. In the early 1980s, industrial research and development grew at an average annual rate of close to 10%. Today, that rate has dramatically declined and other countries have surpassed America in many vital technological arenas. Will spaceflight, one of the critical keys to transportation, communications, and scientific success in the future, be next? This issue is important to the average American.

NASA administrator Daniel Goldin once said, "If we as a nation consciously decide not to go to Mars, our generation will truly achieve a first in human history, we will be the first to stop at a frontier, we will be first to draw a line and say 'this far and no farther' to our children." The biggest benefits of Mars exploration are the benefits of the unknown, benefits of rewards that humans cannot comprehend at this point. For example, when Thomas Jefferson had an extraordinary vision of the future and purchased the Louisiana Territory from France in 1803, many laughed at him for wasting money. Nobody laughs at Jefferson today. Remember, America to Mars, it is about understanding the Earth. ❏

Opposite Page:

This photograph of the Earth was taken by *Apollo 16* astronauts on their way to the Moon in April 1972. Although there is much cloud cover, the United States, most of Mexico, some parts of Central America are clearly visible.

NASA
Discovery

Photo and Art Credits

All photographs supplied by the National Aeronautics and Space Administration belong to the public domain. No copyright is asserted for those photographs. They may be reproduced freely and without permission provided that they are not used to state or imply the endorsement by NASA of a commercial product, process, service, or used in any other manner that might mislead. Other photographs and artwork are copyrighted by their respective private organizations and have been generously provided for use in this book. These photographs and artwork are not in the public domain.

Computer Support Corporation *(Dallas, TX)*
Clipart for various figures
Pages 26–27, 29, 111, 118, 127–128, 131, 133–134, 153, 155–156, 182–183, 196, 216

General Dynamics Corporation *(San Diego, CA)*
Page 24

Lockheed-Martin Corporation *(Denver, CO)*
Page 25, 30

Lowell Observatory *(Flagstaff, AZ)*
Page 268

McDonnell Douglas Corp. *(Huntington Beach, CA)*
Page 22

NASA Goddard Space Flight Center *(Greenbelt, MD)*
(*) Provided by Hubble Space Telescope Science Institute for NASA GSFC
Pages 20, 212*, 233, 236

NASA Jet Propulsion Laboratory *(Pasadena, CA)*
(*) Provided by U.S. Geological Survey for NASA JPL
Pages 6, 21, 72, 108, 222–226, 228, 230–232, 234, 237–243, 265, 266, 269–270, 271, 273–277, 275*, 279, 281, 293, 294–298*, 299–300, 301*, 302–303

NASA Johnson Space Center *(Houston, TX)*
Pages 2–3, 12–13, 16, 34, 36, 38, 41, 44, 46, 50, 53, 55, 59, 62–63, 70, 74, 80, 82, 85, 89, 91, 93–94, 98, 100–101, 103–104, 106, 112–117, 119–120, 122–123, 125–126, 129–130, 135–136, 138, 140–150, 152, 190–191, 195, 198–205, 207–214, 216, 219, 221, 226, 305

NASA Kennedy Space Center *(KSC, FL)*
Pages 5, 28, 159–172, 174, 177–178, 181, 184–185, 220, 264

Trimble Navigation *(Sunnyvale, CA)*
Page 105

Washington Post *(Washington, DC)*
Pages 14–15

Index

This index is designed to quickly access the narrative text and special features. The index is arranged letter by letter. An "f" following the page locator indicates figures and captions.

D

E

F

J

K

L

M

Q

R

S

Colophon

Wayne Lee works at the NASA Jet Propulsion Laboratory in Pasadena, California, where he designs scenarios and orbital trajectories for robotic missions to Mars. He started with the *Mars Global Surveyor* project in 1994 and is now working on the *Mars Sample Return* project, his fifth mission to the Red Planet.

He is currently 30 years old and holds a bachelor of science degree from the University of California at Berkeley in electrical engineering and computer sciences with an emphasis in neural science and bioengineering. In addition, he has a master of science degree from the University of Texas at Austin in aerospace engineering with an emphasis in orbital mechanics and astrodynamics.

Wayne has also been involved in science education. At Berkeley, he was an instructor for an upper-division class in digital electronics. He has also taught space life sciences to undergraduate students at NASA's Kennedy Space Center. In addition, he spends a fair amount of time on the road speaking about Mars missions to college classes and the general public.

The author's hometown is San Diego, California, but he now lives in the smoggy confines of the greater Los Angeles region with his wife Marguerite. In his vanishing spare time, Wayne enjoys sleeping late, taking afternoon siestas, eating out often, midnight food runs, playing the lottery, watching two television sets at once, and anything mischievous done on the spur of the moment.

Wayne invented the idea for this book on 20 March 1992 while sitting on an American Airlines flight bound for NASA's Kennedy Space Center. Before publication with Facts On File, four beta-test versions were released in conjunction with the Texas Space Grant Consortium and the Flat Satellite Society: 1-Alpha (9 June 1992), 2-Alpha (30 December 1992), 3-Alpha (21 June 1993), and 3-Beta (17 January 1994). The 1-Alpha and 2-Alpha versions were originally called the *1992 SLSTP Spaceflight Handbook*. Version 4-Alpha was released by Facts On File in February 1996. This current release has been designated as 5-Alpha and is the first version to be released in color.

To Rise From Earth was produced on a Gateway 2000 computer with 64 megabytes of memory, a 4-gigabyte hard disk, and a Intel Pentium processor running at 90 MHz. Text and layout were generated using *Adobe FrameMaker 5.5*, and graphical illustrations were drawn using *Micrografx Designer 6.0* and *Adobe Illustrator 8.0*. These figures were imported into *FrameMaker* in Encapsulated PostScript (EPS) format. All photographs and other bitmapped graphics were high-resolution drum scanned into EPS and processed with *Adobe Photoshop 5.0*. In total, the original source files for the book occupied approximately 600 megabytes of disk space.

This book uses the Palatino, Helvetica, Helvetica Condensed, Symbol, Zapf Dingbats, Serpentine, Stop, Friz Quadrata, and Albertus Extra Bold family of typefaces. ❏